A CHANGING WORLD

Landscape Series

Volume 8

Series Editors:

Aims & Scope:

Springer's innovative Landscape Series is committed to publishing high quality manuscripts that approach the concept of landscape from a broad range of perspectives. Encouraging contributions on theory development, as well as more applied studies, the series attracts outstanding research from the natural and social sciences, and from the humanities and the arts. It also provides a leading forum for publications from interdisciplinary and transdisciplinary teams.

Drawing on, and synthesising, this multidisciplinary approach the Springer Landscape Series aims to add new and innovative insights into the multidimensional nature of landscapes. Landscapes provide homes and livelihoods to diverse peoples; they house historic – and prehistoric – artefacts; and they comprise complex physical, chemical and biological systems. They are also shaped and governed by human societies who base their existence on the use of the natural resources; people enjoy the aesthetic qualities and recreational facilities of landscapes, and people design new landscapes.

As interested in identifying best practice as it is in progressing landscape theory, the Landscape Series particularly welcomes problem-solving approaches and contributions to landscape management and planning. The ultimate goal is to facilitate both the application of landscape research to practice, and the feed back from practice into research.

A CHANGING WORLD

CHALLENGES FOR LANDSCAPE RESEARCH

Edited by

Felix Kienast

Swiss Federal Research Institute WSL, Birmensdorf, Switzerland

Otto Wildi

Swiss Federal Research Institute WSL, Birmensdorf, Switzerland

Sucharita Ghosh

Swiss Federal Research Institute WSL, Birmensdorf, Switzerland

Springer

A C.I.P. Catalogue record for this book is available from the Library of Congress.

ISBN-10 1-4020-4434-2 (HB)
ISBN-13 978-1-4020-4434-2 (HB)
ISBN-10 1-4020-4436-6 (e-book)
ISBN-13 978-1-4020-4436-6 (e-book)

Published by Springer,
P.O. Box 17, 3300 AA Dordrecht, The Netherlands.

www.springer.com

Cover photograph by Otto Wildi

Printed on acid-free paper

Foreword by the Series Editors

With the Springer Landscape Series we want to provide a much-needed forum for dealing with the complexity of landscape types that globally occur and are studied. It is crucial that the series highlights this profound diversity – both in the landscapes themselves and in the approaches used in their study. Moreover, while the multiplicity of relevant academic disciplines and approaches is characteristic of landscape research, we also aim to provide a place where different knowledge cultures can be synthesized and integrated.

A Changing World. Challenges for Landscape Research is the eighth volume of the series. The editors, Felix Kienast, Otto Wildi and Sucharita Ghosh, have compiled a fascinating collection of chapters dealing with the social, ecological and spatial processes of landscape dynamics. At the beginning of the 21st century society faces the joint consequences of improved access to resources, locations, and information. This change entered the political discussion with the buzzword "globalization". Given the major contemporary technological achievements such as telecommunication and information technology, genetic engineering, traffic and satellite technology, it is likely that the mutual dependence of human activities and services on specific locations is losing importance. Places might become interchangeable and "placelessness" of capital and people may become the rule. At the same time, alienation from the local environment, along with the growing virtual environment that is created by information technology will continue to increase, thereby encouraging people to seek identification with unique, real places. Processes of this kind can be observed all over the world. They have decisive effects on landscape development, zoning regulations, the establishment of large conservation areas and land management schemes.

Landscape research addresses these challenges, where however, existing paradigms and methods have to be extended to adapt to the needs of land managers, politicians and the public for a sustainable land development. In this book, researchers from various disciplines discuss emerging fields within the framework of the "driving forces" of both landscape research and landscape development. Rather than offering a comprehensive overview of all issues relevant to landscape research, the book addresses various contemporary "hot topics", emphasizing scientific, technological and societal trends in these fields.

Toulouse and Aberdeen, October 2006

Henri Décamps
Bärbel Tress
Gunther Tress

Contents

Preface

This book is written by a team of scientists who jointly address conceptual and methodological challenges of landscape research. It stems from a peer review of the Landscape Department at the Swiss Federal Institute of Forest, Snow and Landscape Research (WSL). This peer review, however, did not follow the traditional pathways. Rather than asking the peers to value institutional merits and financial schemes, they were asked to review a predecessor of the present book and to identify future potentials of landscape research. This led to a fruitful future-oriented meeting of peers and reviewees about challenges for landscape research.

We wish to thank the following colleagues for their comments during the review process and their contribution to specific chapters of the present book: Thomas C. Edwards, Utah State University, U.S.A.; Wolfgang Haber, Technische Universität München, Germany; Terry Hartig, Uppsala University, Sweden; Siegfried Heiler, Universität Konstanz, Germany; László Orlóci, University of Western Ontario, Canada; Joachim Saborowski, Universität Göttingen, Germany; Monica Turner, University of Wisconsin-Madison, U.S.A.; Allan Watt, Centre for Ecology and Hydrology, Banchory, UK; Robert Weibel, University of Zürich, Switzerland. In addition more that 25 anonymous reviewers helped in the peer review of the individual articles.

We would also like to thank the WSL staff Jacqueline Annen Gilgien and Ruth Landolt for design and layout and the WSL for financial support to cover part of the production costs. The support of Springer Verlag and the Series Editors of the Landscape Series is gratefully acknowledged.

Change and Transformation: A Synthesis

Felix Kienast*, Otto Wildi and Sucharita Ghosh

Swiss Federal Research Institute WSL, Zürcherstrasse 111, CH-8903 Birmensdorf, Switzerland
* Corresponding author: e-mail: felix.kienast@wsl.ch

Global biophysical and socio-economic changes and technological advances manifest themselves in changing land-use and altering landscape properties and functions. The industrial revolution in the nineteenth century was such an example, followed by the age of almost unlimited mobility starting in the twentieth century. Meanwhile, the last decade of the twentieth century has seen an astonishing development in information technology affecting almost every facet of the society. Easy and nearly unlimited access to computers, satellites and communications systems has also affected the way landscape research is done today. Data are obtained in massive amounts and data mining is now an issue. Also in the biological sciences, modern methods such as molecular genetics have revolutionized our understanding of ecology and evolution and how these interact with the environment. Following modern trends in science, landscape research has become computationally intensive, with strong theoretical components. Now, information is quantified, hypotheses are tested and scientific inference is formal.

Landscape research is an interdisciplinary science. It deals with complex environmental processes at multiple spatial and temporal scales. While the interdisciplinary nature and the focus on space-time processes are shared with other fields as well (cf. spatial ecology or bio-geochemistry), the subject of interest – "landscape" – is unique to landscape research. Popularly, "landscape" is understood as a portion of land or territory that the eye can capture at a single glance. Translated into scientific terms landscape can be considered a fraction of the globe's surface, that has been shaped by natural and human driving forces yielding specific qualities for the life of its inhabitants.

Some of the important events in the course of the history of landscape research include, the promotion of the terms landscape architecture in 1828 (Frederick Law Olmsted, designer of Central Park in the city of New York), landscape ecology in the 1930s, and the founding of the International Association for Landscape Ecology in the 1980s. Today, landscape research is the result of several evolutionary lines that are not contradictory but differ in emphasis. Two of these may be called "European" (which is also represented in the United States and elsewhere), and "American" (which is also common in Australia and Canada). They address different value systems and this is reflected in the diversity of coexisting definitions of landscape research and particularly of landscape ecology. Europeans, with their continent's long history of dense human inhabitation, traditionally envision the landscape to include a strong human component. The term landscape has Latin roots reaching back to the term "regio" which eventually evolved into the old German term "lantscaf". "Scaf" gave rise to the English term "shape" and the german term "schaben" or "schaffen". Thus "Landschaft" means land that was shaped by similar human land-use, and so it is not generally thought to be a natural area sensu stricto that is void of human influence. A detailed linguistic analysis can be found in Haber (2002)[1]. On the other hand, North Americans and Australians often view the landscape to be free of human influences, or else they consider such influences to be of less importance. Just

1 Haber W. 2002. Kulturlandschaft zwischen Bild und Wirklichkeit. Schweiz. Akademie der Geistes- und Sozialwissenschaften, Bern.

F. Kienast, O. Wildi & S. Ghosh (eds.), A Changing World. Challenges for Landscape Research, 1–4.

as the British ecologist Arthur Tansley (1935)[2] spoke of "anthropogenic" plant climaxes, while American ecologist Frederic Clements (1936)[3] focused on "natural" climaxes, so do Forman (American) and Naveh (Isreali) reflect these transatlantic differences in seeing human activities as either external or else integral to landscape research. Within Europe itself, the southern countries that refer to paysage, paesaggio, and paisaje – with etymological origins in pagus (village/town/ country) – seem to envisage an even stronger human presence in shaping land than the northern countries that refer to landshap, landschaft, landscape, and their derivatives. We also recognize that the worldwide spread of landscape-related research after 1980 challenges and complements the "European" and "American" paradigms of landscape research. We see emerging research centers in Asia, South America and Africa that voice their views about landscape values and will contribute to a wider understanding of space, place and changes in time.

This book has three sections that show new avenues for landscape research. These are (1) value systems and sociological aspects; (2) ecological observations, data management and ecological methods to identify processes such as migration and dispersal; and (3) concepts for landscape pattern recognition, statistical analysis of landscape and environmental time series data analysis and dynamic ecological modeling.

Value Systems – Major Drivers of Landscape Dynamics

Value systems determine which landscapes are worth preserving and which goods and services of landscapes shall be used or maintained. While values may form the core components of the most influential action theories, currently there is little empirical knowledge about the role of values in landscape research. This is the starting point of "Value systems: drivers of human-landscape interactions" by Buchecker *et al.* Based on two empirical studies, these authors discuss people-landscape interactions and highlight the potential of value-based landscape research.

A more practical perspective on how planning (e.g., biodiversity action plans) is driven by value systems is in "The role of value systems in biodiversity research" by Duelli *et al.* These authors suggest that a transparent discourse about value systems and corresponding indicators is needed. Rather than attempting to reconcile different value systems, their simultaneous relevance must be recognized while different indicator-sets are developed to account for the diverging objectives.

"The meaning of 'landscape' – an exegesis of Swiss government texts" by Longatti and Dalang presents a semantic analysis of the word "landscape" as it has occurred in a number of Swiss government documents over the last 40 years. The authors highlight how altering social value systems are mirrored in the altered use of the term landscape.

In "Space and place – two aspects of the human-landscape relationship", Hunziker *et al.* identify three recently developed concepts dealing with the human dimension of landscapes. First they elaborate on the concept of perceiving the physical space. In a second phase they compile theories dealing with landscape perceived as place. Finally they discuss the effect of landscapes on psychological restoration.

---

2 Tansley A.G. 1935. The Use and Abuse of Vegetational Concepts and Terms. Ecology 16: 286–289, 303–307.

3 Clements F.E. 1936. Nature and Structure of the Climax. The Journal of Ecology 24: 252–284.

Ecological Observations and Processes

Due to the rapid technological achievements in remote sensing, since the 1990s, a wealth of data on land cover characteristics over large geographical regions have become available. "Modern remote sensing for environmental monitoring of landscape states and trajectories" by Zimmermann *et al.* is an introduction to aspects of remote sensing that are relevant for landscape research. The emphasis of this article is on ecological applications rather than on data-processing. The authors demonstrate a wide range of possibilities for using such data, and show the benefits and the difficulties of combining remotely sensed data with field observations. In "A large-scale, long-term view on collecting and sharing landscape data", Lanz *et al.* discuss accessing data from widely distributed repositories based on open standards and illustrate the important role of metadata for long-term monitoring and data reliability.

Careful interpretation of past land use and land cover helps to reconstruct patterns and processes within historic landscapes. Historical considerations also contribute to public discussions about the past and the future of landscapes. This is presented in the article by Bürgi *et al.* titled "Using the past to understand the present land use and land cover".

Proxy data originating from tree rings provide information on longer term regional and large-scale climate history. In their paper, "On selected issues and challenges in dendroclimatology", Esper *et al.* discuss quantification of climatic signals retained in certain tree ring parameters, and low frequency variations in long-term temperature reconstruction.

Paradigms and theories play important roles in understanding ecological processes. A prominent example is the theory of "island biogeography", already well-known in landscape management, e.g., in reconnecting isolated habitat patches. However, most landscape theories still await confirmation with empirical data. Modern methods, e.g., molecular biology, or satellite imagery, have the potential to rigorously question these paradigms. Testing of paradigms with genetic methods is the concern of the two articles "Integrating population genetics with landscape ecology to infer spatio-temporal processes" by Holderegger *et al.* and "Landscape permeability: from individual dispersal to population persistence" by Suter *et al.* The article by Holderegger *et al.* sets the scene for an emerging field in landscape research: landscape genetics. These authors show how beneficial molecular techniques can be for analyzing migration pattern, dispersal and gene flow. Suter *et al.* use capercaillie (*Tetrao urogallus;* Aves; Tetraonidae) as an example, to illustrate how relating spatial population patterns to landscape structure is limited by the lack of empirical data, and how genetic analysis may help to understand dispersal patterns.

Spatial Pattern Recognition, Time Series Analysis and Dynamic Modeling

This section is about principles, models and methods for quantitative analysis of landscape data. It starts with the article "Identifying and quantifying landscape patterns in space and time" by Bolliger *et al.* This is an overview of various indicators to assess landscape patterns. "Essay on the study of the vegetation process" by Wildi and Orlóci is about governing principles in vegetation analysis. In this essay, the authors discuss nonlinearity, scales, randomness, and other notions such as chaos. To understand why such notions are relevant, consider a chaotic system. In some situations, even very simple deterministic dynamic systems may be chaotic, with a behavior so complex that it mimics randomness. Why is chaos an issue? It is important because it may hamper the prediction of the state of a system, an important concern of ecologists. An emerging conclusion from these two papers is that the analysis of complex landscape data requires highly specialized statistical methods.

A landscape may be viewed as the realization of a space-time stochastic process. In "Statistical analysis of landscape data: space-for-time, probability surfaces and discovering species", Ghosh and Wildi present novel methods for analyzing landscape data in three different contexts. They explain the hypothesis of space for time substitution, nonparametric probability and quantile surface estimation, and the role of self-similarity in extrapolating hyperbolic species-area relations. A second article with rigorous statistical treatment is "Memory, non-stationarity and trend: analysis of environmental time series". The authors Ghosh *et al.* discuss models for changing seasonality, long-memory or slowly decaying autocorrelations, deterministic trend versus stochastic trend-like behavior, non-stationary and non-Gaussian stochastic processes and introduce wavelets and nonparametric curve estimation. Long-term time series observations from a number of regions illustrate the methods.

Handling different scales simultaneously is a key skill for understanding and managing landscapes. This is the topic of "Model up-scaling in landscape research" by Lischke *et al.* It is an overview of up-scaling techniques and considers hierarchy theory as an ideal frame-work for successful up-scaling. Hierarchy theory leads to a general formulation of the up-scaling process, which consists of (a) aggregating source scale variables to target scale variables and (b) deriving the associated target scale model functions. Properly integrating space and time plays a crucial role in predictive modeling. This is shown in the second article by Lischke *et al.* titled "Dynamic spatio-temporal landscape models". The authors claim that modeling at the landscape scale is most effective with the new generation of dynamic regionalized and spatially linked spatio-temporal (SLST) models taking into account both local dynamics and spatial interactions. The authors discuss various SLST models for landscape research.

In conclusion we note an acceleration of progress in landscape research as the methods, the availability of data resources and the awareness of public interests are concerned. It may be a coincidence that this goes parallel to the observed accelerated change of the landscape due to the ongoing globalization of interactions as well as climate- and land use change. The society is expecting solutions to newly emerging problems. We are convinced that our joint contributions from natural and social sciences will be well received by the readers of this book.

Value Systems – Major Drivers of Landscape Dynamics

Value Systems: Drivers of Human-landscape Interactions

Matthias Buchecker*, Susanne Kianicka and Berit Junker

Swiss Federal Research Institute WSL, Zürcherstrasse 111, CH-8903 Birmensdorf, Switzerland
* Corresponding author: e-mail: matthias.buchecker@wsl.ch

Abstract
Value systems are generally acknowledged to be a constitutive element of human life and certainly play a fundamental role in human-landscape interactions. Whereas, however, values form main factors within the most influential action theories, there is only little empirical knowledge about the role of values in landscape research. Based on existing knowledge on the role and nature of people's value systems presented in a first part, the findings of two empirical studies on people-landscape interactions are re-interpreted on the background of two recent value concepts. The insights and difficulties encountered illustrate the potential of value-based landscape research and highlight challenges for future research.

Keywords: values, landscape, value concept, qualitative interviews, survey

F. Kienast, O. Wildi & S. Ghosh (eds.), A Changing World. Challenges for Landscape Research, 7–26.

The Relevance of Values for Landscape Research

What is the importance of human value systems in the context of landscape research? In accordance with the understanding of landscapes as the result of natural and human driving forces, this paper claims that people's values are a fundamental component of the so-called "human dimension" of landscape. Valuation is a constitutive part of human life and behaviour, as humans' actions and reactions are not determined by instincts, but rather subject to (contingent) decisions (Bechmann 1978). A human being without a clear framework to select among existing options to act or judge could probably survive in the wilderness, but he or she would not be able to establish stable relations to others and integrate in groups or a society.

In the context of human-landscape interactions, values are constitutive in two ways. They do not only shape humans' use of their land resources, which is a main driving force for landscape development. They are also a main formative source that humans perceive landscape not just as nature. Values and meanings assigned to landscape structures allow them to combine landscape elements to a consistent landscape picture (Simmel 1993). In return, the examination of the ways people perceive and interpret landscapes can reveal a great deal about their value systems (regarding the environment). Although the fundamental importance of values for landscapes has been recognised in both the scientific and practical context, there is still a lack of empirical knowledge about their role for landscape interactions (Bäuerle 1984; Joas 1997). The scientific analysis of values in landscape contexts bears great potential to deepen the understanding of human-landscape relationships and thus widen the perspective on many fundamental topics of landscape research and land-use, i.e.:

- Socio-economic developments, which represent a central driving force for landscape transformations, go along with alterations of values. Therefore, knowledge of values can contribute to a better understanding of the (human) conditions of such transformations.
- Conflicts with regard to landscape development and management often originate in the circumstance, that the landscape has distinct meanings for different stakeholders. Thus, the assessment and consideration of stakeholders' values can help to prevent conflicts and enhance consensus-finding processes.
- People's perception and valuation of landscapes is a main element of their regional attachment and commitment. Understanding the values assigned by people to their landscape would help to determine visions of landscape developments that are conducive to social well-being.

This paper approaches the interrelation between values and landscapes in several steps: First, a detailed introduction to existing value concepts and theories is given. A short insight into the existing definitions and the values' role within the most influential action theories is followed by the presentation of the most recent value concepts. Here, the focus is laid on the approaches of two authors: Schwartz and Taylor. A next section gives an overview over existing empirical studies on values regarding nature and environment. In the third section findings of two empirical studies on landscape perception and landscape management are analysed and discussed against the background of the value concepts of Schwartz and Taylor. Thereby, the usefulness of the value concepts is critically evaluated, and additionally the assumed situation-specific nature of values is examined in an illustrative sense. Finally, the conclusion summons up the central findings of the paper, addresses the advantages as well as deficiencies of the approaches used for the data analysis and highlights potentials for future research.

The Role and Nature of Values in Theory

The primary difficulty in value research is that although we can understand what the effect of values is, we cannot explain what they are in a psychological sense (Luhmann 1977). Furthermore, there are multiple and varied definitions that have been applied to the term value (Manfredo *et al.* 2004).

The term value has been defined in very different ways between, but also within, the different scientific disciplines. According to the classical economic theory, values were considered as characteristics inherent in goods, e.g. determined by the production costs. Later, they were understood as subjective judgements by economic agents of goods (Friedrichs 1968). Brown (1984) distinguished between values that are *assigned* through the process of evaluation and values that are *held* values as ideals of life.

In social psychology values are widely defined as cognitive controls of behaviour in the sense of “desired values” (Oerter 1970). A definition of values often used in sociology and anthropology was formulated by Kluckhohn (1962: pg. 395): “values are the desirable which influences the selection from available modes and means”. Especially anthropologists emphasise that value systems are specific for each culture (Kohl 1993) and form key mechanisms of collective identity. In the last decades a certain consensus seems to have emerged within the social sciences to view values as the criteria people use to choose between conflicting preferences and by which they justify actions and evaluate people and events (Bäuerle 1984; Schwartz 1992; Taylor 1989). Accordingly, the functions assigned to values by social scientists are mainly the reduction of complexity in human interactions and the facilitation of individual and collective orientation, but also the support of social integration and the promotion of human motivation.

The question of the role of values was raised and discussed at the beginning of the twentieth century by leading theorists such as Weber, Pareto, Durkheim or Simmel. They postulated the key function of values for human interactions – as a reaction to the utilitarian-oriented economical theories (based on Hobbe’s idea of individuals rationally pursuing their purposes).

The recognition of this “convergence” formed the fundament of Parsons’ action theory *(The structure of social action)* which had a formative influence on science and practice for decades despite all criticism (Joas 1997). According to this theory values represent the central element of a cultural system which in turn essentially influences social interaction as a mediating instance. Parsons sees values as cultural defaults which are on the one hand internalised through the individual’s socialisation and on the other hand become efficient in the form of social norms through institutionalisation. From his normative point of view he expects the culturally given values thus to become a moral authority as a prerequisite for a functioning (socially integrated) society. In order to maintain the humans’ contingency of decision, Parson introduced in his theory a second key element. In each situation of action, the individual has to chose between the binary alternatives of five (transcendental) pattern variables of value orientation: self vs. collective orientation, affective vs. affective neutrality, universalism vs. particularism, achievement vs. ascription and specifity vs. diffuseness (Habermas 1981). So the humans’ actions are based on the same value systems, but the actors (and the context) determine the situational limitations of their value systems (Fig. 1). Criticisms refer to the fact that Parsons sees the values as uniformly pre-determined for the whole society, so that in the ideal case societal conflicts are inexistent and a change of values is hardly possible (Joas 1992). Another shortcoming of his theory is the overestimated role of the driver “value” for human action. Thus, he ignored Freud’s insights that decisions are the result of a mediation between two conflicting instances: the drives (“Es”: needs and unconscious motivations) and the values (“Überich”: norms and ideals). Besides that, Freud’s concepts had nevertheless had a strong influence on Parsons.

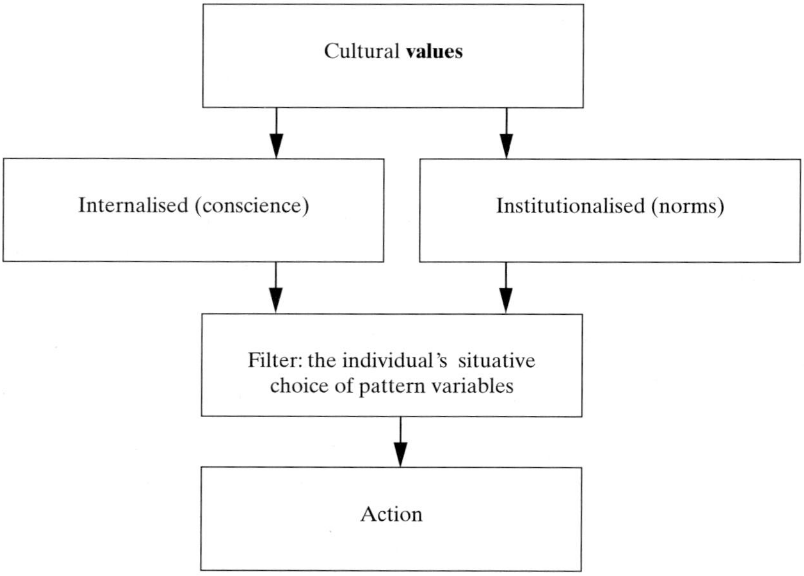

Fig. 1. The role of the values in Parsons' action theory (1953).

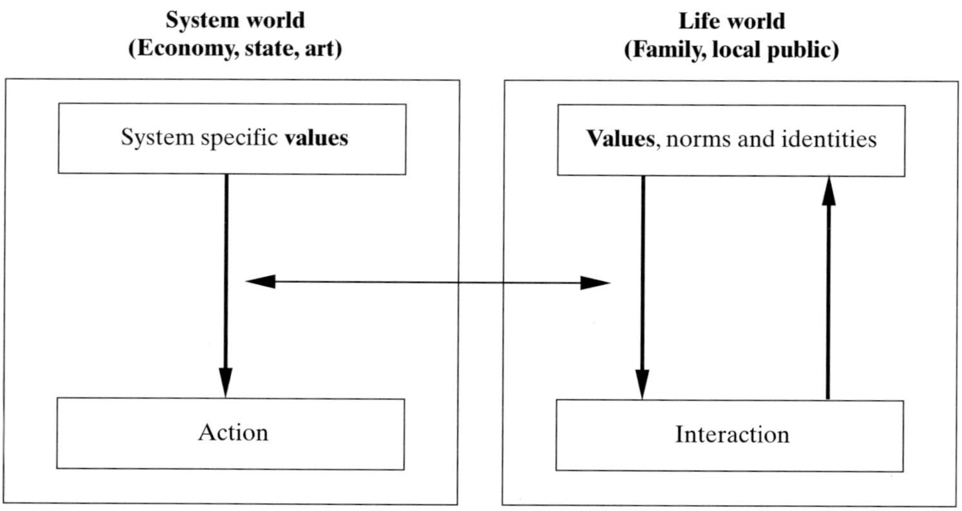

Fig. 2. The role of values according to Habermas' theory of communicative action (1981).

In an effort to overcome the weaknesses of Parsons' action theory, Habermas (1981) elaborated the *Theory of communicative action*. In order to reach his goal he introduced the (new) sphere of *social integration* (the *life world*) in his theory, in which values, norms and identities are reproduced by communicative actions. He contrasted it with the spheres of *functional integration* (system worlds: economy, policy, culture) in which system specific rationalities determine the instrumental actions (Fig. 2). Thereafter a differentiation is needed between people's actions as members of their social group or community (life world) and as agencies of functional roles (system world). In life-world situations, individuals' and groups' value systems influence their actions, whereas in system-world situations, the systems' value systems and thus mainly purpose oriented values determine people's decisions.

As an alternative to these functionalist theories the last decades have seen the emergence of action theories based on "constitutional theories" understanding societal processes as results of the society members' interactions (Joas 1992, pg. 336f). Thereby values are seen as temporarily valid societal defaults that offer the actors an orientation for their actions and, at the same time, are permanently being transformed by the actors through their interactions (e.g. Bourdieu 1979; Giddens 1984) (Fig. 3). According to these theories a value-pluralism does not necessarily threaten the societal order as social integration can also be reached through societal consensus building (Habermas 1981). Conversely, consensus building and (peaceful) societal exchange in general have the effect that the value systems of different groups co-evolve and become more universal – in terms of shared values and the scope of their validity (Mead 1934). This might not only apply to different social groups, but also to interest groups, i.e. members of interest groups might take non purpose-oriented values expressed by affected social groups into account in decision making processes if communication has taken place between these two groups.

A theorist who has dealt with the function and the emergence of values from a constitutional perspective is Charles Taylor (Joas 1997). According to Taylor (1989) values primarily serve as reference points for orientation in life. The *moral topography* offers the individuals a clear framework for action and movement and at the same time criteria to measure the success of their life. He suggests that we understand the values of a society as a (hierarchically and relationally arranged) moral space in which the individuals choose their specific

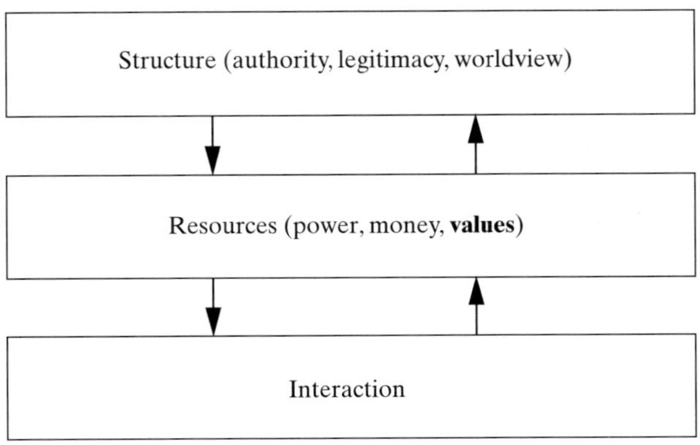

Fig. 3. The role of the values according to the action concepts of contitutional theorists, e.g. the theory of structuring by Giddens (1984): values are produced and reproduced by societal interactions.

Table 1. The ideas of a good life of the Western culture according to Taylor (1989). These ideals are assumed to be the shared basis of the western value systems.

Ideals	Historical origin	Attributes
Naturalism	Scientism	rational, logical
Utilitarianism	Enlightenment	independent, successful
Expressionism	Romanticism	authentic, expressive
Ideal of common life	Luther	responsible, dutiful
Asceticism	Augustin	altruistic, modest
Disengaged rationality	Platon	prudent, equanimous
Heroism	Early manhood	courageous, strong

position and thus establish an individual identity and a framework for action. The most prominent orientation points *(constitutive goods)* form the societies' ideas of "a good life", i.e. ideals of highly respected lifestyles. Such ideals and their attributes are seen as having evolved in the course of the cultural history of each society and sedimented in the society's shared pool of – partly unconscious – moral feelings. Because these feelings are articulated in art and everyday life, these values are constantly being modified – through adaptations to the society's innovations – and further differentiated. For the European culture Taylor mentions the following main ideals of a good life: the ethos of hero (warrior, statesman), detached rationality (wise man), asceticism (holy man), common life (family) and expression (artist) (Tab. 1). According to Taylor, all members of Western societies are oriented to all these ideals, but with varying distance and intensity which together constitute person-specific hierarchies of their value systems. Besides those genuine ideals of a good life, he considers newer ideals of naturalism (determination) and utilitarianism (or value-relativism) as pseudo values behind which value crises might be concealed.

Social psychologists have also developed sophisticated value concepts (e.g. Kellert 1996; Rokeach 1973; Schwartz 1992) that have reached widespread recognition (Manfredo *et al.* 2004). In the most recent of these concepts, Schwartz emphasises – like Taylor – the central role of values for human interactions, in particular for selecting and justifying actions. However, in contrast to Taylor, he deduces the values in a deterministic sense from three universal human requirements: needs of individuals as biological organisms, requisites of coordinated social interaction and welfare-needs of social groups. The value categories he has developed on that basis, however, build on findings of normatively oriented authors such as Parsons and Kluckhohn. Schwartz's concept differentiates between 10 main value categories, called motivational value types, which are arranged around the two bipolar value dimensions "conservation vs. openness to change" and "self-enhancement vs. self-transcendence" (Fig. 4). On the basis of empirical investigations in 20 countries he could corroborate that these value dimensions and categories are universally relevant, while their weights proved to be culturally and individually specific. According to Schwartz, value structures develop along with societal changes, but they are nevertheless expected to remain anchored within the two universal value dimensions.

There are yet only few data-based studies that reveal trends in value shifts (Manfredo *et al.* 2004). Longitudinal studies in Germany have confirmed that in the last decades a value shift towards more openness and more self-enhancement has taken place (Klages 1999). Polling data from North America show a rapid increase in pro-environmental attitudes from the mid-1960s to the early 1970s, a decline in the 1980s, and renewed growth in the 1990s (Dunlap 2002).

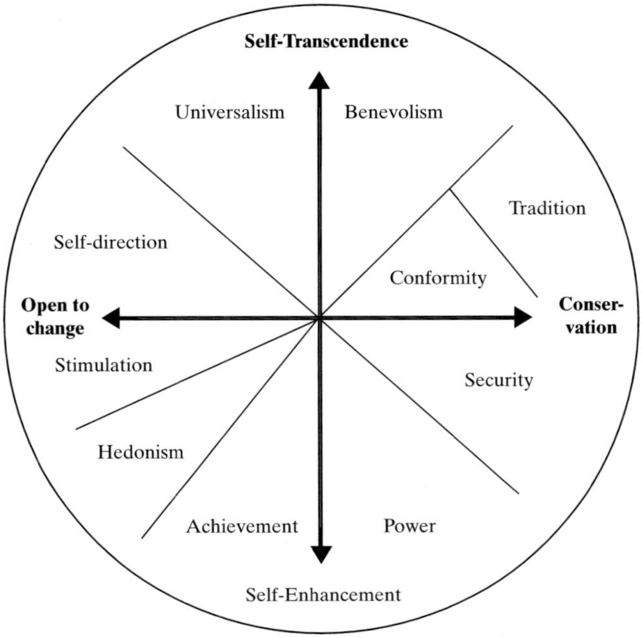

Fig. 4. Theoretical model of relations among motivational types of values, and bipolar value dimensions (Schwartz 1992).

To sum up, there seems to be a consensus among theorists that value structures are universally pre-disposed by the human nature, shaped by the cultural history and open to further development. However, the question is contended whether values have a situation-specific relevance. Whereas Habermas (1981) differentiates in his theory of communicative action between steadily reproduced values valid in life world situations and system-specific values in system world situations, Giddens (1984) and other "constitution" theorists (Joas 1996) see value exchanges through all situations as part of a societal process.

As landscape development takes place in an area of overlap between life world (social meaning) and system world (land-use), it is crucial to understand both the affected people's value systems and their situation specific relevance. In order to illustrate and explore this, we will focus in the empirical part of this chapter on two questions considered in two case studies:

1. How do the values of insiders (own life world) and outsiders (other life world) of an area differ?
2. Which differences in value orientation can be found between representatives of local interest groups (system world) and local residents (life world)?

These questions will be approached by confronting empirical material with the value concepts by Schwartz and Taylor. Before that, we will give a short overview of research findings on values in the context of landscape perception and land management.

Value Systems in Empirical Landscape Research

Although values can be expected to be highly relevant in landscape perception and land management, only few empirical studies focussing on their role exist (Bäuerle 1984; Meyer and Buchecker 2005). This is probably due to the fact that research on people's values in connection with objects is far more complex than investigating people's personal value-orientations in their lives. It is assumed that people's value-orientations also guide the formation of their values regarding parts of the environment (Schultz and Zelezny 1999), but people ascribe values to concrete objects in a creative way during interactions (Touraine 1985). This has methodical consequences: Values held by individuals or groups can be (quantitatively) assessed and weighted on the basis of existing concepts or frameworks of (hypothetical) value categories. Values ascribed to landscape elements and landscape developments, however, can only be (indirectly) interpreted by the qualitative analysis of legal text documents (laws, communal guidelines) or data gained from in-depth interviews because these values are in general not conscious to people.

Systematic qualitative studies that focus on people's values oriented toward landscape developments have not yet been done. A larger number of articles can be found that combine the terms "value" and "landscape" in their titles; without exception however, they do not explicitly address people's deeper values, but rather more superficial attitudes. Some of these studies implicitly give us information about the underlying motives of people's attitudes regarding landscapes (Buchecker 2005; Willis and Campbell 2004; Pursell 1992; Sell and Zube 1986). One of the most substantial contributions in this sense of focusing on people's values expressed in combination with the perception of landscape change was made by Hunziker (1995). He interviewed local residents, tourists and experts on their perception of reafforested areas in an Alpine valley. Thereby he found that the interviewees of all origins judged the different reafforestation scenarios by referring to the same four (value) categories labelled as "tradition", "nature conservation", "profit" and "emotions". This corresponds with the assumption adopted by the value concepts of Schwartz and Taylor that members of the same culture principally share and use the same values. The interviewees, however, differed in the weighing of these value categories and thus judged the development differently. Although general group differences between locals and non-locals could be determined, all interviewees showed ambivalent attitudes towards reafforestation. A subsequent quantitative study on the same topic (Hunziker and Kienast 1999) confirmed that the non-locals put stronger emphasis on the value categories profit and emotion and exhibited higher preference for spontaneously reafforested areas, whereas the locals' value hierarchy with tradition being at the top brought about their preference for open areas.

Bäuerle (1984) used another approach to identify the values relevant for people to judge their local environment. He investigated the residents' leisure-oriented requirements concerning their outdoor area (which he considered as indicators of individually held values) by means of a standardised questionnaire and put the results in relation with requirements for outdoor areas represented in laws and public regulations (formalised value consensus). He found a considerable gap between actually held values and the formerly found value consensus and interpreted this as an expression of a change of values. However, he did not consider the relations of power and influence in the community, which could have contributed to the explanation of this gap. He concluded that therefore spatial planning should be accompanied by the research of spatially oriented values.

Another (indirect) way to assess the role of values for people's landscape interactions could be the measurement of correlations between people's value orientations and their behaviour, attitudes or perceptions. Astonishingly, systematic studies on the (inter-)relations between people's value orientations and their landscape perception do not seem to exist yet.

Some research has been done to measure the interaction between people's value orientations and their (pro-environmental) behaviour, whereby Schwartz's value concept was often used. Karp (1996) for example found that "self-transcendent" and "openness to change" values are strong predictors for pro-environmental behaviour. Other authors using a similar approach have found a limited significance of values as direct predictors of behaviour (Thogersen and Grunert-Beckmann, 1997). According to an empirical study by Corraliza and Berenguer (2000), the predictive power of values for environmental behaviour is only strong in cases of consistency with situational variables (facilitation of behaviour). Norton and Hannon (1997) propose a place-based approach to environmental values hypothesising that the values' behavioural relevance depends on people's relation to a place. Finally, Papadakis (2000) found a strong link between idealistic values ("ökozentriert", i.e. focussing on ecological issues or pro-development) and political predispositions relating to the environment and a weak link with utilitarian values. In the most recognized psychological theory of behaviour, the theory of planned behaviour by Fischbein and Ajzen (1975), values are explicitly included only in the form of the factor "perceived social norms" and do not appear as a specific factor. So it is not astonishing that research findings on the role of values for human behaviour in a more encompassing sense, e.g. landscape relevant behaviour, are not yet available.

To sum up, there seems to be a shared basis of cultural values at least within Western societies, even though the individuals differ in the weighing of the value categories respectively the hierarchy of their value system. These value systems seem to have a (driving or regulating) influence on landscape perception and ecological behaviour. There is, however, a lack of empirical research examining if, in what form and to what extent the context of actions has an influence on the individuals' value-systems (value hierarchies) and their effect on people's actions – as proposed by theorists such as Weber, Parsons, Habermas and Joas.

This issue will be considered below by discussing the results of two empirical studies on people' perceptions of river revitalisation (Study 1) and changes of alpine landscapes (Study 2). There the focus will be on the differences between life-world and system-world situations (the clash that is seen as particularly relevant for landscape conflicts) by investigating these differences on the one hand by comparing people's value-references in different role situations (local residents vs. members of regional interest groups) and on the other hand people's value-references in different place-relations (insiders vs. outsiders).

Study 1: Value System Differences between Interest Groups and the Wider Public

Identifying the residents' attitudes towards river revitalisation

The objective of this project was to identify the local residents' expectations and attitudes towards a planned river revitalisation project of the Thur in Northern Switzerland. For this purpose, data were gathered by conducting qualitative interviews and distributing standardised questionnaires in two neighbouring communities, the urban community of Weinfelden and the (rural) village of Bürglen. Thereby, samples of two target groups were addressed: the representatives of regional interest groups directly involved into the decision making process (N = 24) and the wider public of the two communities (N = 240). Here, we will focus on the two questions of the standardised questionnaire which address a) the meanings people associate with the river Thur and b) the aspects people wish to be improved by the river revitalisation. To answer these questions, respondents were offered a set of answer options each of which they had to rate on a five-point scale. By comparing the two groups' mean ratings of the answer options, the differences of these groups in terms of their perception of the river and their preferences for the river management could be revealed. In order to explore in which sense the "life world" and the "system world" group differed in the values they referred to concerning river perception and river management decisions, the answer options were related to the value categories of the value concepts by Schwartz and Taylor.

The meaning associated with the river Thur

The representatives of the regional interest groups rated the meaning "economic use" clearly higher than the wider public. This was to a less extent also true for the meanings "danger", "achievement", "dynamic", "spectacle" and "source of life" (Fig. 5). Very similar, however, were the ratings both groups attributed to the meanings "part of myself" and "my (collective) home". The meanings "nature" and "recreation area" which received the highest ratings by both groups were much more favoured by the wider public than by the members

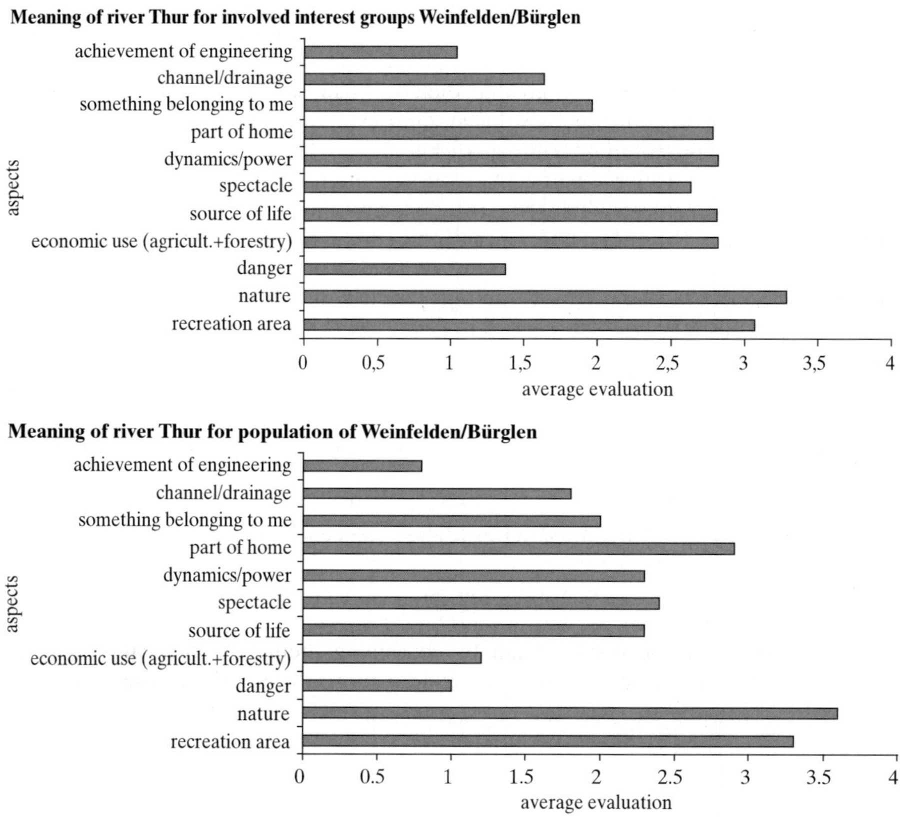

Fig. 5. The meaning of the river Thur as expressed by representants of the regional interest groups directly involved in the decision making process (upper graph) and by the local public (lower graph) (Junker and Buchecker in press).

of interest groups. These results can be interpreted as follows: For the latter the river appears to be at the same time a utility good and a place of private retreat which however has to be achieved by subjugating nature ("danger", "achievement"). For the wider public the river is almost exclusively a place of private retreat and security which should be conserved. Interestingly, however, for both groups the collective reference to the river is stronger than the individual one ("my (collective) home" > "part of myself").

When we consider the ratings of the two groups by drawing on the value concepts by Schwartz and Taylor, we come to the following interpretations: The high rating of the river meanings "naturalness" and "recreation" are strong indicators that the value categories **universalism** ("naturalness") and **self-direction** ("recreation", also "part of mine") of Schwartz's value concept seem to be important for both groups' judgements. The meaning "economic use" highly rated by the interest group can best be assigned to the value categories **security** and **power**, which are to some degree also relevant for the wider pubic ("part of home", "danger "). In any case, the value-orientation **self-enhancement** proved to be least relevant for both groups. According to Taylor's terminology, expressive ("naturalness", "part of me"), common life ("my home") and utilitarian values ("economic use", "recreation", "achievement") stand in the foreground for both groups, whereas older value orientations (hero ethics, disengaged rationality, asceticism) seem to be irrelevant. So according to both value concepts, both groups seem to refer to more or less the same value categories. The two groups differ in the weighing of these values, but we see no principal differences in the role of their value systems, for also the interest group referred to non-purpose-oriented values. Here we have to point to the methodical limitations of these interpretations: As the members of both groups could only choose from a given set of answers – a general feature in standardised surveys – we do not know about any group differences of values that were not presented in the answer options.

Aspects to be improved by river revitalisation

To answer this question, the respondents had to rate the importance of a given set of aspects relevant to revitalisation (Fig. 6). Here the two groups agreed only in the rating of one answer option, the importance of "flood protection" which they both rated astonishingly low. The options "water quality" and "groundwater" were ranked slightly higher by the members of the interest groups. Large differences were found for the options "recreation" and especially "naturalness", much more preferred by the wider public, and the aspect "agriculture", which according to the wider public should be regarded as less important than according to the interest groups' view. Interestingly, the opposite tendency could be observed with the aspect forestry, which is probably seen by both groups as close to nature conservation and thus less profit-oriented. In contrast to the perception of the river's meaning, here the members of the interest groups seem to react predominantly in their role as purpose-oriented users.

When we consider the results again by combining them with the value concepts of Schwartz and Taylor, we can interpret them in a more differentiated way. According to Schwartz's value concept the members of interest groups mainly referred to the value category security ("water quality", "groundwater", "agriculture"), whereas the wider public favoured the categories **self-direction** ("recreation", "leisure") and **universalism** ("naturalness", "forestry"). This seems to confirm the principal differences of the two groups in valuing landscape measures: they mainly relate to values of different value-orientations (self-transcendence vs. conservation). Interestingly, however, from this perspective the members of the interest groups appear to be mainly oriented towards the collective interests,

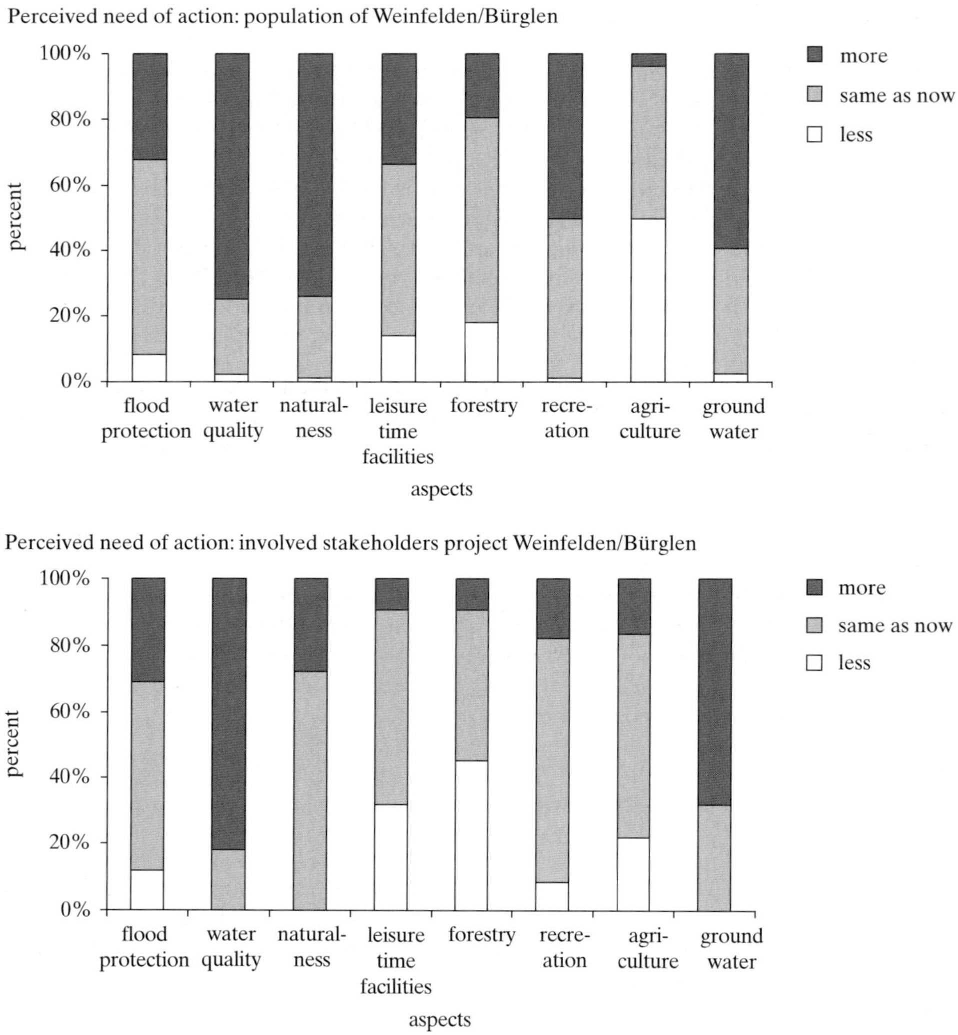

Fig. 6. The importance of aspects to be improved by the river revitalisation of the river Thur ranked by the representants of the regional interest groups directly involved in the decision making process and by the local public (Junker and Buchecker in press).

whereas the wider public is more concerned about their personal needs. Interpretations point to a similar direction when we apply Taylors value concept: Most of the aspects favoured by the interest groups can be attributed to the ideal "common life" in Taylor's terminology, whereas the ones most preferred by the wider public rather belong to the ideal of "expression" and "utilitarianism". So the members of the interest group do not relate to purely purpose-oriented values typical for system-world situations, but rather to traditional (collective) lifeworld-oriented values.

Local landscape projects and values

The members of the interest groups directly involved in the decision making process of the river revitalisation project and the wider public showed much more similar ratings when asked about the river's meanings than when their favoured objectives of the project were addressed. These results suggested that the members of interest groups seem to perceive existing landscape elements more in their role as life-world-oriented residents, whereas they react to imminent landscape changes more as economy-oriented users. These vague interpretations could be sharpened considerably by re-analysing the groups' ratings on the background of the value concepts by Schwartz and Taylor. The use of both value concepts confirmed that both groups refer to the same value categories when they perceive the river whereas they show distinctly different value-orientations when they have to judge the importance of project objectives. Interestingly, however, the interest group do not shut out their life-world values and only admit their purpose-oriented values, but rather (over-) activate more existential values. The wider public, in return, focus more than expected on their individual needs, neglecting to a certain degree the needs of the whole community such as water quality or flood protection. So the observed differences between the two groups concerning the valuation of project objectives cannot be explained with the antagonism between a typical life-world and system-world behaviour, but has rather to be ascribed to a differentiation between a modern life-world view (individual well-being) adopted by the wider public and a traditional life-world view (collective well-being) by the interest groups. As the study took place in a rather rural context (although a small city was included) these interpretations certainly have to be limited to rural contexts.

Study 2: Value Differences between Locals and Tourists

Perceived qualities of Alpine places

The objective of this study was to examine how locals and tourists in Alvaneu, an Alpine community in Eastern Switzerland, perceive the landscape and it's development. The inquiry took the form of a qualitative interview study within which two groups of "theoretically" sampled (Glaser and Strauss 1967) persons were consulted: representatives of the main residential groups and representatives of the region's diverse tourist segments.

The content analysis (Mayring 2000) of the interviews revealed that locals and tourists seem to ascribe quasi the same qualities to the place. Fundamental for both groups were: *quietness, secluded and simultaneously central location, intact landscape, typical architecture, cultural heritage, clear village structure, intact village community, special geographical features, diverse nature, and the mountains.* Furthermore, the analysis of the significance of these fundamental place qualities for each group showed, that for the locals the village community comes on top, whereas for the tourists the physical landscape and village features are of much higher importance.

However, an in-depth analysis of the values standing behind the interviewees' statements revealed basic differences between the ideals of the locals and tourists. For the identification of their different value systems and hierarchies the motivational types suggested by Schwartz and Taylor's pool of (culture specific) "ideas of a good" life prove to be a very useful – if not indispensable – tool. Below, value differences will be illustrated by focussing on two qualities of the place, i.e. (1) *intact landscape and diverse nature* and (2) *view of the place and cultural heritage.*

Intact landscape and diverse nature

The majority of representatives of both groups characterised Alvaneu as a "unique" place. They justified this with the place's intact landscape and richness of natural features. However, the attribute "intact" turned out to be used ambiguously, on the one hand referring to the well-kept landscape and on the other hand to wild and untouched parts. Still, the values found behind the perception of landscape and nature of locals and tourists were incongruous:

The locals seem to be very proud of their "pristine place", as they call it. With the emphasis on the intact and diverse qualities of the place the locals primarily tried to distinguish their place – and thus themselves – from rather urban places and those cultures. Their positive view of the place's uniqueness was not even overshadowed by its deficiencies in infrastructure and comparably difficult economic situation. Thus, the unspoilt character of their place in comparison to other (urban) places was valued higher than qualities related to the means of existence: To have "more green space" and an "undamaged nature" is more important, than to have all the infrastructure that is available in cities. Also, they are convinced that their place is attractive for recreation seekers (outsiders). According to this, these qualities are closely linked to the local's feelings of social belonging and their view of the character of their own group and thus their group identity. According to Schwartz's motivational types these qualities address **security** (intact environment) and **achievement** (pristine place), however not on an individual level, but in a collective sense as a social group. From Taylor's culture-historical perspective the assessed need for distinction can be related to the centuries of resistance against urban expansion in rural areas. It can also be seen as resulting from the locals' adoption of the widespread nostalgic (urban) ideal of unspoilt Alpine nature and culture, which the locals needed in a self-reflexive sense to positively distinguish their social identity from the economically superior urbanized parts of the country.

In the case of the tourists, the attribution of the characteristics intactness and diversity also served as a means of distinction. But their fascination of the place primarily emanates from the numerous possibilities for individual recreation and adventure there, in contrast to the situation in their everyday world. They particularly like the fact that they can "just step out into nature" and "feel close to nature". Also important to them is the fact that the landscape appears to be partly "untouched by humans" und thus "still wild". They appreciate the high (ecological) variety and possibilities of unrestricted exploration and discovery that the local landscape offers. Thus, they named these characteristics to distinguish their stay in the place on the one hand from the hustle and bustle and on the other hand from the monotony of their everyday lives. Therefore, their statements reveal particular individual needs related to deficits in their everyday life world. Referring to Schwartz' concept (see Fig. 4), the value types **self-direction** (unrestricted exploration of the environment), **universalism** (closeness to nature) and **stimulation** (ecological variety, possibilities of discovery) stand here in the foreground. Viewed from Taylor's culture-historical perspective (Tab. 1), these needs can be attributed to the outsiders' utilitarian ideals (landscape as object of utility for outdoor activities) but also to romantic-expressive ideals (outdoor activities as a way to achieve a unity with nature). Furthermore, naturalistic ideals (ecological variety) are contained.

Thus, the locals address markers of their group identity with these characteristics, while the tourists primarily refer to their personal compensation needs.

View of the place (local scenery) and cultural heritage

Both, the local people as well as the tourists for the most part really enjoyed "the village as such". Decisive for this liking was the fact that the village still has a traditional structure, with an old centre and a couple of "typical" elements of cultural heritage (e.g. old farmhouses, stables, wells, baking and washing houses *[pastregls]* and a church as well as chapel). But again, an in-depth analysis revealed value differences between the two groups:

For the locals it is important that built landscape elements "suit the place" in the sense that they fulfil a particular purpose. Thereby, the function of an element mostly seems to outweigh its appearance. The purpose of a building is judged on the basis of how well it serves the place and the local community as a whole. The appearance in turn, is determined by the local building regulations, which in Alvaneu demand a steep roof, particular size, orientation, and material (timber and stone). Furthermore essential for the locals is that appearance and function of built landscape elements correspond with each other. Alterations of old buildings and the construction of new buildings are, however, not considered as a threat, but as absolutely adequate and necessary innovations. Such changes demonstrate that the place undergoes a developmental process and does not stagnate. Still, elements of (built) local cultural heritage are generally appreciated, but only if they are well kept and still used. If not, they represent decay, which is completely undesirable for the two locals. However, generally most important is, that the beautiful view of the place is not being ruined through any kind of developments of the built landscape. The locals' emphasis on the settlement structure and cultural heritage can be brought in connection with Schwartz's value-types **security** (not ruined view of the place), **conformity** (correspondence of appearance and function) and **tradition** (continuity of the local architectural style, preservation of cultural heritage). However, Schwartz's value concept does not adequately explain the purpose-orientation of the locals; value categories such as functional (between hedonistic and achievement) and autonomy (between security and power) seem to be lacking in this concept. From Taylor's point of view these characteristics relate to the ideal of everyday life, particularly in a collective sense, i.e. a village successfully struggling for its existence. Built elements are viewed as positive, if they express a strong social cohesion (observed rules, uniform style, shared past) crucial for survival and indicate a prosperous (self-determined) development (conservation of cultural heritage, local innovations).

In contrast, the tourists are mostly very strict in their judgement on whether built elements suit a place or not. In their descriptions of the place, tourists predominantly spoke about nostalgic motives and thereby expressed demands like "a traditional village structure" with "a clear centre" and "a church in the centre". Thus, for them it is predominantly the appearance of an element that matters. For example, many of them consider new large chalets to be disturbing, and some of them even generally dislike new buildings. The tourists particularly appreciate those built elements, which correspond with their image of an authentic mountain village. As far as the function is concerned, they generally approve of buildings if they are directly of personal use to them – or, if they at least do not get in the way of their needs. Therefore, the statements of the tourists about the "building structure" and "cultural heritage" also reveal Schwartz's value types **security** (clear village structure) and **tradition** (elements of authentic Alpine culture) however, in a much more abstract sense, i.e. closer to the value type **universalism**. The widespread desire among tourists for nostalgic features is also expressed in a nearly religious desire for a clear order and thus more sense of life. According to Taylor's approach, the tourist's view can be ascribed to the romantic ideal which is characterised on the one hand by being against progress (nostalgia) and on the other hand by its reverence to self-expression (authenticity).

To sum up, the locals address life-world ideals in a very basic sense, whereas the tourists reveal idealistic and ideological values and are looking for a good life far beyond their everyday life.

Tourist places and values

For the most part the residents and tourists of Alvaneu share their view about the most important qualities of the place. Consequently, their conflict potential regarding the development of the place seems rather marginal at first sight. But even though most of the named qualities are connected, the analysis revealed considerable differences in the values between these judgements of insiders and outsiders. According to Schwartz's value concept, for the local residents the value types **tradition, benevolence** and **security** stand in the foreground. For the tourists the value types **self-direction** and **universalism** are the most significant ones for their valuation of their tourist place. The value types **stimulation** and **hedonism** (individual recreation and pleasure) are only addressed by tourists – even if they are only of marginal importance. To summarize, the large basis of shared values originating from the same cultural sources indicates a good potential for a mutual understanding. The strong position of values, which could be labelled as "conservative" in the locals' value systems apparently stands in opposition to the values signifying self-transcendence (self-direction, universalism) prevailing with tourists. From this perspective, the potential of conflicts between locals and tourists appears to be considerable.

From the point of view of Taylor's value concept the values shaping the tourists' appreciation of the place can mainly be assigned to romantic ideals and to a lesser degree to utilitarian and rational ideals. For the locals' valuations of their place, Taylor's value categories do not match very well. Certainly his most appropriate category for the locals' judgements is the ideal of the common life. But the locals' ideals of a good life seem to incorporate characteristics not necessarily associated with that ideal such as conformity and social coherence which originate from historical experiences of rural culture. It seems that Taylor's value hierarchy is specific to an urban society and not appropriate for an examination of a rural community, a well known problem which goes back to the distinction between society *(Gesellschaft)* and community *(Gemeinschaft)* made by Tönnies in 1887 (1988). From a rural cultural-historical perspective the locals' values could be related to the feudal system in which the village community was collectively responsible for delivering the taxes, so that mutual control (conformity) as well as mutual help (coherence) was existential. Pfister (1997), however, argues that conservative circles artificially generated the ideal of rural harmony in early modernity in order to promote their anti-socialist ideology. This raises the question whether the rural ideals have developed as a counter-culture to the urban culture, or whether they are part of an urban (anti-progressive) culture integrated into the rural culture.

But also this culture-historical perspective based on Taylor's value concept reveals, that the criteria seemingly shared by locals and tourists can be traced back to ideals of completely different origin. These ascertained differences between the two groups' value systems are likely to raise to the surface in discussions about the future landscape development and contain quite a potential for conflicts of interests.

Insights into the Role of Value Systems for Human-landscape Interactions

Within the theories of value research there is a consensus that:

- the values serve people as criteria in order to choose between alternatives/opportunities of action as well as to judge persons and events;

- the value systems of people of the same cultural background include the same value categories;
- the weighing of the values respectively their hierarchical order in the value systems is group-specific as well as individually different and – thus (values are) a constitutive element of individual and group identities;
- the societal priorities of values and the group of shared values (also new values) can alter in the course of societal changes and developments;
- the value systems of persons can turn out to be distinct in different (role) situations.

The analysis of the empirical data illustrated in more detail in which way different (role) situations (of people) influence the values underlying the perception of landscapes as well as the attitudes towards landscape changes. It became apparent that local residents in situations of personal economic interest (system-world situation) and local residents in situations of pure residential interests (life-world situation) in fact share most of the values in their value systems. In the rating of the meanings attributed to the river landscape by these two groups, both groups related to the same value categories. And among the aspects to be improved by the river revitalisation, the members of the interest groups proved to be even more concerned about collective needs and security than the other residents. So, the content of people's value systems does not alter in different situations; the only thing that changes is the hierarchical positions of particular values. Members of interest groups do not base their land-use decisions purely on purpose-oriented values; though at least in rural contexts they might prefer traditional (collective-oriented) values to modern (individual-oriented) values.

The picture that emerged from the comparison between people in residential situations (life world) and people in tourist situations (system world) is slightly different: Both groups seem to value the landscape on the basis of nearly the same criteria, often even using the same wording in their judgements. But a closer examination revealed that locals and tourists associated the same criteria with different meanings and with values that partly even belong to opposite value orientations (Fig. 5). These differences are certainly related to the fact that for the tourists the landscape has primarily a recreational function. In contrast, the function of social integration, which the landscape typically has for people in life-world situations is supposed to be only marginally important to them. From this point of view the broad congruence of the criteria underlying the two groups' perceptions of the landscape is really surprising. Strikingly, social integration seems to be to some extent an issue for the tourists whose actual socio-cultural groups exist far from the places where they seek recreation, i.e. they also seem to be partly in a life-world situation, although an imagined one (nostalgia, no real integration intended). The differences in the value orientations and value systems relevant for locals' and tourists' perceptions of Alpine landscapes are partly due to the different character of their need for integration, but partly also due to the different socio-cultural background, i.e. the differences are caused by a mixture of situational and cultural factors.

According to the societal theories by Weber, Parsons and Habermas, the value systems of individuals are not necessarily in all situations drivers or rather regulators of their interaction with the environment. For example in affective or existential situations, people's reactions often contradict their value systems. Most of the landscape-relevant interactions, however, take place in contexts were the value systems are supposed to be highly relevant and thus represent so-called purpose or value oriented actions. As our case studies illustrated, value systems are in fact powerful drivers for landscape perception. Purpose-oriented values (related to the economic use of the landscape) not belonging to the value system in Taylor's sense also play an important role, although not a dominant one. In situations, however, where decisions for landscape management are made, these "non-genuine" utilitarian values become more dominant, at least among the interest groups. The wider public does not modify its value system for that situation in that way. But whereas the wider public's value systems of today tend to neglect conservative and more collective-oriented values, the interest groups at least in rural contexts mainly focus in such situations on collective-oriented utilitarian values.

To sum up, these findings suggest that the inclusion of all interest groups as well as the wider public in decision-making processes concerning landscape development is necessary. For this guarantees that all values at stake will be taken into account. Preferably, the negotiation process should focus on the participants' values (not their interests or claims). Because at least interest groups and the wider public within a local area have the same value systems, the shared values promise to provide a good basis for consensus building. This is to a certain extent also true for landscape conflicts between outsider and insider groups, e.g. in tourist areas. As shown above, the seemingly similar landscape criteria used by these groups can differ here considerably in its underlying values. Thus, in the context of consensus building it is particularly important to put much emphasis on the identification and articulation of the basic values relevant for these groups in order to avoid a consensus purely on the level of words. The public articulation of shared or conflicting values is according to Taylor (1989) (in general) valuable. This is an aspect of communication that is often neglected, although this is an opportunity for a society or a group to renew and confirm its value orientations.

Implications for a Future Value-based Landscape Research

Research on the impact of individuals' and groups' value systems on landscape perception and landscape management has so far been neglected (Meier and Buchecker 2005). A main reason for this is the difficulty that people's values and value systems are not directly accessible as they are to a far extent unconscious or at least not discursively present to them. For the examination of value systems, theoretical concepts suggesting value categories are needed. Such concepts have been elaborated in the last decades (e.g. by Rokeach, Schwartz, Taylor), but they have not yet been systematically tested and reflected for their capacity to help identifying the value systems of individuals or groups in landscape relevant situations. In this chapter we have tried to apply two of these concepts to empirical data gathered in two case studies. The results show that a) both concepts can contribute in a complementary way to identify the values behind people's landscape perceptions, and b) that not all value-orientations are covered by these concepts.

In the value spectrum proposed by Schwartz there is a lack of value categories covering utilitarian (purpose-oriented) and romantic (expressive) ideals. Taylor's value concept – which at least claims validity for the Western culture – in turn, does not contain value categories typical for rural ideals such as tradition and conformity. In general, the Schwartz's conservative value-orientation **(tradition, conformity, security)**, which is also highly ranked in Western culture, is not represented in Taylor's. Thus, both concepts require an extension, but offer a good basis for mutual completion: Taylor's approach is more adequate to openly confront the question of societal value orientations, whereas the approach of Schwartz certainly offers the advantage of better operationalization.

Theoretical and empirical research is needed to develop and test a really comprehensive framework qualified to identify and map people's (situational) value systems. On the basis of such frameworks, adequate methodical approaches for qualitative and quantitative research of the values' role in people-landscape interactions will have to be developed. A better understanding of this complex interrelation can only be achieved with an improved and internationally shared methodology.

References

Bäuerle G. 1984. Der Freiraum als räumliches Wertobjekt: Ansatz zu einer räumlichen Werttheorie. München: Minerva-Publikation.

Bechmann A. 1978. Nutzwertanalyse, Bewertungstheorie und Planung. Bern.

Bourdieu P. 1979. Entwurf einer Theorie der Praxis. Frankfurt: Suhrkamp Verlag.

Brown T.C. 1984. The concept of value in resource allocation. Land Economics, 60 (3): 231–246.

Buchecker M. 2005. Public place as a resource of social interaction. In: Turner P. and Davenport E. (eds.). Spaces, Spaciality and Technology. The Kluwer International Series on Computer Supported Cooperative Work, Vol. 5., Heidelberg: Springer-Verlag, pp. 79–96.

Corraliza J.A. and Berenguer J. 2000. Environmental values, beliefs, and actions – A situational approach. Environment and Behavior 32 (6): 832–848.

Dunlap R.E. 2002. An enduring concern: Light stays green for environmental protection. Public Perspective 13: 10–14.

Fishbein M. and Ajzen L. 1975. Belief, attitude, intention an behaviour: An Introduction to the theory and research. Addison Wesley.

Friedrichs J. 1968. Werte und soziales Handeln. Ein Beitrag zur soziologischen Theorie. In: Veröffentlichungen der Akademie für Wirtschaft und Politik. Hamburg.

Giddens A. 1984. The constitution of society: outline of the theory of structuration. Cambridge: Polity Press.

Glaser B.G. and Strauss A.L. 1967. The discovery of grounded theory: strategies for qualitative research. New York: de Gruyter. 271 p.

Habermas J. 1981. Theorie des kommunikativen Handelns. Frankfurt a. M.: Suhrkamp Verlag.

Hunziker M. 1995. The spontaneous reafforestation in abandoned agricultural lands: perception and aesthetic assessment by locals and tourists. Landscape and Urban Planning 31: 399–410.

Hunziker M. and Kienast F. 1999. Impacts of changing agricultural activities on scenic beauty – a prototype of an automated rapid assessment technique. Landscape Ecology 14: 161–176.

Joas H. 1992. Die Entstehung der Werte. Frankfurt am Main: Suhrkamp.

Joas H. 1996. Die Kreativität des Handelns. Frankfurt am Main: Suhrkamp.

Joas H. 1997. Die Entstehung der Werte. Frankfurt am Main: Suhrkamp. 320 p.

Junker B., Baumeler M., Debrunner R., Nigg P., Poncini C. and Zschokke M. 2003. Wie sieht die Bevölkerung aus Weinfelden und Bürglen ihre Thur? natur+mensch 44 (5): 4–7.

Junker B. and Buchecker M. in press. Social science contributions to the participatory planning of water systems-results from Swiss case studies. In: Castelletti A. and Soncini-Sessa R. (eds.). Topics on System Analysis and Integrated Water Resources Management. Elsevier, Oxford.

Karp D.G. 1996: Values and their effect on pro-environmental behavior. Environment and behavior 28 (1): 111–133.

Klages H. 1999. Wertewandel und bürgerschaftliches Engagement an der Schwelle zum 21. Jahrhundert. Speyerer Forschungsberichte 193. Speyer: Forschungsinstitut für öffentliche Verwaltung.

Kellert S.R. 1996. The value of life: biological diversity and human society. Island Press: Washington.

Kluckhohn C. 1962. Values and value-orientations in the theory of action: an exploration in definition and classification. In: T. Parsons, and E. Shils (eds.), toward a general theory of action, Cambridge: Harward University Press. S. 388–433.

Kohl K.-H. 1993. Ethnologie – die Wissenschaft vom kulturell Fremden. München: Beck.

Luhmann N. 1977: Zweckbegriff und Systemrationalität. Über die Funktion von Zwecken in sozialen Systemen. Tübingen.

Mayring P. 2000. (7. ed.). Qualitative Inhaltsanalyse: Grundlagen und Techniken. Weinheim. Deutscher Studienverlag.

Manfredo M.J., Teel T.L. and Bright A.D. 2004. Application of the concepts of values and attitudes in human dimensions of natural resources research. In: M.J. Manfredo *et al.* (eds.), society and natural resources. A sumary of knowledge, prepared for the 10th International Symposium on Society and Resource Management.

Mead H. 1934. Mind, Self, and Society. Chicago: Chicago University Press.

Meier Ch. and Buchecker M. 2005. Soziokulturelle Aspekte der Landschaftsentwicklung. Grundlagen für das Projekt Landschaft 2020 des BUWAL. HSW-Schriftenreihe Umwelt und Natürliche Ressourcen, Nr. 1.

Norton B.G. and Hannon B. 1997. Environmental values: A place-based theory. Environmental Ethics 19 (3): 227–245.

Oerter R. 1970. Struktur und Wandlungen von Werthaltungen. München.

Papadakis E. 2000. Environmental values and political action. Journal of Sociology 36 (1): 81–97.

Parsons T. 1951. The social system. New York: The Free Press.

Pfister Ch. 1997. Landschaftsveränderung und Identitätsverlust: Akzentverschiebungen in der Modernisierungskritik von der Jahrhundertwende bis um 1970. traverse 2: 49–68.

Purcell A.T. 1992. Abstract and specific physical attributes and the experience of landscape. Journal of Environmental Management 34: 159–177.

Rokeach M. 1973. The nature of human values. New York: Free Press.

Schwartz H. 1992. Universals in the content and structure of values: theoretical advances and empirical tests in 20 countries. Advances in experimental Social Psychology 25: 1–65.

Sell J.L. and Zube E.H. 1986. Perception of and response to environmental change. Journal of Architectural and Planning Research 1986/3: 33–54.

Schultz P.W. and Zelezny L. 1999. Values as predictors of environmental attitudes: evidence for consistency across 14 countries. Journal of Environmental Psychology 19: 255–265.

Simmel G. 1993. Das Individuum und die Freiheit. Frankfurt am Main: Fischer Taschenbuch Verlag.

Taylor C. 1989. Sources of the self: The making of a modern identity. Cambridge (Mass.): Harvard University Press.

Thogerson J. and Grunert-Beckmann S.C. 1997. Values and attitude formation towards emerging attitude objects: From recycling to general waste minimizing behavior. Advances in Consumer Research 24: 182–189.

Tönnies F. 1988 (8. ed.). Gemeinschaft und Gesellschaft: Grundbegriffe der reinen Soziologie. Darmstadt: Wissenschaftliche Buchgesellschaft.

Touraine A. 1985. An Introduction to the Study on Social Movements. Social Research 52: 749–787.

Willis S. and Campbell H. 2004. The chestnut economy: The praxis of neo-peasantry in rural France. Sociologia Ruralis 44 (3): 317–331.

The Role of Value Systems in Biodiversity Research

Peter Duelli*, Priska Baur, Matthias Buchecker, Felix Gugerli, Rolf Holderegger and Thomas Wohlgemuth

Swiss Federal Research Institute WSL, Zürcherstrasse 111, CH-8903 Birmensdorf, Switzerland
* Corresponding author: email: peter.duelli@wsl.ch

Abstract
Landscape research needs to consider physical features and processes, as well as human preferences. Anthropocentric value systems rule the world of perception and valuation of landscape quality, but different stakeholders have different motivations for planning, managing or protecting landscapes. The potential for conceptual or practical conflicts arising from such differing value systems is illustrated by the diverse perception of biodiversity as a landscape quality of growing importance. Spending public money to conserve or enhance biodiversity in cultivated landscapes means different things to different people. Hence, it is crucial to select pertinent indicators for measuring the success or failure of environmental schemes. In biodiversity evaluation, the main value clash is between indicators for species conservation (rare and threatened species) and ecological resilience or ecosystem functions (species richness). Similar conflicts are discussed for ecological concepts and theories concerning landscape quality. In order to avoid the image of conflicting expert opinions, landscape research should be aware of the relevant underlying value systems and declare them while going public.

Keywords: evaluation, perception, indicators, biodiversity, stakeholders

F. Kienast, O. Wildi & S. Ghosh (eds.), A Changing World. Challenges for Landscape Research, 27–34.

Introduction

In the context of landscape management, proponents of nature conservation often deplore the fact that their prime value system does not manifest itself in market prices and thus gets lower priority when conflicting with other anthropocentric values such as economic profits. This is just one example to illustrate how value systems may differ among stakeholders because each of them may use and value different properties within a landscape.

All people orient themselves by value systems. In the context of landscape ecology, the basic importance of value systems is reflected in the spectrum of motivations and convictions of stakeholders, be they politicians, landscape managers and scientists, or consumers and taxpayers. However, stakeholders leave their own normal motives unreflected and are in particular unaware of the motivations and value systems of their conflict partners. They are inclined to project their own value systems onto others, or, even further, to assign to their individual values system a general prescriptive meaning. Similarly, researchers are often not conscious about their motivations and the extent to which they are influenced in their daily work by their own individual value system. Altogether, this can lead to misunderstandings and conflicts.

The objectives of this chapter are to discuss the role of value systems in landscape research by means of illustrative examples, and to draw conclusions on how to develop future directives, explicitly taking into account the importance of value systems. We think this to be an important prerequisite to mitigate misunderstandings and conflicts concerning subjects and results of landscape research.

The Role of Value Systems in Human Interactions

Value systems offer the individual a point of orientation to interpret life's meanings (Joas 1997). Shared value systems allow people to co-operate and to build collective identities. In multi-fractioned societies, building a consensus requires a mutual understanding of the various value systems involved (Godschalk and Paterson 1999). Value systems are hierarchical and cultural, but also individually specific (Taylor 1989). Hypergoods, such as cultural concepts of individual identity or of what makes a good life, also influence the hierarchy of value systems. Value systems are not static; they develop during an individual's life as well as with societal changes (Breakwell 1986; Klages 1999). To add to the complexity, single individuals as well as interest groups can simultaneously have competing value systems. Value systems are important, but they are not sufficient to understand and explain decision making. The latter are also influenced by extrinsic factors such as financial perspectives or reputation. In the economic view of human behaviour, action is determined by expected benefits and costs, as well as preferences and constraints (Frey 1992). While preferences relate to value systems, which could also be interpreted as intrinsic incentives, constraints apply to extrinsic incentives, but also social values and norms (Fishbein and Ajzen, 1975). Thus, to understand and explain human behaviour, both have to be considered, intrinsic motivations and value systems as well as extrinsic incentives. Together with extrinsic incentives, value systems motivate and regulate actions and social interactions. Extrinsic incentives may have the strongest influence on decisions and practical measures affecting a landscape, both at an individual or political level. In practice, ethical values tend to be overruled by economic considerations. Consequently, the amount of financial support for the investigation and enhancement of non-vital values in landscapes is low in comparison to other economically more promising fields. As the economic perspectives, and societal perspectives in general, are inherently anthropocentric, we should also take into account anthropocentric value systems in landscape research as well as for conservation or restoration planning and management schemes. Most important seems to be collaborative planning that strives for activating the potential of local cooperation and, in particular, for integrating all the values of the local society into the planning process (Godschalk and Paterson 1999).

In a tentative classification of the value systems in landscape ecology, we distinguish between primary values (essential or vital needs) and secondary values (non-essential or non-vital needs), whereby primary values are in general rated higher than secondary ones. Examples of vital values are health, security, or food supply, while we consider ecological integrity, nature conservation, cultural heritage, mobility, and recreation to be important, but not of vital importance. Biologically speaking, a decrease in primary, vital values would negatively affect individual Darwinian fitness, while many people deprived of secondary, non-vital values seem to do quite well.

Value Systems and Paradigms in Landscape Research

When theories and concepts seem to be corroborated by empirical results or numerous observations, they tend to become paradigms among stakeholders and practitioners. It is an important scientific task, also in landscape ecology, to rigorously question these paradigms (Kuhn 1970).

The landscape can be seen from different perspectives, and its quality can therefore be evaluated in different ways. Value systems – either independent or competing – are involved when indicators measure structural diversity, connectivity, rareness, wilderness, or cultural heritage. Similarly, insider and outsider perspectives will have different motivations and goals. The appreciation of a landscape changes with individual knowledge and experience. Specific values apply for individuals, groups, and cultures. A top-down scientific approach will yield different results than a bottom-up approach to valuation with a focus on social aspects or economic willingness-to-pay for a special landscape quality. For social scientists, landscape quality refers mainly to people's perception, which is only one aspect of the quality of life. In the natural sciences, the focus is on the quality of life for plants and animals, including human inhabitants or temporary users.

Many strongly debated issues in landscape research and planning can be interpreted in view of competing or questioned paradigms, based on different value systems. There is a broad spectrum of value-based conflicts, ranging from differing personal interests of stakeholders to differing interpretations among experts on the applicability of ecological concepts and theories.

Conflicting Value Systems and Indicators for Biodiversity

In this paragraph, we aim at demonstrating the difficulties to develop quantitative indicators of landscape quality. Thereby we focus on biodiversity, a landscape quality, which today is of principal importance in landscape research (Nobis and Wohlgemuth 2004).

Gaston and Spicer (2004) distinguish, with respect to biodiversity, between direct-use values (e.g. food, medicine, industrial materials, biological control), indirect-use values (e.g. ecosystem functions), and non-use values (e.g. option value, intrinsic value). Here we use the term "value" in a strictly anthropocentric way: Even with so-called intrinsic or biocentric values, it is basically the people who want to prevent species from going extinct, not the organisms themselves. The various stakeholders in a landscape have different motivations for preserving or enhancing biodiversity. Their value systems are tuned to different aspects or entities of biodiversity. The dilemma with conflicting value systems becomes apparent when quantifiable indicators are to be defined for monitoring and assessing states and trends of biodiversity (Duelli and Obrist 2003b). Today, the most urgent and most important indicators of biodiversity are those needed by administrators for agriculture, forestry, and nature conservation. But given the various value systems among stakeholders, what are the pertinent biodiversity indicators at a national or international level? How could we ever prove that, e.g. the European goal of "halting the loss of biodiversity by the year 2010" (EC 2001) has been met or not?

With the following examples we show in which sense contradictory concepts of biodiversity are based on differing value systems, and we illustrate the need for a strong appreciation of anthropocentric value systems in landscape research.

Agriculture: In the European Union, about 2.4 billion Euros are spent yearly for agri-environmental schemes to promote ecological compensation in agricultural regions (Kleijn and Sutherland 2003). These schemes more or less explicitly include the goal to increase biodiversity in depleted landscapes. But what aspect of biodiversity are these schemes referring to? It is a revealing experience to ask stakeholders in agriculture, nature conservation, landscape planning, or tourism about their motivations to protect or enhance biodiversity. For agricultural ecosystems in industrialised countries, the motivations can be roughly grouped into three major value systems (Duelli and Obrist 2003a):

1. Species conservation: The motivation here is primarily ethical and socio-cultural. The fascination for all things that are rare or endangered is a basic human trait. Biodiversity can also be compared with art: many people suffer emotionally and intellectually if deprived of any form of art. The same is true when people are faced with the loss of an enigmatic plant or animal species threatened by extinction.
2. Ecological resilience: The motivation is primarily ecological, based on the paradigm of the "balance of nature" (Pimm 1991). It is linked to the concept of sustainability, stating that more species can fill more ecological niches, and that more genetic variation provides a better insurance against extinctions due to rapid environmental change. The fervent debate on biodiversity and ecosystem functioning (Naeem and Li 1997; Cropp and Gabric 2002; Loreau 2004) will go on, as long as the conflicting value systems of species conservation and ecological resilience are not disentangled.
3. Ecosystem services: The motivation is both ecological and economic. The basic ecological reasoning is the same as with ecological resilience, but clearly focused on the economic benefit of particular ecosystem functions. Examples are biological control (preventing pest outbreaks and the use of pesticides in agriculture and forestry) or pollination. Rare and threatened species have high appeal in conservation politics, but they are usually of negligible ecological importance. The idea of species redundancy (a species can take over the "role" of another one) is relevant for the value of ecosystem functioning (Janzen 1998) but unthinkable for conservationists.

Other motivations for safegarding agricultural biodiversity are prospecting for genetic resources for medicine, pharmacology, cosmetics, protecting and promoting cultural heritage and cultivated breeds, or the sense of place (Hampicke 1991; Gustafson 2000). In cultivated landscapes, many stakeholders are mainly interested in generating income for farmers when dealing with biodiversity and agri-environmental schemes (Baur 2003). However, when justifying their claim they rely on the value systems of others.

Sustainable development: The goal of nature conservation is to maintain nature (natural or cultural) as such, independent of their utilitarian aspects and use values (Pfister 1997). In contrast, sustainable development can be viewed as a social project and a regulative idea that aims at integrating ecological, economic and social dimensions of human development

(Minsch 1997). Positive tradeoffs between conflicting goals, e.g. between nature conservation and economic development, are judged to be necessary in order to promote an integral and enduring human well-being, including ecological responsibility (WCED 1987). From the point of view of sustainable development, conserving nature is not a goal in itself but a means for ensuring opportunities of human development, today and tomorrow. From this perspective, it follows, however, that in a poor country an ecologically valuable landscape might be sacrificed to economic development, similarly to what had happened in most industrial countries.

Wilderness: The term wilderness today means different things to different people, but it is always linked with naturalness (allowing natural processes), unmanaged nature (no visible human interference), and "authenticity" (Schnitzler and Borlea 1998). Whether secondary nature in formerly cultivated areas can be called wilderness is a matter of debate (Cris 2004). Wilderness areas have a very high appeal for eco-tourism and adventurous recreational activities (Bennett 1994; Bauer 2005). Wilderness can be seen as one aspect of biodiversity, but it may neither correlate with other aspects of biodiversity such as species richness, nor with other values such as ecosystem services or species conservation. Depending on the aim of a nature reserve, it should either remain untouched (wilderness natural dynamics), or managed according to a specific goal and reserved for public use (education, recreation, tourism).

In addition to the above examples of obvious differences between stakeholder interests there are examples where value systems and goals seem to match, but the means and way cause disagreement. Ecological or sociological concepts and theories are often at the base of these conflicts. Furthermore, most of the arguments on biodiversity values are influenced by spatial scale: A species threatened in one country may be common in another; a high local species richness (alpha-diversity) has a different value than a high regional or national diversity (beta- or gamma-diversity).

Fragmentation: The theory of island biogeography (MacArthur and Wilson 1967) has been used to explain the negative effects of fragmentation (Simberloff 1982; Jedicke 1994). Accordingly, a fragmented landscape looses species because the individual habitat islands are small and isolated from each other. On the other hand, according to the mosaic concept (Duelli 1997), a fragmented landscape harbours more species because of higher diversity of habitat types and more habitat heterogeneity, often due to increasing human influence (Korneck *et al.* 1998; Wohlgemuth *et al.* 2002). Depending on the value systems of the stakeholders it might be more important to maintain or enhance local species richness (ecological resilience) than to prevent the loss of a few particularly fragmentation-sensitive species (species conservation). In abundant habitats, where ecological considerations prevail, species richness might be more important, while in rare habitats, such as raised bogs in central Europe, the protection of a characteristic though poor species composition is a prime motivation.

Restoration ecology: In many countries and particular regions, planting new hedgerows or restoring rivers is subsidized with public money because it helps a variety of threatened species to survive in ecologically depleted landscapes (conservation value; Jedicke 1994) and enhances e.g. beneficial insects (biological control value; Zwölfer and Stechmann 1989). However, in the case of planting hedgerows in areas where hedgerows never constituted a traditional landscape element it is refused (value of cultural heritage; Marschall and Bruns 2002). In addition, planting hedgerows can be detrimental for curlews or skylarks that depend on open areas, but they may offer new nesting sites for other bird species (species conservation value).

Is More Always Better?

The term biodiversity as a concept has a positive connotation in the public, whereas in natural sciences biodiversity is treated as a value-free environmental quality, where the components and aspects can be quantified in a way similar to temperature or humidity for describing the climate. With climatic factors, however, we are aware that some like it hot, others cold, while with biodiversity we tend to focus on those value-specific aspects, where more is always better. To take up the above example of the notoriously species-poor raised bogs: Protecting biodiversity here means the conservation of the few species characteristic for raised bogs. In this case, more biodiversity equals more specialist species. Similarly, a higher species richness of lichens in natural forests (high naturalness value) goes along with a lower species richness of carabid beetles, which are predators and thrive in moderately disturbed cultivated areas (low biocontrol value). For biodiversity evaluation, the declared choice of a value-specific indicator allows to keep up the positive connotation of "more is always better".

Conclusions

The above examples illustrate the importance of clarifying the value systems of different stakeholders, including researchers. Value systems are fundamental when it comes to applying scientific knowledge to practice, but also much earlier, when research questions are formulated. The science of landscape ecology should consider and analyse value systems much more carefully. Subconscious value systems must be made explicit, because they also influence the results of scientific research, e.g. the choice of indicators to measure landscape quality. Therefore, we need to develop scientific tools to assess value systems. We must endeavour to learn how values evolve and expand, and what hinders their integration. For instance, differences between urban and rural attitudes can be explained by different value systems. Future research should further provide methods for assigning all indicators proposed for measuring the qualitative and quantitative landscape properties to one or several value systems. In any case, the resulting compromises will, in practice, depend on political and economic interests rather then on scientific grounds – which again is a matter of value systems.

References

Bauer N. 2005. Attitudes towards wilderness and public demands on wilderness areas. In: Kowarik I., Starfinger U. and Oggendorf P. (eds.), Wild forests in the city. Postindustrial urban landscapes of tomorrow. Berlin, Springer: pp. 47–66.

Baur P. 2003. Milch und Blumen – Schritte auf dem Weg zur Professionalisierung von ökologischen Leistungen durch die Landwirtschaft. In R. Oppermann H.U. and Guyer (eds.), Artenreiches Grünland bewerten und fördern – MEKA und ÖQV in der Praxis. Stuttgart, Verlag Eugen Ulmer: pp. 160–171.

Bennett D. 1994. The unique contribution of wilderness to values of nature. Natural Areas Journal 14: 203–208.

Breakwell G.M. 1986. Coping with threatened identity. London, Methuen.

Crist E. 2004. Against the social construction of nature and wilderness. Environmental Ethics 26: 5–24.

Cropp R. and Gabric A. 2002. Ecosystem adaptation: Do ecosystems maximize resilience? Ecology 83: 2019–2020.

de Waard R.S. 2002. Planning with SpaceUseForms in the region of Breda-Tilburg. Habiforum 2002: 1101.
Duelli P. 1997. Biodiversity evaluation in agricultural landscapes: an approach at two different scales. Agriculture, Ecosystems and Environment 62: 81–91.
Duelli P. and Obrist M.K. 2003a. Biodiversity indicators: the choice of values and measures. Agriculture, Ecosystems and Environment 98: 87–98.
Duelli P. and Obrist M.K. 2003b. Monitoring biodiversity in agricultural landscapes: the dilemma of conflicting value systems and indicators. In: Tack J., Branquart E., Caudron A. and Segers H. (eds.): Scientific tools for biodiversity conservation: monitoring, modeling and experiments. Brussels: pp. 170–174.
EC 2001. Environment 2010: our future, our choice. EC, Brussels.
Fishbein M. and Ajzen L. 1975. Belief, attitude, intention and behaviour: An Introduction to the theory and research. Reading MA: Addison Wesley.
Frey B.S. 1992. Economics as a science of human behaviour: Towards a new social science paradigm. Boston, Kluwer.
Gaston K.J. and Spicer J.I. 2004. Biodiversity: An introduction. Malden, Blackwell.
Godschalk D.R. and Paterson R.G. 1999. Collaborative conflict management comes of age. Journal of architectoral and planning research 16: 91–95.
Gustafson P. 2000. Meaning of place: Everyday experience and theoretical conceptualization. Journal of Environmental Psychology 21: 5–16.
Hampicke U. 1991. Naturschutzökonomie. Stuttgart, Verlag Eugen Ulmer.
Janzen D. 1998. Gardenification of wildland nature and the human footprint. Science 279: 1312–1313.
Jedicke E. 1994. Biotopverbund. Grundlagen und Massnahmen einer neuen Naturschutzstrategie. Stuttgart, Ulmer.
Joas H. 1997. Die Entstehung der Werte. Frankfurt, Suhrkamp.
Klages H. 1999. Wertewandel und bürgerliches Engagement an der Schwelle zum 21. Jahrhundert. Speyerer Forschungsberichte: 193.
Kleijn D. and Sutherland W.J. 2003. How effective are European agri-environment schemes in conserving and promoting biodiversity? Journal of Applied Ecology 40: 947–969.
Korneck D., Schnittler M., Klingenstein F., Ludwig G., Takla M., Bohn U. and May R. 1998. Warum verarmt unsere Flora? Auswertung der Roten Liste der Farn- und Blütenpflanzen Deutschlands. Schriftenreihe für Vegetationskunde 29: 299–444.
Kuhn T.S. 1970. Logic of discovery or the psychology of research? In: Lakatos I.M. (ed), Criticism and the growth of knowledge. Cambridge, Cambridge University Press.
Loreau M. 2004: Does functional redundancy exist? Oikos 104: 606–611.
MacArthur R.H. and Wilson E.O. 1967. The Theory of Island Biogeography. Princeton, Princeton University Press.
Marschall I. and Bruns D. 2002. Functional change and idealisation of hedges in the agricultural landscape. Naturschutz und Landschaftsplanung 34: 113–118.
Minsch J. 1997. Nachhaltigkeit und institutionelle Innovationen. In: K. Rennings, O. Hohmeyer (eds.) Nachhaltigkeit. Baden-Baden, Nomos: 297–329.
Naeem S. and Li S. 1997. Biodiversity enhances ecosystem reliability. Nature 390: 507–509.
Nobis M. and Wohlgemuth T. 2004. Trend words in ecological core journals over the last 25 years (1978–2002). Oikos 106: 411–421.
Pfister C. 1997. Landschaftsveränderung und Identitätsverlust. Traverse 2: 49–67.
Pimm S.L. 1991. The balance of nature? Ecological issues in the conservation of species and communities. Chicago, University of Chicago Press.
Schnitzler A. and Borlea F. 1998. Lessons from natural forests as keys for sustainable management and improvement of naturalness in managed broadleaved forests. Forest Ecology and Management 10: 293–303.
Simberloff D.S. 1982. Refuge design and island biogeographic theory: Effects of fragmentation. American Naturalist 120: 41–50.
Taylor C. 1989. Sources of the self: The making of a modern identity. Cambridge (Mass.), Harvard University Press.

WCED, 1987: Our common future. Oxford, Oxford University Press.

Wohlgemuth T., Bürgi M., Scheidegger C. and Schütz M. 2002. Dominance reduction of species through disturbance – a proposed management principle for central European forests. Forest Ecology and Management 166: 1–15.

Zwölfer H. and Stechmann D.H. 1989. Struktur und Funktion von Hecken in tierökologischer Sicht. Verhandlungen der Gesellschaft für Ökologie 17: 643–656.

The Meaning of "Landscape" – An Exegesis of Swiss Government Texts

Peter Longatti* and Thomas Dalang

Swiss Federal Research Institute WSL, Zürcherstrasse 111, CH-8903 Birmensdorf, Switzerland
* Corresponding author: e-mail: peter.longatti@wsl.ch

Abstract
This little linguistic study presents a semantic analysis of the word 'landscape' as it occurs in a number of government texts of the last 40 years that are fundamental for nature and landscape protection in Switzerland. The use of the word 'landscape' in each text mirrors the political development of the concept, which in turn is a response to the outcome of implemented policies.

We first observe a shift from the original 'picture' concept of landscape as a visual experience and at the same time an emotional experience, landscape as home for people, to the concept of landscape as a physical location where biological processes take place.

Then there is a shift back to the social and psychological aspects of the landscape focussing on people's needs and activities in the landscape so that the invisible aspects of the landscape become the dominant features and the idea of the picture finally disappears.

Keywords: BLN, cultural heritage, European Landscape Convention, identity, inventories, local scenery, mire landscapes, natural monuments, nature protection, Regional Landscape Parks, Swiss Landscape Concept, wetland habitats

F. Kienast, O. Wildi & S. Ghosh (eds.), A Changing World. Challenges for Landscape Research, 35–46.

"When I use a word," Humpty Dumpty said in rather a scornful tone, "it means just what I choose it to mean – neither more nor less."
"The question is," said Alice, "whether you can make words mean so many different things."
"The question is," said Humpty Dumpty, "which is to be master – that's all."
Lewis Carroll, Through the Looking Glass

How do we know what people mean when they use the term 'landscape'? Consulting a dictionary is not very helpful because the specific meaning can only be derived from the context. Thus we use the hermeneutic approach which is based on the principle that we cannot understand the parts of a text until we understand the whole and vice versa. This means that we need to look at how the word 'landscape' is related to other terms in a particular text. From these relations we derive the meaning or implied meanings of the word. In legal texts such as the ones dealt with in this study we have the additional benefit that the term must actually be defined exactly when it comes to the implementation of the law or regulation. When an inventory had to be drawn up listing all the mire landscapes of national importance and these landscapes had to be delimited on a map, the surveyors had to have a set of rules with which they could decide where to draw the border of each landscape.

In the context of the later texts, where no implementation has taken place yet, there are no such hard and fast rules. Many scientific interpretations of 'landscape' are collected there, but we cannot be sure how the term 'landscape' will eventually be interpreted out in the field. We can guess that just like in the case of the mire landscapes (see below) some aspects will be dropped for practical reasons and the designated objects (if any are designated at all) will not correspond to the original concept put forward by the scientists.

A Picture of a Scenery: The Dictionaries

When we look up the word 'landscape' in a dictionary, we become aware that the meaning has two aspects: one is visual (i.e. the idea of a picture) and the other is physical or geographical (i.e. a part of the land). See for example the Merriam Webster OnLine entry for 'landscape': "1a: a picture representing a view of natural inland scenery b: the art of depicting such scenery 2a: the landforms of a region in the aggregate b: a portion of territory that can be viewed at one time from one place)."

The Oxford Reference Online points to the history of the word 'landscape':

"4. landscape: Initially introduced into English with reference to paintings and particular ways of seeing the world from the Dutch landskap, the idea of the landscape has gradually expanded to embrace what is perhaps better described as the countryside. (From The Concise Oxford Dictionary of Archaeology in History)" [quote from the Webpage]

The same shift from the artwork to the content of the painting and on to the visual reality that calls for being depicted can also be found in the German language. See for example the entry for the word 'Landschaft' in Brüder Grimms Deutsches Wörterbuch: " ... der weg an sich war meistentheils schlecht und steinig, doch zeigte uns jeder schritt eine landschaft, die eines gemähldes werth gewesen wäre..." (every step showed us a landscape, that would have been worthy of a painting [our translation])

Thus 'landscape' was an aesthetic unit that could be captured on a piece of canvas. Landscapes are still strongly associated with painting even today when visual communication is dominated by other media.

Nature and Cultural Heritage: The Swiss Constitution

The passages in the Swiss constitution relating to the protection of landscapes are the legislative basis for the other documents. Faced with the ever expanding postwar building boom threatening to encroach on all the hitherto untouched areas, an article was added to the constitution in 1962 in efforts to contain this development. It declared the protection of nature and cultural heritage as a task of the Confederation.

> *"Der Bund hat in Erfüllung seiner Aufgaben das heimatliche Landschafts- und Ortsbild, geschichtliche Stätten sowie Natur- und Kulturdenkmäler zu schonen und, wo das allgemeine Interesse überwiegt, ungeschmälert zu erhalten." (old version 1962, BV62)*
> In fulfilling its tasks, the Confederation shall treat delicately "the familiar image of the landscapes and localities we live in", historical sites as well as natural and cultural monuments while preserving them completely where public interest prevails. (our own fastidious translation)

When the constitution was revised in 1999, some minor changes were also made in this article:

> *"Der Bund nimmt bei der Erfüllung seiner Aufgaben Rücksicht auf die Anliegen des Natur- und Heimatschutzes. Er schont Landschaften, Ortsbilder, geschichtliche Stätten sowie Natur- und Kulturdenkmäler; er erhält sie ungeschmälert, wenn das öffentliche Interesse es gebietet." (new version 1999, BV99)*
> In fulfilling its tasks, the Confederation shall take into consideration the objectives of the protection of nature and cultural heritage. Landscapes, local sceneries, historical sites as well as natural and cultural monuments are to be treated delicately; they are to be preserved in their entirety if public interest so requires. (our own fastidious translation)

Three modifications can be made out in the new text:

1. before the passage requiring the delicate treatment of landscapes etc. (hat zu schonen) "protection of nature and cultural heritage" (Natur- und Heimatschutz) was introduced, and which the Confederation has to consider.
2. The word "heimatlich" was omitted: the 'feeling at home' aspect was left out.
3. "Landschaftsbild" (image of a landscape, scenery) was reduced to "Landschaften"(landscapes).

Although these modifications seem rather subtle (and are lost in the official English translation) they nevertheless indicate significant shifts:

1. A shift from the purely visual to the factual. It is no longer explicitly pictures of landscapes but only physical landscapes that are the object of preservation. The picture aspect is however maintained by the following word "Ortsbilder" (local sceneries).
2. A shift from subjective experience and emotional attitudes contained in the word "heimatlich" (a sense of home/feeling at home) to objective qualities as "Natur- und Heimatschutz" (protection of nature and cultural heritage) which refer to specific sites of national importance as we shall see later. The word "Heimat" (home) is still there but in a different context that changes its connotations: landscape is no longer seen as the home of people but as an inherited asset which could be lost and has to be safeguarded.
3. A shift from the life of the inhabitants to the visual state of the past as a reference point for quality. The omission of "heimatlich" (feeling at home) puts more focus on the historical aspects: it follows naturally that landscape elements must not be tampered with but

"preserved in their entirety" (erhält sie ungeschmälert). Economic activities and other human pursuits in the landscape are not considered and there are no references to human well-being (which rings in the word "heimatlich"), nor are landscape changes and modifications mentioned at all. Landscapes appear as static units and human needs such as restoration of physical and emotional health, social interaction, and identification are not connected with them. Natural processes, dynamics of ecosystems that bring change and sometimes deterioration are not mentioned either. The reference point for preservation efforts in the past, there are no goals for developments other that maintaining former qualities. The constitution looks after things that have been there for a long time and have become accepted as natural and cultural assets.

Beautiful Scenery: The Federal Inventory of Landscapes and Natural Monuments (BLN)

As we have seen, the Constitution only states a general committment to the protection of landscapes and careful use of land resources. Objects of national importance that are to be protected are mentioned in Art. 78:

> "3 Er kann Bestrebungen des Natur- und Heimatschutzes unterstützen und Objekte von gesamtschweizerischer Bedeutung vertraglich oder durch Enteignung erwerben oder sichern."
> It may support efforts towards the protection of nature and cultural heritage, and may, by contract or by expropriation, acquire or secure objects of national importance. (official translation, BV99e)

Drawing up an inventory of the sites worthy of protection as an instrument for implementing protective measures was the logical consequence. The first inventory was the BLN. It was derived from an existing inventory named KLN, which had been compiled by 3 NGOs in order to coordinate their protective efforts. The BLN inventory today contains some 160 objects, which were selected and delimited by experts relying on their experience. It is a collection of very different sites varying from individual features such as a single large erratic block of less than a hectare to a whole mountain range of 500 square kilometres.

The purpose of the inventory is the complete preservation of these objects of national importance. The word is "ungeschmälert" meaning 'entirely' or 'unreduced' and indicates a defensive attitude: the entire territory of these sites must be kept free from any intrusions and there is no room for compromise. In order to be included in the inventory an object must meet the following criteria:

- It must be untouched by urban sprawl: the landscape should show few changes and be used mostly in near-to-nature (i.e. traditional) ways. Urbanized areas are explicitly excluded as a different inventory was drawn up (and enacted in 1981) for urban scenery and heritage sites: the Bundesinventar der schützenswerten Ortsbilder der Schweiz (ISOS).
- It must have aesthetic and cultural value: a landscape must be beautiful, unique, scientifically and historically interesting, it should contain relics of former cultural landscapes. In other words, it should be attractive to educated people, historians and scientists in particular.
- The most important landscape types should all be represented in the inventory and the objects should also be distributed evenly across the country.
- See the definition of this last category:

> *"Typlandschaften: Meist naturnah geprägte Kulturlandschaften, die für eine Landesgegend besonders kennzeichnende Oberflächenformen, kulturgeschichtliche Merkmale sowie für Fauna und Flora wichtige Lebensräume enthalten."*
> Landscape types: mostly cultural landscapes of near-to-nature character, with features that are typical of the region, historical characteristics and important habitats for fauna and flora. (our own translation, BNL)

In this inventory the reference point for quality clearly lies in the past. Traditional land use is the only tolerable form of interference with an otherwise natural environment. The envisioned function of these objects is to provide an edifying background for recreation (mainly for educated people) and opportunities to enjoy natural beauty. The landscape is for visitors only. The goal of this inventory is very similar to that of a museum: a collection of valuable objects of cultural and scientific interest is to be safeguarded against loss and degradation. Science study is encouraged, but modifications are not allowed. The objects were not selected on any scientific basis or method but just by expertise and intuition.

Focussing on Habitats: The Swiss Mire Landscapes Inventory

The Inventory of the Swiss Mire Landscapes of Particular Beauty and National Importance is mainly focussed on habitats. The value of a landscape selected for the inventory was mainly based on biological aspects; thus we might consider this inventory the peak of the 'biologistic' approach to landscape. During its implementation the weaknesses of this approach became evident to the administration. This inventory came about against the will of the government as the result of a referendum held in 1987. The objective of the referendum was to stop the army from converting one of the largest and most intact mire landscapes, located near Rothenthurm in the canton Schwyz, into a training ground with barracks and other infrastructure. The supporters of the referendum argued that the originally widespread mire habitats had been reduced to such an extent, that the last remnants had to be defended at all costs with no compromises.

The government saw the point of protecting habitats but for the landscape in question they were willing to make another exception for the sake of national defense. The support of the army in the population was weaker than the government had anticipated and the referendum was accepted in the vote and the strict protection of mire landscapes had to be implemented. Thus one specific type of natural landscape (i.e. characterized by mire habitats) was under strict protection, while all other types were not.

Actually, wetland habitats were finally protected by separate inventories: Raised bogs and fenlands where each is listed in a separate inventory: the Federal Inventory of Raised and Transitional Bogs of National Importance and the Federal Inventory of Fenlands of National Importance, respectively. Based on a different law (NHG - Federal Act on Wildlife, Countryside and National Heritage Protection of 1966) these had been compiled before the mire landscapes inventory and were relatively straightforward: a number of plant communities could be named and these could also be delineated in the field by trained biologists. The objects (i.e. wetland areas) proposed for the bog and the fenland inventories were delineated, evaluated and selected according to strictly scientific standards.

Protected bogs and fenlands did not cause any serious political problems; there was not much opposition because the protected areas were in general quite small, mostly found in remote areas and were not very valuable to farmers. Apart from the historical period when peat was cut, bogs were always treated as waste land of inferior quality where only hardy cattle could be kept. Bogs that are left alone just grow and their characteristics get more pronounced. Fenlands are a bit more problematic because they depend to some degree on traditional farming, i.e. on reed and grass cutting for litter production. If this form of human intervention stops, bushes will take over and the open areas will eventually be overgrown by trees and the typical vegetation disappears. So protecting the fenlands also implies some measure of human intervention and the farmers have to be encouraged to continue or even resume traditional methods or else someone else will have to do a cut every now and then.

Now the mire landscapes inventory is a more delicate matter because it deals not just with some habitats but with quite large areas where wetlands are a dominant feature but with large tracts of other types of land inbetween. A whole valley may be inside the perimeter of an object and so the protection applies to all of the land of a farmer or even several. The people who found themselves included in the inventory were naturally very sensitive about the protection goals and the measures: to some it looked as if they had been condemned to traditional farming with no alternative. At the given scale of the mire landscapes, the economic situation of the inhabitants, whose collaboration was also necessary to achieve the protection goals, became an issue. It was no longer just a question of preserving a natural state of the past.

A mire landscape had to have the typical wetland elements and other natural features. Man-made elements were also allowed as long as they were related to traditional farming and typical for mire landscapes. Modern elements were admitted if they served the purpose of protection indirectly by allowing the farmers to go on using the land and make a living. So for example a new road to a remote alp was acceptable if it allowed the farmer to go and cut the grass with modern machines, while on the other hand it might have violated the requirement of leaving the landscape intact and preserving the traditional look of the scenery. This left much room for debate and negotiations. Modern elements that did not benefit farming were strictly probibited. See for example the Rothenthurm amendment to the federal constitution:

> *"Moore und Moorlandschaften von besonderer Schönheit und von nationaler Bedeutung sind Schutzobjekte. Es dürfen darin weder Anlagen gebaut noch Bodenveränderungen irgendwelcher Art vorgenommen werden. Ausgenommen sind Einrichtungen, die der Aufrechterhaltung des Schutzzweckes und der bisherigen landwirtschaftlichen Nutzung dienen." (BV 99)*
> Mires and mire landscapes of particular beauty and national importance are protected areas. The construction of any kind of building or installation whatsoever, and any operations changing soil structure are strictly prohibited. Excepted are operations and installations necessary for the maintenance of the near-natural landscape and existing agricultural use. (Translation by Grünig, A. 1994)

The following definition was used to select and delimit the objects of the inventory

> *"Eine Moorlandschaft ist eine Landschaft, welche vom Mooraspekt dominiert wird. Sie muss schön und naturnah sein und in der Regel Weite, landschaftliche Einheit und Abgeschlossenheit aufweisen, darf also keine willkürlichen Landschaftsausschnitt darstellen"*
> A mire landscape is a landscape dominated by mire aspects. It must be beautiful and in a near-to-nature state. Generally, there should be large open space and the landscape should be a clearly confined unity, not just any piece of territory. (our translation)
> (Schlussbericht EDI/BUWAL, 1991)

This old visual priciple 'what you see from a vantage point' lead to troubles in the implementation. As a consequence of the idea that any economic activity other than traditional farming is a deterioration in a mire landscape, there were many political sqabbles where villages and individual landowners tried to get their land excluded from the object perimeter. To reach a compromise, the principle of the visual delimitation had to be abandoned. It became evident that invisible realities, such as ownership of land, had to be taken into consideration, and changes were made accordingly. So now a mire landscape may end at a property boundary, which makes no sense to the simple admirer of scenery.

The aesthetic aspect of the landscape was very important although it leaned a bit on scientific insights and modern attitudes. Traditionally, bogs and fens had not been considered beautiful for a long time – very much like mountains before alpinism. Mires were unpleasant places hostile to human beings where you would not go if you could avoid it. (Even the biologists who had to survey the vegetation found the mire habitats, where they got soaking wet or mercilessly burnt by the sun, a rather demanding work place.) So, many of the locals failed to see the beauty of a mire landscape and could not understand why these of all things should be protected.

The protection goals were to maintain traditional scenery, attractive for tourists; to preserve the mire habitats and other natural landscape elements, attractive to naturalists; to create a network of natural habitats for the benefit of endangered species; and to support traditional farming as an essential part of the aesthetic value of the landscape.

The point of reference for quality was on the one hand in the past but on the other hand the biological concept of an undisturbed habitat: the biological richness of the cultural landscape. Thus the inventory calls for regeneration of those habitats that are found in a poor state.

Providing an Economic Perspective: Regional Landscape Parks

After the shortcomings of the mire landscapes inventory and a critical analysis of the state of protected sites in Switzerland in an OECD report (WPEP 2000, p. 240), the legislators were looking for a new instrument to protect landscapes and natural habitats and fostering the local economy at the same time. It had become increasingly clear that abandoned areas did not automatically meet the protection goals such as species diversity and richness of structures. The landscapes would become monotonous, accommodate fewer species, lose a lot of their character, and some habitats would disappear altogether. It was realised that in many cases the cultural aspects are the distinguishing part of a landscape and these depend on human activities. Consequently the protection efforts must consider the economic situation of the local people and offer some economic perspectives for them.

The concept of regional landscape parks, which was developed internationally (UNESCO biosphere reserves 1995), serves just this purpose and it is included in the revision of the NHG law (Federal Act on Wildlife, Countryside and National Heritage Protection of 1966) which is not yet completed. The UNESCO's definition of a biosphere reserve in a nutshell:

> "Biosphere Reserves are areas of terrestrial and coastal ecosystems promoting solutions to reconcile the conservation of biodiversity with its sustainable use. They are internationally recognized, nominated by national governments and remain under sovereign jurisdiction of the states where they are located. Biosphere reserves serve in some ways as 'living laboratories' for testing out and demonstrating integrated management of land, water and biodiversity. Each biosphere reserve is intended to fulfil three basic functions, which are complementary and mutually reinforcing:
>
> – A conservation function – to contribute to the conservation of landscapes, ecosystems, species and genetic variation;

- A development function – to foster economic and human development which is socio-culturally and ecologically sustainable;
- A logistic function – to provide support for research, monitoring, education and information exchange related to local, national and global issues of conservation and development."

(UNESCO 1995)

A minimum size of 100 square kilometres is suggested for such parks and the establishment would be a bottom-up process. This means that the people of the region would have to take the initiative. The parks would get the status of a UNESCO biosphere reserve. The bottom-up process, however, leaves the question open of how a consistent approach can be achieved across the whole country. Along with natural aspects, features of cultural heritage would be essential for the identity of the park and sustainability would be another requirement.

The protection goals are to maintain and increase the quality of the natural environment and the cultural heritage, but also to develop a sustainable economy and to improve the quality of life of the inhabitants. This is to be achieved by creating a label for marketing tourism and local products but also by monitoring the development of the environment and doing research on the people-landscape relationship. Research is seen as an essential part of the protection effort.

While the mire landscapes inventory was almost totally focussed on the past with only biological restoration and regeneration in mind, the landscape parks are at least partly oriented towards the future: instead of just preserving some attractive parts, the effort goes towards creating an identity, which of course would be based on existing natural and heritage features, but the point is the development; and the reference for quality is not the past state of the landscape but the economic perspective for the local inhabitants. Landscape is seen as a place where economic processes occur and change is an objective and not just a danger to avert.

The Comprehensive View: The Swiss Landscape Concept

The regional landscape parks are an instrument to counteract the economic decline of relatively remote rural areas, but they are not suitable for the periurban areas where most people live and work. In 1997 the Federal Government introduced the Swiss Landscape Concept as an instrument for spatial planning. It is not another inventory of some more valuable sites but it lists the overall targets for nature and landscape protection.

The declared aim is to establish a long-term partnership between land users and landscape protectors and sustainability is the overall objective for natural and cultural landscapes.

> *"Es [i.e. the Swiss landscape concept]*
> *– strebt eine nachhaltige Entwicklung der Landschaft an.*
> *– fördert den Dialog zwischen Nutzern und Schützern in der Landschaft."*
> (sustainable development and dialogue between landscape users and protectors)

It also lists all the aspects of landscape that are considered:

> *"Lebensraum: Auch alle anderen Arten haben Ansprüche und Recht auf Leben in ihren Lebensräumen.*
> (habitats for other species)

Naturraum: Nur noch wenige Gebiete sind von menschlicher Tätigkeit und Nutzung unbeeinflusst geblieben. Gerade deshalb ist die noch bestehende Dynamik in diesen Naturlandschaften sehr bedeutend.
(unhindered dynamics of natural processes)
Kulturraum: Landschaft umfasst Stadt und Land. Sie ist das geschichtliche Gedächtnis für unsere Gesellschaft.
(town and countryside as cultural heritage)
Wirtschaftsraum: Die Landschaft bildet seit jeher und auch in Zukunft die Grundlage des Wirtschaftens.
(fundamental resource for economic activities)
Erlebnisraum: Im Erlebnisraum suchen wir Begegnungen in Natur und mit Kultur, sinnliche Erfahrungen, Erholung oder Herausforderung und Abenteuer.
(opportunities for experiences of nature and culture, health restoration and adventure)
Wahrnehmung und Bewertung: Das Bild der Landschaft, das wir durch alle unsere Sinne wahrnehmen, ist entscheidend geprägt durch unser kulturelles Bewusstsein.
(perception of landscape is determined by our cultural awareness)
Identifikationsraum: Die Verbundenheit zu einer vertrauten Umgebung ist wesentlich, dass wir wieder vermehrt Verantwortung für das Wohnumfeld übernehmen.
(identification with a familiar environment and responsibility)
Zeugin der Erdgeschichte: Geotope sind Landschaftsteile, welche die Millionen Jahre alte Geschichte unserer Erde, des Lebens und des Klimas besonders typisch und anschaulich belegen. Sie sind aus ökologischer, touristischer, pädagogischer und wissenschaftlicher Sicht sehr wichtig. Ihre Zerstörung ist unwiederbringlich.
(natural heritage)
Gemeineigentum: Interessenausgleich zwischen dem Charakter der Landschaft als öffentliches Gut und dem privaten Eigentum. Der zweifache Zugriff auf denselben Raum erfordert Verhandlungen und Konsensfindung.
(dual character of landscapes as public good and private property makes consensus finding necessary)
International: ...müssen die einzelnen Regionen und Länder für ihre ökologisch und kulturell wertvollen Landschaften ihre Verantwortung übernehmen.
(responsibility for ecological and cultural values in an international context)"
(LKS 1998)

Different development targets are set for natural and cultural landscapes. The features that have to be protected are naturalness, identity, diversity, beauty and uniqueness. The general protection targets are to minimize intervention in the landscape by careful use of remaining land resources and to improve the ecological quality of intensively used land. The following more specific management targets are listed: to allow dynamic processes, to improve the water sites, to maintain and connect habitats for flora and fauna, to keep historical and cultural aspects of a landscape visible, to maintain a suitable environment for historical sites and monuments, to strengthen people's attachment to nature, landscape and cultural heritage, to maintain the regenerative capacity of renewable resources, to reduce the expansion of urbanized areas into the landscape, to preserve and create transitional zones between used land and natural habitats, to create different levels of usage and to improve the quality of intensively used landscapes (in particular the areas where many people live). For all this the fundamental connection between land use and preservation is regarded as essential.

One focus of the term landscape is still on habitat protection, but the attention is also on the land between the habitats. The ecosystem approach has made it clear that individual habitats cannot exist in isolation, that they need to be connected and that they are affected by dynamic processes that occur in the landscape. Another reason for a holistic view of the landscape is the recognition of human needs regarding the landscape.These needs, however, are still mainly the needs of nature loving and culturally interested people, who are occasional visitors to a landscape, and not the needs of the inhabitants who might even work there.

Landscape still mostly stands for natural landscapes in contrast to urban areas, the countryside aspect still dominates. The requirement that "history and significance of the landscape should be readable" points to museum-like qualities and the overall attitude towards change and development is still defensive: changes and new uses of land that lead to new building activities are regarded as consumption of landscape resources which has to be minimized. Landscape is seen as a non-renewable resource and land use as an irreversible process, although one might argue that a landscape never disappears it just changes its look. The idea of regeneration is however mentioned in connection with intensively used landscapes. This could be taken for a sign of a more pro-active attitude.

Centring on Human Well-being: The European Landscape Convention

Switzerland is planning to sign the European Landscape Convention of the Council of Europe in 2005. The convention already has a certain influence on administrative practices and is therefore worth looking at although it is not a product of the Swiss administration.

The European Landscape Convention is another concept and thus not very explicit about details, however it illustrates the official view of how the subject of 'landscape' is held today across Europe.

In Chapter I of the Convention under General provisions, Article 1 – Definitions, we find the following definition of landscape:

> "A 'Landscape' means an area, as perceived by people, whose character is the result of the action and interaction of natural and/or human factors;"

In the preamble to the convention the specific aspects are spelled out in more detail:

> "The landscape …
> … has an important public interest role in the cultural, ecological, environmental and social fields, and constitutes a **resource favourable to economic activity** and whose protection, management and planning can contribute to job creation;
> … contributes to the **formation of local cultures** and
> … is a **basic component of the European natural and cultural heritage,** contributing to **human well-being** and consolidation of the **European identity**;
> … is an important part of the **quality of** life for people everywhere: in **urban areas and in the countryside**, in degraded areas as well as in areas of high quality, in areas recognised as being of outstanding beauty as well as **everyday areas**;
> … **is a key element of individual and social well-being** and
> … its protection, management and planning entail rights and responsibilities for everyone" [our own emphasis]
> (Preamble to the European Landscape Convention, ELC 2000)

The purpose is safeguarding the common heritage by providing a unified approach. Because there are other conventions dealing with protection of species and habitats, this one clearly focusses on the other aspects of landscape. Landscape is defined as something perceived by people and formed by the interaction of natural and and human factors – no longer a habitat here and a historical site there that have to stay as static as possible. Landscape has become dynamic.

The focus is on the quality of life in urban as well as in rural settings. Tourist attractions, everyday landscapes in periurban areas, beautiful and degraded parts of the landscape are all included and seen in their cultural, ecological, economic and social aspects.

Landscape is an element of local as well as European culture, it forms the identity of the people and is generally an essential factor for their well-being. It is also an economic factor and thus a resource that has to be managed for optimum use. As perceptions and psychological needs are evidently subjective aspects of landscapes, objectives of development cannot be established without including people in the decision making process and the measures have to be based on consensus or else the targets will not be reached.

With such a multitude of approaches and disciplines involved it is clear that the objectives of landscape development are no longer plain to see even for experts. To define objectives and quality standards that can be measured has become a complex interdisciplinary undertaking and no single discipline can come up with an answer for everybody. The European Landscape Convention therefore calls for training programmes and research to assess landscape qualities, to provide procedures to establish the different needs of the people at large and indiviual stakeholders, to provide methods for consensus finding and to introduce instruments for planning and managing landscape development. It is no longer possible to state in a few sentences what must be protected and how to go about it. The convention lists the range of aspects where solutions must be found.

With the focus on local and European identity and the well-being of the inhabitants, we are back at the original concept of 'Heimat' (homeland) of our first example, the early version of the Swiss constitution. But now it is more than familiar sights and cultural heritage: ultimately the subject is the identity of the people and thus every aspect of their daily lives. The countryside aspect of 'landscape' and the idea of the painting have become marginal while invisible aspects have taken over: the pictures in the minds of people, the feelings that go along with these, the interactions that take place and the roles people play. The social sciences are now firmly established in the business of landscape protection.

References

Government documents

[BLN] Bundesinventar der Landschaften und Naturdenkmäler
URL: http://www.umwelt-schweiz.ch/buwal/de/fachgebiete/fg_land/bln/objekte/index.html

[BV62] Bundesverfassung der Schweizerischen Eidgenossenschaft of 19 Mai 1874: Art. 24sexies, amendment accepted in the vote of 22.6.1962.

[BV99] Bundesverfassung der Schweizerischen Eidgenossenschaft of 18. April 1999. SR 101.
URL http://www.admin.ch/ch/d/sr/c101.

[BV99e] Federal Constitution of the Swiss Confederation of 18 April 1999.
URL http://www.admin.ch/ch/index.

[WPEP 2000] OECD Working Party on Environmenntal Performance 2000: OECD Environmental Performance Reviews (1st Cycle) Conclusions & Recommendations 32 Countries (1993-2000)
URL http://www.oecd.org/dataoecd//19/56/2432829.pdf

[ELC 2000] Council of Europe. European Landscape Convention. Florence, 20 October 2000.
URL http://conventions.coe.int/Treaty/en/Treaties/Html/176.htm.

[LKS 1998] Swiss Landscape Concept. Order number 310.137eng.
URL http://www.landschaftskonzept.ch/d/_start.htm

[UNESCO 95] The Seville strategy for biosphere reserves.
URL http://www.unesco.org/mab/nutshell.htm.

[Schlussbericht 1991] EDI /BUWAL (ed) 1991: Inventar der Moorlandschaften von besonderer Schönheit und von nationaler Bedeutung. Schlussbericht. Bern. 209 pp.

Literature

Grünig A. (ed.) 1994: Mires and Man. Mire Conservation in a Densely Populated Country – the Swiss Experience. Birmensdorf, Swiss Federal Institute for Forest Snow and Landscape Research. 415 pp.

Space and Place – Two Aspects of the Human-landscape Relationship

Marcel Hunziker[1*], Matthias Buchecker[1] and Terry Hartig[2]

[1] Swiss Federal Research Institute WSL, Zürcherstrasse 111, CH-8903 Birmensdorf, Switzerland
[2] Uppsala University, Institute for Housing and Urban Research, S-80129 Gävle, Sweden
[*] Corresponding author: e-mail: marcel.hunziker@wsl.ch

Abstract
Studies of the "human dimension" of landscapes have become increasingly important in landscape research because of the roles that humans play either as causes of ecological alterations or as legitimate users of the landscape. An important use of landscapes is as a physical "space" for living but also as a "place" with its meanings and contributions to societal identity. In this chapter, we present some of the key theories of landscape experience and empirical research related to those theories. They are grouped around three concepts: First, we survey theories dealing with landscapes perceived as a physical space, covering topics such as environmental preference and the evolutionary basis of the psychological processes through which preferences arise. Secondly, we summarize some of the theories dealing with landscape perceived as place. Here we discuss concepts such as "sense of place" and "place identity". We emphasize that place identity is a particular element contributing to sense of place. Thirdly, we discuss theory and research concerning the role of landscapes for psychological restoration, which bridges the approaches that treat landscape as space and those which treat it as place. In the conclusion, we provide some suggestions for further integrative work.

Keywords: landscape, preference, sense of place, psychological restoration, theories, literature review

F. Kienast, O. Wildi & S. Ghosh (eds.), A Changing World. Challenges for Landscape Research, 47–62.

Introduction

Landscape research consists not only of ecological research but also of social science research. The latter, often called "human-dimension research", deals with the multi-faceted interrelationship between landscape and society or individuals. This social aspect of landscape research has become increasingly important during recent years and it will become even more important in the future. There are two main reasons for this:

- First, in a comprehensive understanding of landscape ecological systems, humans are seen to play an important role in the system. Humans have so far been treated mainly as a cause of disturbances in natural systems, but more and more humans are also recognized as legitimate users of the system, particularly as "receivers" of material goods such as agricultural and forestry products and immaterial goods such as psychological restoration and (visual) information. Thus, from a basic scientific point of view, human-dimension research is needed for a comprehensive understanding of the socio-ecological systems that manifest themselves in landscapes. This includes the investigation of the above-mentioned "receiving" of goods by the system element "human being" (Nassauer 1997). This aspect of landscape research is still somewhat neglected, and there is a need to strengthen research efforts in this respect.
- Secondly, sustainable development involves more than matters of ecological balance. It aims at long-term ensuring of material and immaterial needs of the population. To this end, these needs must be investigated. Since the human needs that constitute the social aspect of sustainability remain underrepresented in sustainability research, more effort must be made in the future to better understand them. Knowledge of people's needs, including the reasons for these needs, is a prerequisite for designing nature conservation and landscape planning measures that can be accepted by the public and, thus, have a chance of succeeding in the long run (Hunziker *et al.* 2001; Luz 1993; Stoll 1999; Schenk 2000). Landscape planning and nature conservation measures that conflict with people's needs will face opposition. Even when such measures are in line with people's needs, educational and other interventions may be required to foster acceptance of planning and conservation measures. Knowledge about people's needs can support the design of such interventions.

As landscape research and sustainability research increasingly incorporate the human dimension, we are faced with the complexity of the human character. Each human is, simultaneously, a biological organism; a person with a unique set of capabilities, experiences, and aspirations; a social being acting within various roles in various groups; and a carrier of culture (e.g. Bourassa 1991). The complexity of the human condition finds expression in the experience of landscape, which is that component of human-dimension research on which we intend to focus in this chapter. Our intention here is to discuss some well-known and frequently used theories of landscape experience and some of the empirical research guided by or related to those theories. In doing so, we want to further the incorporation of the human dimension, and to help landscape-planning and nature-conservation practitioners develop successful strategies and measures.

Because humans are at the same time biological and social beings, one should not be surprised that the numerous theories dealing with landscape experience differ remarkably in the way they treat biological versus social determinants of landscape experience. In this chapter some of these theories are highlighted together with approaches that bridge between the two perspectives. Finally we suggest further integrative work.

To cope with the complexity of landscape experience, we find it useful to refer to two modes of landscape perception, one as space and one as place. The two modes receive widely differing weights depending on our biological inheritance and our psycho-social-cultural background. In the space mode, people perceive the landscape primarily in terms of their biological needs; that is, they focus on the (instrumental) use of the landscape. In the place mode, however, people perceive the landscape primarily in terms of self-reflection (experiences, achievements) and social integration (values, norms, symbols, meanings). This is a long standing distinction. For example, Simmel (1993) differentiated in a similar way in his "philosophy of landscape" between the animals' drive-defined perception of space and the humans' perception of landscape, which he described as a creative act. Thus, when individuals or groups become familiar with a particular space and link it with their cultural values, social meanings and personal experiences, it becomes a place for them (Tuan 1977). In other words, personal, social and cultural processes of appropriation superimpose a layer of meaning on space (Altman and Low 1992) and thus transform it into place.

Review of Theoretical and Empirical Literature

Theories about the human-landscape relationship can be roughly divided into two major groups. The theories in one group primarily focus on the relationships between universal, mostly physical characteristics of landscape and evaluative judgments such as preference. In these theories, landscape is considered as space. The theories in the second group primarily focus on the cultural and group specific meanings of the landscape through which space becomes transformed into place. Thus, the two major groups of theories on the human-landscape relationship can be defined in terms of their primary focus on space vs. place. In the following, we discuss the best known theories in each of the two groups.

Theories and studies regarding landscape perceived as space

The best-known theories focus on landscape as space and build on assumptions about the survival needs of early, prehistoric humans regarding their environment. The perceptual capabilities and predispositions, which evolved to meet these survival needs are assumed to still function as an "inborn" basis of the human-landscape relationship. In modern humans, however, these perceptual capabilities and predispositions may not function so much as a necessary aid to survival, though they still find expression on the "psychological" level of landscape preference. Even today, then, according to these theories, the best liked landscapes tend to be those which would have helped to satisfy the survival needs of primitive humans due to their special spatial characteristics.

One such theory, the savannah theory of Orians (1980, 1986), puts substantial weight on the fact that the first humans lived in the African savannah. Orians supported his theory with several observations: first, the European explorers of North America preferred for their first settlements savannah-like landscapes with groups of trees, views onto lakes and rivers, and vista points from which one could oversee the whole region (Orians 1980). Shephard (1969) made a similar observation about the settlers of New Zealand. Secondly, it is argued that, in countries around the world, people tend to arrange the cultural landscapes similarly to that of the natural savannah landscape. That is, many cultural landscapes represent a mosaic of open grassland and groups of trees (Orians 1980). And thirdly, savannah-like landscapes occur in many paintings (Smith 1989).

The literature on landscape perception includes various empirical tests of the savannah theory. For example, Balling and Falk (1982) found that savannah landscapes were highly preferred over other landscapes, especially dense forest and desert landscapes. Moreover, the savannah landscapes received particularly high preference ratings from the young children in their sample (ages 8 and 11). The children also rated the savannah significantly higher than other, also positively judged landscapes, whereas there were no significant differences between the judgments of the savannah and the other most-preferred landscapes when given by older persons. Balling and Falk (1982) interpreted this result with regard to the low grade of socialization of children, which they claimed made it easier to express an "inborn" biological-instinctive reaction. Lyons (1983) however, argued that the savannah preference of children might be caused by the fact that the savannah is most similar to those landscapes where children normally play, in parks with meadows and groups of trees. This interpretation treats the savannah preference as a product of social norms rather than biological rules. However, one can argue that there is also a reason for constructing parks in the manner described, which in turn might support again the savannah-theory.

Appleton (1975, 1996) based his prospect-refuge theory on the need of primitive humans for shelter and for keeping close watch over their surroundings. It differs from the savannah theory in that it is restricted to what Appleton considered the most important of the primitive human's survival needs, that of "seeing without being seen". He justifies this restriction (Appleton 1975: 73) with the argument that fulfilling the need for shelter and surveillance of the surroundings is an intermediate step for fulfilling the other basic needs.

Various attempts have been made to test the validity of prospect-refuge theory. In particular, differences between the genders have been studied in this regard (e.g. Nasar 1988). For example, Hull and Stewart (1995) found that the men and women in their sample focused on different things when moving through a landscape. Some authors, however, consider such differences as indicators of differing social rules (Balling and Falk 1982; Bernaldez *et al.* 1987; Lyons 1983; Strumse 1996). More empirical evidence in support of Appleton's theory has been reported by Clamp and Powell (1982), Woodcock (1982), Abello and Bernaldez (1986), Mealey and Theis (1995) and Hägerhäll (2000), but still other authors have concluded that their results did not offer support for the theory (e.g. Klopp and Mealey 1998).

The information processing theory of Kaplan and Kaplan (1989) assumes that those landscapes are preferred which stimulated and facilitated the primitive human's acquisition and rapid processing of information and thus promoted the development and differentiation of a capacity for planning action in the environment. This theory analyzes landscape perception in terms of complexity, mystery, coherence and legibility (Kaplan and Kaplan 1989: 52ff). Complexity and mystery relate to the need to gather information, while coherence and legibility relate to the need to make sense of the information gathered. These informational characteristics of the perceived environment can also be ordered along a temporal continuum: complexity and coherence refer to immediately available and interpretable information, whereas mystery and legibility refer to the possibility for gaining more information and yet maintaining orientation as one moves further into the landscape. Various empirical studies have examined the influence of one or all of these four characteristics on preferences for scenes (e.g. Gimblett 1990; Strumse 1994a,b; Coeterier 1996; Van den Berg *et al.* 1998). It was commonly found that one or more but not all of the informational characteristics positively predicted preference for scenes of widely varying kinds (e.g. Herzog 1989). Still other studies have found negative correlations between the informational characteristics and preference (Gimblett 1990). In a recent meta-analysis, Stamps (2004) directed attention to the heterogeneity of findings produced by the empirical work with the theory, which is the most extensively tested of the psychological theories on landscape preference. Stamps also suggested some methodological solutions and theoretical revisions to address the issue of non-reproducibility. Herzog and Leverich (2003; see also Herzog and Kropscott 2004) have addressed the non-reproducibility issue with specific regard to legibility.

A final theory of interest here has guided research, which directly examined a presumed functional correlate of landscape preference, namely, restoration from psycho-physiological stress. Ulrich's (1983; Ulrich *et al.* 1991) psycho-evolutionary model of affective and aesthetic response to environments has a number of basic assumptions in common with the theories described above. It assumes that rapid-onset affective responses to certain visual configurations in the environment had adaptive value over the course of human evolution, and that people today remain biologically prepared to prefer those configurations. The affective response is assumed to be elicited by environmental "preferenda", which are features or stimulus characteristics of the environment whose vague nature may preclude cognitive judgments but which still suffice for eliciting generalized affect. Ulrich (1983) assumes three basic kinds of preferenda in natural environments: gross structural aspects of settings, gross depth properties that require little inference, and general classes of environmental content. More specifically, affective reactions are evoked by a scene's complexity, focality (degree to which it contains a focal point or an area that attracts the observer's attention), depth, and ground surface texture. Threatening features, deflected vistas, and water also may work in drawing out an initial reaction. In this specification of environmental features, one sees points of correspondence between Ulrich's model and the theoretical analyses of Orians, Appleton and the Kaplans.

The model of Ulrich (1983) also refers to the initial motivating state of the person on encountering the landscape. If the person is experiencing a high level of arousal, then an initial affective response of interest and liking may open the door to a process of restoration. Experimental work by Ulrich has documented differences in restoration from acute demands (an exam, a horrifying film) with measures of emotion and physiology (e.g. Ulrich 1979; Ulrich *et al.* 1991) under different environmental conditions, presented with photographic or video simulations. This work has encouraged direct empirical assessments of relations between landscape preference and psychological restoration as a functional outcome (e.g. van den Berg *et al.* 2003), an issue that we will return to later in this chapter when we discuss connections between theories about landscape as space and theories about landscape as place. We now turn to that latter group of theories and related empirical research.

Theories and studies regarding landscape perceived as place

Transforming spaces into places is existential activity, as through the creation of places people visualise, memorise and thus stabilise constitutive human goods such as the sense of belonging, social integration, purposes that give meaning to life (values) and the sense of self (Williams *et al.* 1992). Sense of place is perhaps the most general concept which describes the relationship between people and their (local) spatial settings, subsuming other concepts such as place attachment, place identity and place dependence (Jorgensen and Stedman 2001). Place attachment is described as a positive emotional bond that develops between groups or individuals and their environment (Altman and Low 1992; Korpela 1989). Place dependence refers to how well a setting serves goal achievement given an existing range of alternatives (Jorgensen and Stedman 2001; see also Stokols and Shumaker 1981). Finally, place identity represents those aspects of self identity which involve and are reflected by the environment and its social and personal meanings (Buchecker 2005; Korpela 1989; Twigger-Ross and Uzzell 1996; Proshansky *et al.* 1983).

A large amount of research on sense of place has been conducted over the last several decades. Much energy has been invested in differentiating and operationalising the diverse dimensions or aspects of sense of place. Some authors maintain that these attempts have so far essentially failed (Pretty *et al.* 2003; Jorgenson and Steadman 2001). The failure can be attributed to the strong linkages among the diverse aspects of sense of place. This view resonates with Relph's (1976) recommendation to use sense of place as a tool for integration. Such an integrative approach encourages the researcher to take all aspects of people-place interaction into consideration, but it also bears the risk of asking far too much of him or her. Therefore, Hummon (1992) suggested differentiating sense of place into two "functional" dimensions: a) a cognitive dimension of sense of place which helps people to understand the place and thus allows them to establish an external orientation, and b) an emotional dimension which offers information on one's relationship to places and thus enables individuals to build up an internal (self-referent) orientation. According to Graumann (1983) a cognitive understanding of the place is a precondition (but not a sufficient condition) for establishing a relation to a place.

Especially in the last years, the cognitive dimension of sense of place has largely been neglected. Important older contributions to this research field stem from geographical "mental map" studies (e.g. Krüger 1987), which consider place knowledge and place representation, and studies in the ecological psychology tradition (Barker 1968; Fuhrer 1990), which focus on place-related behavioural rules. According to mental map studies, landmarks and clear borders support people's efforts to establish a clear representation of their places. Behavior setting studies show that people's meaningful interactions most commonly occur in settings in which they can connect with sufficiently clear behavioural rules (Proshansky *et al.* 1983; Buchecker 2005). This supports appropriation in that people know how to act within their environment in a secure way and thus build up a personal relationship to it.

In contrast to the cognitive dimension, the emotional dimension of sense of place has been a main issue not only in environmental psychology, but also in anthropology in the last years. However, not all aspects of sense of place have been equally embraced (Manzo 2003). Place research has focused in the last decades on people's favourite (or special) places and settings, and in particular those within their residential area. Empirical studies have found that informal meeting places (Oldenburg 1989), places symbolizing collective belonging (Buchecker 2005), places used in childhood, places frequented during leisure activities, and natural settings outside of the closer residential area often have particularly high emotional significance to a local residential population (see also Korpela *et al.* 2001). These places offer people opportunities to individually or collectively appropriate them.

Another considerable amount of recent place research has concentrated on the influence of time spent in a place on the people-place relationship. For example, Hay (1998) showed that as the amount of time spent in a place increases, the relationship to the place, and in particular the attachment, intensifies and becomes deeper (from "aesthetic experience" to "part of place") as well as more comprehensive (from special place to area-wide). Manzo (2003), however, emphasized that a more extended sense of place does not necessarily mean that the relationship has a better or more positive quality; sense of place might also be connected with negative or ambivalent emotions (Cooper 1995; Relph 1976), and it may also entail too much structure (Buchecker 2005). Thus, places to which we feel committed can also seem oppressive and imprisoning (Tuan 1974), and unknown and personally meaningless places can bring relief and new perspectives. The lack of attention to this ambivalent character of sense of place constitutes a shortcoming of recent research, according to Manzo (2003).

A less-studied emotional aspect of the sense of place involves people's relationships to groups and the relationships that hold between groups (Manzo 2003). Pratt (1984) emphasised the problematic character of this aspect in showing that the sense of people's rootedness and belonging is often obtained by the (symbolic) exclusion of others from that place. Similarly, Waitt (2000) found in her empirical study that the preservation of places often implies a preference for one group's cultural heritage over that of another group. In agreement with Dixon and Durrheim (2000) it can be concluded that while personal preferences and experiences influence people's relationships with places, these personal preferences can themselves be seen as products of a larger context.

In comparison to the long tradition of research on the causes of sense of place, the research on the consequences of sense of place is still in its initial phase. A starting point for this research was formulated by Greider and Garkovich (1994). When a person or a social group transforms space to place through direct experiences and interactions, it becomes part of the person's or group's "self". This may bring about a sense of responsibility for that place, as its loss or damage threatens the group's or person's self-identity (Breakwell 1986). A correlation between sense of place and a sense of responsibility for or even commitment to the given place has often been hypothesized (e.g. Buchecker *et al.* 2003; Volker 1997; Falk and Kilpatrick 2000), but the presence and magnitude of this correlation have not yet been sufficiently studied. There is, however, some structured empirical evidence (and much anecdotal evidence) that strong forms of sense of place representing unique ties between people and place are correlated with feelings of intense caring for the locale. Eisenhauer *et al.* (2000) showed that connections with 'special' places with particular meanings incorporate sentiments that go beyond judgements about utility. Such places cannot be substituted by other sites with similar attributes. A strong sense of place can therefore provoke people to react with high levels of concern about management practices (Schroeder 1992; Williams *et al.* 1992). For example, Syme *et al.* (1993) could show that in the context of wetland preservation, environmental concern – which is supposed to be closely linked to place attachment – is a motivating factor for involvement in nature preservation.

Sense of place also contributes indirectly to pro-environmental behaviour, as it is an important factor for social capital, which facilitates collective action for mutual benefit (Woolcock 1998). Falk and Kilpatrick (2000) found that social capital results from interactions that draw on (local) knowledge resources and identity resources. Sense of place not only contributes to both of these resources, but also fosters social interaction (Buchecker 2005).

The concept of sense of place encompasses an extremely broad area of inquiry. Arguably, the more specific concept of place identity is better suited for use as an analytical tool for understanding people-place relations, as it has a more well-elaborated grounding in psychological theory. The account in the following section will focus on this concept. As we will subsequently show, together with the psychological restoration concept, the place identity concept offers an opening to the integration of space- and place-focused theories of landscape experience.

Place identity as a particular element in sense of place

Place identity is not to be understood as a sub-aspect of sense of place, but rather as a specific perspective on people-place relations, namely, a self-reflective perspective.

According to Proshansky *et al.* (1983), place serves as an external memory for people's place-related aspects of their self-identity, called place identity. The function of place-identity is to regulate (stabilize and develop) people's self-identity (Fuhrer and Kaiser 1994). This regulating function of place for people's identity is crucial, because self-identity is a very unstable and at the same time existential cognitive construct constituted by social interactions and thus threatened by external changes (in relationships, resources) or internal changes (in confidence, anxieties) (Breakwell 1986). Places, and especially residential places, are suited to serve as external memories of people's place-related identity because they form the sceneries of people's (everyday) social interactions.

According to Graumann (1983), people's (social) identity is connected to place by the process of identification which unfolds in three steps: (1) identifying one's environment, (2) being identified by the others in the environment, and (3) identifying oneself with one's environment (or a part of it). In a further stage, more active forms of identification can take place by appropriating a place, that is, by leaving physical or social traces there (Weichhart 1990). As soon as a place reminds an individual of the main features of his or her identity (social belonging and qualities, individual abilities and qualities, cultural values), that individual can re-build and thus regulate his or her identity, which is necessary after even slight set-backs in everyday life. And as identity development is a life-long process, a person may not feel well at a place unless he or she can periodically re-appropriate the place, which allows that person to update and develop his or her identity (Fuhrer and Kaiser 1994). Thus, individuals can establish a place identity in places which are characterised by continuity yet at the same time offer them sufficient opportunities for appropriating the settings and leaving individual and collective traces there.

These requirements are in agreement with the identity process theory advanced by Breakwell (1986; Twigger-Ross and Uzzell 1996), according to which the regulation and development of identity is, at least in our Western culture, founded on the principles 'continuity', 'self efficacy', 'distinctiveness' (both associated to appropriation) and 'self esteem'.

An empirical study by Fuhrer and Kaiser (1994) showed that people escaped from their private homes if they could not succeed in regulating their identity in them, trying instead to compensate in more distant places. Similarly, Röllin and Preibisch (1993) found that with the increase of urbanisation, local residents increasingly stayed away from their residential area and withdrew in their leisure time either into the privacy of their homes or into distant recreation areas. This raises the question of the impact of urbanisation and modernisation on two crucial aspects of place quality: place identity and the regulation of identity, in particular in residential areas. These areas have a special importance for identity formation for two reasons: they normally are the place of first socialisation, and residents commonly have a relatively greater degree of control within their home area, which is a precondition for an active spatial identification.

In the following, the interrelations among social change, place identity, and landscape perception will be considered on the basis of two qualitative studies. The deeper aim of this account is to provide an understanding of the current development of place-space relationship. For this end, we will contrast a cross-spatial comparison between two Swiss communities differing in their degree of urbanisation (Buchecker 2005) with a cross-temporal comparison within a relocation project in England (Speller *et al.* 2002).

The analysis of in-depth interviews in the two Swiss communities showed that the residents strongly identified themselves with their community, regardless of the degree of urbanisation. Within the given village (defined as the main settlement of the community), the residents mainly referred to symbols of collective belonging in expressing their place identities. However, whereas the residents of the less urbanised community thereby focused their identification on collective elements such as the village structure or the communal water catchment, those of the more urbanised community mainly referred to the more abstract idea of the (lost) village community. In spite of their strong identification, the residents also associated the village with feelings of restriction and imprisonment, and they missed having opportunities to individually appropriate the village in both a physical sense (e.g. far-reaching building restrictions) and a social sense (e.g. lack of informal meeting places). Strict traditional rules and norms seemed to allow the residents little room to leave individual traces in the village; only children and unadjusted adolescents could establish a personal relationship to places within the scope of social control. Consequently, most of the residents could only regulate and develop the social and collective aspects of their identity within the village, and not the individual aspects. As these aspects are especially vulnerable ones, the residents have had to regulate these aspects elsewhere.

Almost all of the interviewees admitted that they actively avoided their respective village in their leisure time and tried to get to natural areas as fast and often as possible (and whenever possible by car or bicycle). The residents of the less urbanised community usually frequented the natural areas within their community, while the residents of the more urbanised community escaped to recreation areas outside of their closer region.

This seeming difference is also reflected in the relation between the residents and their communities' close-to-nature areas (defined as areas which are at least partly ruled by natural dynamics). Whereas the residents of the less urbanised community were very enthusiastic about these areas and could name many places to which they felt connected, in particular in a personal way, the interviewed residents of the more urbanised community found it difficult to indicate pleasant places there and complained about the omnipresent noise. Also, in the less urbanised community, only a few residents referred to places in nearby natural areas which they had physically appropriated (e.g. fire-places or tree huts). More residents there mentioned places reminding them of special (social and personal) experiences. But most of the residents referred to places they felt attracted to because of their beauty. Seel (1991) has argued that objects are in general experienced as aesthetically pleasing if they correspond with the observer's values or/and if the observer perceives it as a work of art and is thus animated to imaginative activity (i.e. to virtually shape the object). Speculatively, then, aesthetic experiences might be seen as an abstract form of appropriation and identification and may thus serve to compensate for or substitute the lack of active individual appropriation. Aesthetic experiences allow the residents and tourists to regulate their individual identity, but in a general or indirect way.

The findings of the qualitative research suggest that residential areas of peri-urban regions are (increasingly) split in two separated spheres in terms of identity regulation: the village as the sphere of collective identity and the close-to-nature areas (nearby or in more distant recreation areas) as the sphere of individual identity. As additional evidence for the existence of these two poles of identification, it seemed that the residents perceived and valued these areas with conflicting criteria. The criteria applied to areas in the village were harmony, orderliness, familiarity and serviceability. In contrast, those applied to the nearby natural areas were variety, beauty, surprise, silence and secludedness.

These fundamental differences between the spheres in terms of perception and evaluation per se may bring about conflicts concerning landscape development, as there seems to be little consensus among the residents (and especially between residents and visitors) about the exact location of these spheres. The schism of the residential area into two different spheres in terms of identification and identity regulation may complicate the residents' integration of individual and social identity and thus challenge the place's orientation-giving function (Hay 1998).

In the above-mentioned English study, the residents' relation to their old and new village was studied at five points in time during a relocation process (Speller *et al.* 2002). Before the relocation of the village, the residents identified themselves almost exclusively in a collective way with their old community. People there did not seem to feel the need to make the outside of their house different from the houses of the others. During the process of planning the new village, however, the residents began to verbalise the desire for future distinctiveness of their new houses. When they moved to their new houses they enjoyed individualising them, but at the same time they also started to miss the sense of community and complained about the lack of interaction. However, the desire for individual identification in this case appeared to be stronger than the desire for social belonging. The sudden change of preference from collective to individual identification with the village suggests that unwritten laws or norms had been inscribed into the structures of the old village, inhibiting its residents from expressing individuality within the old village. And with the disappearance of these structures, the residents started to express some seemingly repressed individual desires.

The two studies suggest that urbanisation and modernisation are accompanied by an increased desire for individual identification, whereas the old village structures maintain norms and meanings of the traditional collectively oriented society and thus inhibit individual appropriation and identification within the village. If this ban is not lifted by a fundamental change, such as a relocation of the village, then the residents may feel the need to regulate their individual identity outside of social control, in the transit sphere between place and space. This tendency seems to erode social interaction and thus contribute to a steadily proceeding though perhaps unwitting alienation from the village community, resulting in turn in the spread of dormitory villages and an increased demand for nature (abstract individual identity) as well as nostalgia (abstract social identity). Thereby the place-space dichotomy (self-regulated orientation vs. freedom) dissolves and ends in a new dichotomy of a place-space mixture (value-congruent areas) on the one hand and alienated space (value-free functional areas) on the other hand.

Toward Integrating the Concepts of Space and Place

To cope with the complexity in the experience of landscape, we grouped theories about landscape perception into those focused on space and those focused on place. The distinction between space- versus place-focused theories corresponds with the amount of attention that each accords to biological inheritance versus social-cultural background. We do not argue, however, that the distinction is a sharp one. Although the space-focused theories adopt assumptions about an evolutionary basis for rapid affective reactions to spatial and other features of the physical environment, they also acknowledge that the initial affective response can subsequently be modified by personal experience with the place and by cultural background (Kaplan and Kaplan 1989; Ulrich 1983, 1993). For their part, place-focused theories about the experience of landscape may not acknowledge the specific concerns of the space-focused theories, or they may discount the importance of specific evolutionary assumptions about the basis of landscape preferences. Ultimately, however, they cannot deny the fact of innate bases for the experience of landscape. After all, at the most fundamental level, the human perceptual apparatus as it has evolved allows us to "see" electromagnetic radiation within a certain range of wavelengths, to "hear" vibrations in the air within a certain range of frequencies, and so on. Thus, the issue in integrating space-focused and place-focused theories is not one of whether the basis for such an integration exists, but rather one of where to build solid and useful bridges between them.

We mentioned earlier that Ulrich's (1983) model of aesthetic response to the landscape helps to form a bridge between those theories concerned primarily with space and those concerned primarily with place. His theoretical account described restoration from stress as an extension of the initial affective response to particular configurations perceived in the landscape, a response that started from spontaneously emerging feelings of interest and liking. In postulating an immediate functional value for landscape preferences, his account resembles the other space-focused theories that we have overviewed here. In its reference to restoration, however, it suggests a particular approach to theoretical integration.

The approach builds on the concept of restorative quality and it appeals to us for three reasons. First, current theories about the restorative qualities of person-environment transactions are extensions of theories concerned with landscape preferences. We have already stated that this is the case for Ulrich's psycho-evolutionary model. It holds as well for the attention restoration theory of Kaplan and Kaplan (1989), and Kaplan (1995), which has roots in their information-processing model of preference. These theories place the process of responding to the immediate landscape in relation to what had come before as well as in relation to what would follow. That a particular scene could support restoration of physiological arousal to more moderate levels as described in Ulrich's model assumes that the person had come to that scene from some other situation in which arousal had been elevated. Further, the model sees adaptive value in restoration in that the person may thus be better prepared to deal with what comes next. Thus, the concept of restoration draws our attention to recurrent changes and how particular kinds of preferred landscape experiences figure in those changes. This in turn leads us to consider how the different theories of landscape experience are arrayed along dimensions of the duration and significance of different forms or modes of landscape experience, from momentary aesthetic and affective responses to scenery to periodic restorative experiences to long-standing place attachments and place identity.

Second, and consequently, reference to environmental restorativeness also opens into a discussion of the development and maintenance of place identity. A developing line of research situates the purposive use of restorative environments in a larger context of ongoing self-regulation, and in doing so sheds additional light on the restoration-preference connection. Some of this research activity departed from Korpela's (1992) observations about adolescents' descriptions of experiences in their favourite (i.e. most preferred) places. His subjects often referred to some need for restoration on going into their favourite places and to changes that are characteristic of restoration while in those places. It appeared that the favourite place served an environmental strategy of self-regulation; by affording a restorative experience, the favourite place helped the person fulfil functional principles that are thought to guide self-regulation. That the place served the person in this way was seen as a basis for the person's liking for and attachment to the place (see also Korpela and Hartig 1996; Korpela *et al.* 2001; Korpela *et al.* 2002).

Finally, reference to environmental restorativeness can also inform a discussion about some of the practical consequences of sense of place. Hartig *et al.* (2001) have discussed how the perception of restorative quality in a non-spectacular natural environment is associated with ecological behaviour. They argued that people will seek to protect not only the specific natural places that they rely on for psychologically stabilizing experiences like restoration, but also other places like them. Moreover, they argued that people will seek to protect natural places through mundane activities that have rather indirect effects, such as recycling or driving less, as well as through activities that are dedicated specifically to preservation, such as activism and voting for legislation that creates new nature preserves. Their work thus builds on a theme common in the literature on environmentalism. Numerous prominent figures in the environmental movement have in their personal accounts described how some strongly felt emotional bond to the natural world motivated and sustained their activism (e.g. Fox 1985).

Conclusion

Studies of the "human dimension" of landscapes have become important in landscape research and will become increasingly so because of the roles that humans play in the landscape as sources of ecological impacts and as legitimate users of the landscape. A comprehensive understanding of the landscape requires a clear understanding of the character and the function of the human-landscape relationship. In a first step it is helpful to recognise that landscape experience can be differentiated into two modes, as place and as space. These experiential modes enable humans to fulfil different basic human needs: recreative and aesthetical activities and restoration on the one hand, regulation of identity and representation of meanings (values, norms, experiences) on the other hand. As we asserted at the outset, knowledge of people's needs is a prerequisite for designing nature conservation and landscape planning measures that are acceptable to the public. Our discussion here of frequently used theories of landscape experience and of empirical research related to those theories should give a sense of the complexity of human needs connected with the experience of landscape. In this context it is, however, important to also see the complementary character of space and place as modes of landscape experience. The functions of space may over the life of an individual become intertwined with the functions of place. So far, relations between space and place functions have not yet been researched in a systematic form.

Concluding here, we wish to share a few ideas about directions for further integrative work. Clearly, given our foregoing discussion, the role of restorative experience in the development of place preferences and place attachments deserves further attention. In particular, the reciprocal character of those relations has received little attention. Having seen that restorative experiences can figure in the development of place preferences, we would like to know whether those who have strong preferences for particular kinds of places, such as natural areas, show more effective restoration when they enter such a place, even one that they have not visited before.

Little research has been done on the interaction between place identity (or sense of place) and landscape preference (though again see the work of Korpela and colleagues). According to the results presented above, it might be hypothesised that people who cannot regulate their individual identity within their everyday surrounding have a stronger preference for natural elements. Conversely, it would be worth studying whether people whose preferences do not match the predictions of the space-related evolutionary theories also show special characteristics in terms of their place identity.

Beyond reference to restorative experience as a bridging phenomenon, as we have proposed here, there are few attempts to integrate across a larger part of the above-discussed plurality of (overlapping) predictors of landscape preference. More comprehensive models would be helpful, however, when the landscape preferences of different parts of the population have to be determined. When such models are established theoretically, there is a need for testing their validity empirically, particularly when transfer into the practice of landscape planning and nature conservation is foreseen.

We discussed above the role of duration for landscape experience, assuming landscape as something static. Yet, of course, landscape changes over time. However, there is a considerable research gap regarding systematic analysis of the judgements of temporal landscape change (though see the work of Zube and Sell, e.g. Sell and Zube 1986; Zube *et al.* 1982). Here questions arise such as the effect of the rate of change or the significance of the symbolic meaning of landscape elements affected by change: do people get used to the changed landscape when the changes are small and slow enough? Are there any quantitative or qualitative (symbolic) thresholds of change where even slow and small changes lead to a loss of preference and thus to reactance? These gaps in the scientific work around landscape experience need to be bridged: one of the main challenges for planning and conservation involves these slow, small and thus less apparent changes in the landscape. Working in this field would not only mean doing integrative research regarding the space and place aspects of the landscape. Integration would go beyond social science and include other fields of landscape ecology, which also try to deal with small, slow changes and thresholds. A useful precedent on which such work could build is the limits-of-acceptable-change approach within wilderness management in the USA (e.g. Stankey *et al.* 1985).

Another unresolved problem is the scale-dependency of landscape preferences. Some landscape changes might be accepted when considered from large distance, but rejected when details can be observed. This issue should be further investigated in order to improve the reliability of generalisations. It would, in addition, provide a further possibility for integration of the social and natural science approaches in landscape research.

Scale-dependence has not only a spatial but also a social dimension: Whereas certain developments are well accepted by the majority of the population of a whole nation, they may be objected to by the inhabitants of the region where they take place or by those social groups otherwise affected by them. Further investigations and theoretical considerations are necessary to find out which types of landscape change are universally preferred or not, and which types of change are principally to be evaluated from a local or regional perspective. Here again, integrative consideration of the space and place aspect of landscape is necessary – and possible.

Finally, we observe that there are still numerous challenging research questions within each single field of landscape research. However, striving towards integration of landscape-experience research and, furthermore, towards integration of social and natural science work regarding landscape might not only be challenging, but fruitful and significant for practical work like landscape planning and nature conservation – and thus for the reality out there.

References

Abello R.P. and Bernaldez F.G. 1986. Landscape preference and personality. Landscape and Urban Planning 13: 19–28.

Altman I. and Low S. 1992. Place attachment. New York: Plenum Press.

Appleton J. 1975 (2nd revised edition, 1995). The experience of landscape. New York, Wiley.

Balling J.D. and Falk J.H. 1982. Development of visual preference for natural environments. Environment and Behavior 14, 1: 5–28.

Barker R.G. 1968. Ecological psychology: Concepts and methods for studying the environment of human behavior. Stanford, CA: Stanford University Press.

Bernaldez F.G., Gallardo D. and Abello R.P. 1987. Children's landscape preferences: From rejection to attraction. Journal of Environmental Psychology 7, 2: 169–176.

Bourassa S.C. 1991. The aesthetics of landscape. London and New York: Belhaven Press.

Breakwell G.M. 1986. Coping with threatened identity. London: Methuen.

Buchecker M., Hunziker M. and Kienast F. 2003. Participatory landscape development – overcoming social barriers to public involvement. Landscape and Urban Planning 64: 29–46.

Buchecker M. 2005. Public place as a resource of social interaction. In: Turner P. and Davenport E. (eds.). Space, spatiality and technology. Dordrecht: Kluwer.

Clamp P.P. and Powell M. 1982. Prospect-refuge theory under test. Landscape Research 7: 7–8.

Coeterier J.F. 1996. Dominant attributes in the perception and evaluation of the Dutch landscape. Landscape and Urban Planning 34: 27–44.

Cooper M.C. 1995. House as a mirror of self: Exploring the deeper meaning of home. Berkeley, CA: Conari Press.

Dixon J. and Durrheim K. 2000. Displacing place identity: A discursive approach to locating self and other. British Journal of Social Psychology 39: 27–44.

Eisenhauer B.W., Krannich R.S. and Blahna D.J. 2000. Attachments to special places on public lands: An analysis of activities, reason for attachments, and community connections. Society and Natural Resources 13: 421–441.

Falk I. and Kilpatrick S. 2000. What is social capital? A study of interaction in a rural community. European Society for Rural Sociology 40, 1: 87–110.

Fox S. 1985. The American conservation movement: John Muir and his legacy. Madison: University of Wisconsin Press.

Fuhrer U. and Kaiser F. 1994. Multilokales Wohnen. Bern: Verlag Hans Huber.

Fuhrer U. 1990. Bridging the ecological gap: Behaviour settings as interfaces. Environment and Behaviour 22, 4: 518–537.

Gimblett H.R. 1990. Environmental cognition: The prediction of preference in rural Indiana. Journal of Architectural and Planning Research 7: 222–234.

Graumann C.F. 1983. On multiple Identities. International Social Science Journal 35: 309–321.
Greider T. and Garkovich L. 1994. Landscapes: The social construction of nature and the environment. Rural Sociology 59, 1: 1–24.
Hägerhäll C.M. 2000. Clustering predictors of landscape preference in the traditional Swedish cultural landscape: prospect-refuge, mystery, age and management. Journal of Environmental Psychology 20, 1: 83–90.
Hartig T., Kaiser F.G. and Bowler P.A. 2001. Psychological restoration in nature as a positive motivation for ecological behavior. Environment and Behavior 33: 590–607.
Hay R. 1998. Sense of place in developmental context. Journal of Environmental Psychology 18: 5–29.
Herzog T.R. and Kropscott L.S. 2004. Legibility, mystery, and visual access as predictors of preference and perceived danger in forest settings without pathways. Environment and Behavior 36, 5: 659–677.
Herzog T.R. and Leverich O.L. 2003. Searching for legibility. Environment and Behavior, 35, 4: 459–477.
Herzog T.R. 1989. A cognitive analysis of preference for urban nature. Journal of Environmental Psychology 9: 27–43.
Hull R.B. and Stewart W.P. 1995. The landscape encountered and experienced while hiking. Environment and Behavior 27: 404–426.
Hummon D.M. 1992. Community attachment: Local sentiment and sense of place. In: Altman I. and Low S.M. (eds.). Human behavior and environments: Advances in theory and research. Volume 12: 253–278. New York: Plenum Press.
Hunziker M., Hoffmann C. and Wild S. 2001. Die Akzeptanz von Raubtieren, Gründe und Hintergründe – Ergebnisse einer repräsentativen Umfrage in der Schweiz. Forest Snow and Landscape Research 76, 1–2: 301–326.
Jorgensen B.S. and Stedman R.C. 2001. Sense of place as an attitude: Lakeshore owners attitudes toward their properties. Journal of Environmental Psychology 21: 233–248.
Kaplan R. and Kaplan S. 1989. The experience of nature. A psychological perspective. Cambridge University Press. 340 pp.
Kaplan S. 1995. The restorative benefits of nature – towards an integrative framework. Journal of Environmental Quality 15: 169–182.
Klopp B. and Mealey L. 1998. Experimental mood manipulation does not induce change in preference for natural landscapes. Human Nature 9, 4: 391–399.
Korpela K.M. 1989. Place-identity as a product of environmental self-regulation. Journal of Environmental Psychology 9: 241–256.
Korpela K.M. 1992. Adoloescents' Favourite Places and environmental self-regulation. Journal of Environmental Psychology 12: 249–258.
Korpela K.M., Hartig, T., Kaiser F.G. and Fuhrer U. 2001. Restorative experience and self-regulation in favorite places. Environment and Behavior 33: 572–589.
Korpela K.M., Kyttä M. and Hartig T. 2002. Restorative experience, self-regulation, and children's place preferences. Journal of Environmental Psychology 22: 387–398.
Korpela K. and Hartig T. 1996. Restorative qualities of favorite places. Journal of Environmental Psychology 16: 221–233.
Krüger R. 1987. Wie räumlich ist die Heimat. Geographische Zeitschrift 75, 3: 160–177.
Lee C. and Hummon D.M. 1993. A place to call home. The Sociological Quarterly 34, 1: 111–131.
Low S.M. and Altman I. 1992. Place attachment: A conceptual inquiry. In: Altman I. and Low S.M. (eds.). Place attachment. pp. 1–12. New York: Plenum Press.
Luz F. 1993. Zur Akzeptanz landschaftsplanerischer Projekte. Europäische Hochschulschriften. Kiel: Peter Lang.
Lyons E. 1983. Demographic correlates of landscape preference. Environment and Behavior 15, 4: 487–511.
Manzo L.C. 2003. Beyond house and haven: Toward a revisioning of emotional relationships with places. Journal of Environmental Psychology 23: 47–61.
Mealey L. and Theis P. 1995. The relationship between mood and preferences among natural landscapes – an evolutionary perspective. Ethology and Sociobiology 16: 247–256.

Nasar J.L. 1988. Environmental aesthetics: Theory, research, and applications. New York: Cambridge University Press. 521 pp.

Nassauer J.I. (ed.) 1997. Placing nature – culture and landscape ecology. Washington, DC: Island Press.

Oldenburg R. 1989. The great good place. New York: Paragon House.

Orians G.H. 1980. Habitat selection: General theory and applications to human behavior. In: Lockard J.S. (ed.). The evolution of human social behavior. pp. 49–66. New York: Elsevier.

Orians G.H. 1986. An ecological evolutionary approach to landscape aesthetics. In: Penning-Rowsell E.C. and Lowenthal D. (eds.). Landscape meanings and values. London: Allen and Unwin.

Pratt M.B. 1984. Identity: skin, blood, heart. In: Bulkin E., Pratt M.B. and Smith B. (eds.). Yours in Struggle: Three Feminist Perspectives on Anti-semitism and Racism, pp. 27–81, New York: Long Haul Press.

Pretty G.H., Chipuer H.M. and Bramston P. 2003. Sense of place amongst adolescents and adults in two rural Australian towns: The discriminating features of place attachment, sense of place and place dependence in relation to place identity. Journal of Environmental Psychology 23: 273–287.

Proshansky H., Fabian A.K. and Kaminoff R. 1983. Place-identity: Physical world socialization of the self. Journal of Environmental Psychology 3: 57–83.

Relph E. 1976. Place and placelessness. London: Pion.

Röllin P. and Preibisch M. 1993. Vertrautes wird fremd Fremdes vertraut. Kulturelle Vielfalt und nationale Identität, Nationales Forschungsprogramm 21. Basel: Helbing and Lichtenhahn.

Schenk A. 2000. Relevante Faktoren der Akzeptanz von Natur- und Landschaftsschutzmassnahmen? Ergebnisse qualitativer Fallstudien. Publikation der Ostschweizerischen Geographischen Gesellschaft Neue Folge, Heft 5.

Schroeder H. 1992. The spiritual aspect of nature: A perspective from depth psychology. In: Stoep V. (ed.). Proceedings of the 1991 Northeastern Recreation Research Symposium, pp. 25–30.

Seel M. 1991. Eine Ästhetik der Natur. Frankfurt a. M.: Suhrkamp. 388 pp.

Sell J.L. and Zube E.H. 1986. Perception of and response to environmental change. Journal of Architectural and Planning Research 3: 33–54.

Shephard P. 1969. English reaction to the New Zealand landscape before 1850. Pacific Viewpoint Monograph 4.

Simmel G. 1993. Das Individuum und die Freiheit. Fischer Taschenbuch Verlag, Frankfurt am Main.

Smith B. 1989. European vision and the South Pacific. Melbourne: Oxford University Press.

Speller G., Lyons E. and Twigger-Ross C. 2002. A community in transition: The relationship between spatial change and identity processes. Social Psychological Review 4, 2: 39–58.

Stamps A.E. 2004. Mystery, complexity, legibility and coherence: A meta-analysis. Journal of Environmental Psychology 24: 1–16.

Stankey G.H., Cole D.N., Lucas R.C., Petersen M.E. and Frissell S.S. 1985. The Limits of Acceptable Change (LAC) system for wilderness planning. Gen. Tech. Report INT-176, USDA Forest Service Intermountain Forest and Range Experiment Station, Ogden, UT.

Stokols D. and Shumaker S.A. 1981. People in places: A transactional view of settings. In: Harvey J.H. (ed.). Cognition, social behavior, and the environment, pp. 441–488.

Stoll S. 1999. Akzeptanzprobleme bei der Ausweisung von Grossschutzgebieten. Diss. Univ. Berlin.

Strumse E. 1994a. Environmental attributes and the prediction of visual preferences for agrarian landscapes in western Norway. Journal of Environmental Psychology 14: 293–303.

Strumse E. 1994b. Perceptual dimensions in the visual preferences for agrarian landscapes in western Norway. Journal of Environmental Psychology 14: 281–292.

Strumse E. 1996. Demographic differences in the visual preferences for agrarian landscapes in western Norway. Journal of Environmental Psychology 16, 1: 17–31.

Syme G.J., Beven C. and Sumner N. R. 1993. Motivation for reported involvement in local wetland preservation. Environment and Behavior 25, 5: 586–606.

Tuan Y.F. 1977. Space and place. London: Arnold.

Twigger-Ross C.L. and Uzzell D.L. 1996. Place and identity processes. Journal of Environmental Psychology 16: 205–220.

Ulrich R.S. 1979. Visual landscapes and psychological well-being. Landscape Research 4: 17–23.

Ulrich R.S. 1983. Aesthetic and affective response to natural environment. In: Altman I. and Wohlwill J.F. (eds.). Behavior and the natural environment. Plenum Press, New York.

Ulrich R.S., Simons R., Losito B.D., Fiorito E., Miles M.A. and Zelson M. 1991. Stress recovery during exposure to natural and urban environments. Journal of Environmental Psychology 11: 201–230.

Ulrich R.S. 1993. Biophilia, biophobia, and natural landscapes. In: Kellert S. and Wilson E.O. (eds.). The biophilia hypothesis. pp. 73–137. Washington, DC: Island Press.

Van den Berg A.E., Koole S.L. and Van der Wulp N.Y. 2003. Environmental preference and restoration: (How) are they related? Journal of Environmental Psychology 23: 135–146.

Van den Berg A.E., Vlek C.A. and Coeterier J.F. 1998. Group differences in the aesthetic evaluation of nature development plans: A multilevel approach. Journal of Environmental Psychology 18: 141–157.

Volker K. 1997. Local commitment for sustainable rural landscape development. Agriculture, Ecosystems and Environment 63: 107–120.

Waitt G. 2000. Consuming heritage: Perceived historical authenticity. Annals of Tourism Research 27, 4: 835–862.

Weichhart P. 1990. Raumbezogene Identität. Stuttgart: Franz Steiner Verlag.

Williams D.R., Patterson M.E. and Roggenbuck J.W. 1992. Beyond the commodity metaphor: Examining emotional and symbolic attachment to place. Leisure Sciences 14, 1: 29–46.

Woodcock D.M. 1982. A functionalist approach to environmental preference. Ann Arbor: Ph.D. Dissertation, University of Michigan.

Woolcock M. 1998. Social capital and economic development: Toward a theoretical synthesis and policy framework. Theory and Society 27: 151–208.

Zube E.H., Sell J.L. and Taylor J.G. 1982. Landscape perception – research, application and theory. Landscape Planning 9: 1–33.

Ecological Observations and Processes

Modern Remote Sensing for Environmental Monitoring of Landscape States and Trajectories

Niklaus E. Zimmermann[1*], Robert A. Washington-Allen[2], Robert D. Ramsey[2], Michael E. Schaepman[3], Lukas Mathys[1], Benjamin Kötz[4], Mathias Kneubühler[4] and Thomas C. Edwards[5]

[1] Swiss Federal Research Institute WSL, Zürcherstrasse 111, CH-8903 Birmensdorf, Switzerland
[2] Department of Wildland Resources, Utah State University, Logan, Utah 84322-5230, U.S.A.
[3] Centre for Geo-Information, Wageningen University and Research Centre, Droevendaalsesteeg 3, NL-6708 PB Wageningen, The Netherlands
[4] Remote Sensing Laboratories, Department of Geography, University of Zürich, Winterthurerstr. 190, CH-8057 Zürich, Switzerland
[5] USGS Utah Cooperative Fish and Wildlife Research Unit, College of Natural Resources, Utah State University, Logan, UT 84322-5290, U.S.A.
[*] Corresponding author: e-mail: niklaus.zimmermann@wsl.ch

Abstract
Contemporary and emerging remote sensing technologies, combined with biophysical first principles and modern spatial statistics allow for novel landscapes analyses at a range of spatial and temporal scales. In the past, supervised or un-supervised classification methods and the development of indices of landscape degradation and other derived products based on multi-spectral imagery of various resolutions has become a standard. Biophysical indices, such as leaf area index, fraction of photosynthetically-active radiation, phytomass or canopy chemistry, can be derived from the spectral properties of satellite imagery. Indices of changes in landscape composition and structure can be measured from the thematic maps originating from remotely-sensed imagery. Additionally, 30-year or longer time series of historical remote sensing archives (Landsat, AVHRR) allow retrospective studies of the historical range of variability and the trajectories of both landscape elements and biophysical properties.

A trade-off exists between high spatial and high temporal resolution when comparing platforms. Development of new, improved sensors and analysis techniques, such as sub-pixel classifications resulting in the development of continuous fields for formerly discrete classes, has reduced this trade-off. High spectral resolution and multiple view angles even enhance the potential for accurate retrieval of variables such as Albedo and chlorophyll concentration. Thus, powerful monitoring tools for land use/cover change detection are arising from such analyses. They can lead to an improved understanding of landscape states and processes. Finally, this evolution allows for mapping and monitoring of new landscape features that were not much used to date.

Keywords: leaf area index, satellite imagery, temporal resolution, continuous fields

F. Kienast, O. Wildi & S. Ghosh (eds.), A Changing World. Challenges for Landscape Research, 65–91.

Introduction to Remote Sensing for Landscape Oriented Ecologists

Landscape research is increasingly making use of remotely sensed data for analysis of pattern, gradients, and trajectories of landscape elements and properties. Here, we give an introduction to aspects of remote sensing relevant for landscape research – with emphasis on ecological applications rather than on the processing of data. Instead, we demonstrate the breadth of possibilities for using such data, and the promises and difficulties arising from combining remotely sensed sets with landscape and field data.

We identify three main purposes for integrating satellite or airborne imagery into landscape oriented ecological research: (a) pattern recognition for classification or mapping of landscape and other surface objects, (b) develop, map, or visualize continuous representation of surface properties, and (c) monitoring landscape trajectories and detecting change of both discrete or continuous features at the landscape scale. While historically the first two aspects were rather well distinguished, we nowadays see both aspects mixed in an array of applications.

Mapping landscape pattern through classification is a powerful tool for management and for habitat related analyses. Landscape ecology has advanced significantly from simple remote sensing through improved abilities to characterize pattern, analyze the formation and change of pattern through time, and evaluating its implications for populations, communities, and ecosystem processes (Turner *et al.* 2001). However, landscape-scale ecological research is often also focused on the analysis of gradients and how gradual changes of important variables in space influence the response and behavior of species (Austin 2002; Austin and Smith 1989; Collins *et al.* 1993), ecosystems (Chapin *et al.* 2002; Likens *et al.* 1977; Pomeroy and Albert 1988), or urban growth (Herold *et al.* 2003; Klostermann 1999). Ecological theory is thus highly influenced by such analyses and the development of statistics tightly follows these requirements. Thus both the classification of images and the derivation of indices and continuous fields from the same imagery can contribute significantly to the advancement of landscape research in an ecological context.

Remote sensing principles

Remote sensing has emerged as a comparably new tool in landscape ecological research, enhancing the recognition of pattern, the analysis of the state and the trajectories of landscapes, as well as the recognition of landscape properties. In combination with Geographical Information Systems (GIS), biophysical first principles, and modern spatial analysis methods it allows for a powerful retrieval and interpretation of pattern and processes of landscapes at a range of spatial and temporal scales. Here we briefly introduce the most important characteristics of remote sensing for ecologists working at a landscape scale.

Remote sensing is founded on the following principles: a) images of the earth (land and sea) surface and of the atmosphere are sampled from a rather large (remote) distance, usually from aircraft or from satellites, and b) recorded as the reflected electromagnetic radiation of the Earth's surfaces and objects (Fig. 1). This is done either passively by recording the ground reflected light of the solar illumination (in the visible, thermal or microwave range), or actively by recording reflectance and emissions from a source that has been projected from the same platform to the earth surface.

Remote sensing platforms most widely used in landscape ecology are based on passive systems, notably scanning the visible and the near infrared part of the spectrum (e.g. 400–1000 nm). The usefulness of detecting and recording reflected radiation originates from the fact that objects on the ground reflect and/or absorb incoming radiation selectively

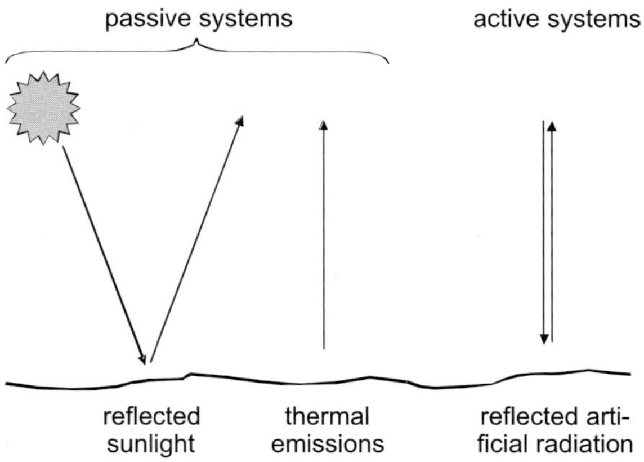

Fig. 1. Examples of the flow of radiation and remote scanning of reflections and emissions (adapted from Albertz (2001)).

(depending on the surface properties), and thus they reflect the incoming energy at variable intensities along a range of wavelengths. A comparison of incoming vs. reflected intensity of the radiation allows us: (a) to distinguish objects on the ground, and (b) to draw conclusions on the nature of respective objects.

Green leaves, for example, do not reflect sunlight at equal intensity across the irradiated spectral range, but mainly reflect the green and near-infrared wavelengths. Red and blue are absorbed and used to generate energy-rich biochemical compounds for photosynthetic processes. Consequently, a leaf appears to be green because red and blue colors are absorbed by the leaf. The high reflection of infrared and the low reflection of red wavelength have led to the development of various vegetation indices that allow us to map vegetation greenness on the Earth's surface (Huete *et al.* 1997; Payero *et al.* 2004). While remote sensing is used in a wide range of applications and research fields such as atmospheric sciences or oceanography we mostly concentrate on applications related to the ecology of terrestrial landscapes.

Most digital remote sensing instruments record electromagnetic energy along a specific scale by collecting photons, converting them to electrons, and saving them as digital numbers. The raw image has no true Earth projected coordinates, nor an inherent metric coordinate system. Intensive processing is required before such images can be used for further analysis. Processing involves geo-registering the images on the Earth's surface in a defined projection, and translating the digital numbers into meaningful physical units through calibration. Following an atmospheric correction step, the radiance will be converted to ground reflectance by removing the effect of atmospheric properties, such as trace gasses (e.g. ozone, water vapor) and aerosols. The so-called ground or surface reflectance is a directly comparable unit that can be used for multi-temporal and cross-sensor analysis (Jensen 2004). As this reflects an ideal case, a large variety of relative atmospheric correction approaches has been established to cope with this normalization procedure.

Not all applications require the same level of processing. Some vegetation indices may be derived from original brightness values, and not percent reflectance, since they are based on ratios rather than on real data values. Still, as a rule of thumb, we expect better and repeatable results if images are carefully processed with respect to registration and radiometric correction (Schlaepfer and Richter 2002).

A history of platforms

Remote sensing originated with the first aerial photographs taken from balloons and then migrating to powered aircraft (Cohen and Goward 2004). These photographs were visually analyzed and interpreted for a variety of applications. This approach was accelerated during the First World War and matured during the Second World War. Thereafter, remote sensing became standard for spying activities during the Cold War period, which required fine spatial resolution for a successful interpretation and recognition of identified targets on the ground.

With the launch of the ERTS-1 satellite (later renamed Landsat), a new era of civilian remote sensing came into existence. The Multispectral Scanner (MSS), which recorded reflected electromagnetic energy within 4 distinct "bands," or regions of the spectrum, allowed researchers to map vegetation distributions. These initial images covered comparably large areas (185 x 185 km) at an 80 m resolution, covering the non-polar regions of the globe every 18 days. With the launch of the 4th Landsat platform, the MSS was supplemented with the Thematic Mapper (TM), which was better calibrated and recorded energy within 7 bands of the electromagnetic spectrum ranging from visible to the thermal wavelengths. The TM had a 30 m spatial resolution and a 16-day revisit period. The current Landsat 7 platform supports the Enhanced Thematic Mapper with increased calibration and the addition of a 15 m panchromatic (b/w) band. These platforms have provided a near-continuous record of the Earth's surface for 32 years, and is likely to continue well into the foreseeable future.

Another operational civilian sensor is the NOAA satellite platform Advanced Very High Resolution Radiometer (AVHRR), which scans the entire earth surface every day at a spatial resolution of 1.1 km, and then is calibrated to means of 4 and later 5 spectral bands.

The selection of spectral bands measuring surface reflectance is critical for the development of ecological applications, although some satellites were originally developed for other purposes. In general, new multi- to hyper-spectral sensors for space and aircrafts are now developed with a placement of bands along the spectral axis from visible light through near to thermal infrared and onto microwaves to primarily detect different properties and characteristics on the solid surface, on water and atmosphere (see e.g. Ustin *et al.* 2004).

An array of sensors aboard satellite and aircraft now exist that can be distinguished with respect to spatial, temporal, and spectral (radiometric) resolution. Depending on their characteristics, they are designed to meet specific application related requirements. Two platforms carrying several different sensors relevant to landscape level ecology have recently been launched: EOS TERRA (and its sister platform AQUA) launched by NASA, and ENVISAT launched by the European Space Agency (ESA). The products from NASA are available for download on the internet either for free or for a small charge[1]. As is the case for the Moderate Resolution Imaging Spectroradiometer (MODIS) sensor data, some imaging products are fully post-processed by science teams, and thus ready to use for landscape level research.

For regional scale applications of many different kinds the Landsat (MMS/TM) sensors still hold a dominant role, as is obvious from the variety of applications that have been developed. Cohen and Goward (2004) identified 4 major reasons for this: (a) the data are comparably affordable, (b) long (+30 years) time series now exist, (c) the spatial resolution of the images is ideal for regional management (+/– 30 m), and (d) most of the important discriminating spectral domains for landscape research in the visible (VIS), the near infrared (NIR), and in the shortwave-infrared (SWIR) range, are available. This makes the TM

[1] See e.g.: http://edcimswww.cr.usgs.gov/pub/imswelcome/

sensor an ideal source for pattern recognition, gradient mapping and land change detection. In fact, the sensor concept of Landsat (and thus of many follow-up sensors, such as SPOT, IRS, or ASTER, and also the earlier NOAA-AVHRR) originated from the recognition that photosynthesis does not use the full light spectrum, but only a fraction of it, and that red and NIR bands discriminate differing vegetation patterns. Many of the ecologically relevant indices used today are based on Landsat data. These indices include the "Tasseled Cap", with brighteness, greenness, and (later with TM) wetness values, calculated from the Landsat images. In addition, many different vegetation, soil and geologic indices were developed from Landsat. These were mostly based on a simple ratio (NIR/RED), and the Normalized Difference Vegetation Index (NDVI = (NIR+RED)/(NIR-RED), both of which express the amount of green leaf mass, specific leaf area, or chlorophyll content.

The NOAA AVHRR – originally developed for meteorological purposes – was first used by Compton J. Tucker of the U.S. Geological Survey to explore it's applicability for ecological applications. He found that the NDVI concept could – with modifications – well be applied to AVHRR (e.g. Tucker *et al.* 1985). A second sensor was thus now available, which was complementary to Landsat with similar advantages, such as: (a) it is free of charge, (b) long (+30 years) time series exist, (c) the grain size of the images is ideal for continental to global scientific analyses (~1.1 km), and (d) most of the important discriminating spectral domains for landscape research in the visible (VIS) and in the NIR range are available. The lower spatial and radiometric resolution is compensated by the much higher (daily) temporal resolution. Because of this high temporal resolution, the AVHRR sensor was instrumental in the identification of vegetation and other surface patterns, and in the detection of global level climate change effects (Zhou *et al.* 2001) by linking these pattern with climate variables (Nemani *et al.* 2003).

Pattern Recognition and Image Classification

Classification of remotely sensed imagery with the purpose of obtaining thematic maps of the landscape is one of the main activities related to landscape oriented remote sensing. Supervised (with prior knowledge of the classes to be obtained) and unsupervised (based on statistical clustering without prior knowledge of the classes to be obtained) classification have both been used. In applications relevant to landscape ecological research, both supervised and unsupervised classifications are appropriate, but they differ with respect to the goal. The first approach aims at classifying an image based on a predefined classification scheme. Such efforts are usually appropriate if a pre-defined schema is necessary to solve associated scientific or management questions. The second approach is used mainly as an exploratory analysis tool in order to objectively find pattern at the landscape scale. By this, finer nuances of the landscape pattern may be found that are not known prior to the analysis. The principle task resulting from unsupervised classifications is to associate the obtained classes with meaningful ecological descriptions. Both approaches have long been used to generate vegetation maps from single- to multi-scene images (Homer *et al.* 1997) for the purpose of mapping, or for the derivation of additional ecological or conservation information (Scott *et al.* 1993). While classifications using a single scene may be comparably easy to achieve, doing the same from multi-scene mosaicked imagery (or even more so when combining scenes through time) presents many additional difficulties.

The "Gap Analysis Program" (GAP) within the US is a good example of classification efforts at landscape scales. This program is founded on the recognition that a species-by-species or threat-by-threat approach to protect individual wildlife species or to maintain species richness is not a promising approach (e.g. Burley 1988; Scott *et al.* 1993; Scott and

Jennings 1998). As an alternative, the GAP program attempts to: (a) provide an overview of habitat resources across large spatial domains (classification of land cover at a state or regional level), (b) evaluate the potential distribution of animals (and plants) based on mapping suitable habitats, and (c) identify gaps in the protection of biodiversity at state, regional, and national scales. Thus, the classification is meant to support management purposes, which is often the case in classification exercises. Edwards *et al.* (1996) tested the usefulness of the GAP approach for a variety of species in National Parks in Utah using the land cover classification from Homer *et al.* (1997). They concluded that the GAP-based modeling process seems robust enough to provide a reasonably high level of accuracy for use in conservation planning at the ecoregion level.

Many different classification approaches are in use for the GAP program, where scientists usually have to face the problem of classifying complex vegetation pattern across comparably large spatial scales, which requires the simultaneous processing of a large number of Landsat scenes. Most often, though, scenes are first mosaicked together and corrected for atmosphere effects, either directly using e.g. the ATCOR method (Richter 1990, 1998), or indirectly, using pseudo-invariant features (Hall *et al.* 1991, Jensen 2004). An initial unsupervised classification generates spectral classes for which field data is collected to discriminate vegetation types in a post-classification stage that share similar spectral, but dissimilar ecological properties (Homer *et al.* 1997). Statistical rules are then applied in order to optimize this post-classification processing. Often, the field work and the subsequent model refinement is stratified by ecoregion (Homer *et al.* 1997, He *et al.* 1998; Lunetta *et al.* 2002). Such stratifications by ecologically meaningful parameters tend to improve classification accuracies (De Bruin and Gorte 2000).

Other classification efforts have concentrated on the global level, resulting in a series of products with various detail and (dis-)advantages. One of the first attempts based on an international effort was the International Geosphere and Biosphere Programme (IGBP) land cover map based on NOAA-AVHRR data (Belward *et al.* 1999; Loveland and Belward 1997; Townshend *et al.* 1994). Other examples include the PELCOM land cover map (Mucher *et al.* 2000) or the UMD global land cover map (Hansen *et al.* 2000), both based on AVHRR. Nowadays, such global land cover maps have become a standard product under the MODIS data processing (Friedl *et al.* 2002). The products are used primarily as inputs into various global simulation models related to atmosphere and vegetation dynamics (Sellers *et al.* 1997; Townshend *et al.* 1994). While such pre-defined classification schemes are appropriate for many purposes, the mixing of pure land cover classes within pixels due to the continuous variation found in the landscape (Foody and Hill 1996), and due to the mixed nature of land cover at coarse (≈1 km) spatial resolution (Schowengerdt 1996), may cause problems. A more continuous representation of the proportional fraction of individual cover classes per pixel may serve as an alternative to the thematic classification approach.

Image classification of vegetation and other ecologically relevant features (e.g. dominant species, habitat types, cover) at a landscape scale is common. In most cases, spectral information is the predominant source to map such features, though – as discussed in the GAP approach – ancillary information such as ecoregions, or topographic information is often added, especially in post-processing of unsupervised classifications. Rarely though, such classification efforts make use of ecological theory, the niche concept or the wide array of tools and techniques often used in predictive habitat modeling (see Guisan and Zimmermann 2000 for a review). While the latter approaches often lack the spatially explicit recognition of pattern (which is one of the key advantages of using remotely sensed imagery), they make the best use of the links between environmental drivers (climate, soils, etc.) and the resulting habitat distribution pattern. Such links are not only founded on statistical procedures, but also on ecological theory (see e.g. Austin 2002; Austin and Smith 1989; Collins *et al.* 1993;

Guisan and Zimmermann 2000) for linking such concepts with predictive habitat modeling. We argue here that image classification – especially when carried out on multi-scene imagery across large spatial domains – could significantly profit from links to ecological theories, and from including ecologically relevant drivers such as heat sum, moisture balance, or global radiation for improving image classifications. On the other hand, predictive habitat modeling can profit equally much from the fact that remote sensing allows recognizing and map successional stages, rather than simply the environmental potential of species or habitats.

Characterization and Mapping of Continuous Gradients

One of the most powerful approaches in ecological research is to map gradients first and classify later (if necessary), meaning that many aspects of nature are hidden if objects are classified first, and then described in terms of properties later. Thus, properties and characteristics of landscapes or ecosystems should better be described directly as gradients that own various behaviors in space and time. Two advantages arise from this approach; (a) any property can be mapped and monitored at its proper spatial and temporal scale and spatial dimension, and (b) the detection of change (of a property) is directly observable from multi-temporal image analyses, and not restricted to the change of classes. One example for this is the monitoring of forest resources using thematic change detection procedures. In this situation, binary maps are generated to delineate forests from non-forests, and change detection is based on the analysis of subsequently derived forest/non-forest maps. This inherits an initial problem, i.e. what is considered a forest. Thus, training data inevitably represent a classification scheme. Once calibrated, we can only detect change if a pixel in a classified image switches between forest and non-forest. Gradual changes are not observable. If pixels are comparably large and if the thematic classification of an image includes many cover classes, then change detection becomes even more difficult and error prone. This is due to the mixed nature of the landscape classified, with few pure pixels available only (Foody and Hill 1996; Schowengerdt 1996).

One solution to this is to monitor the forest or tree cover fraction directly. Represented as a continuous layer, we can monitor the change in forest resources more directly, for finer gradual change. We need only to consider the uncertainty of the continuous map as a source of error for change detection. The approach described above has resulted in the concept of land cover and vegetation continuous fields (DeFries *et al.* 1999; Hansen *et al.* 2002, 2003) at the global scale as an alternative to a fixed classification scheme (Belward *et al.* 1999; Friedl *et al.* 2002; Loveland and Belward 1997; Townshend *et al.* 1994).

In general, the representation of continuous fields relies on three basic approaches, though they are often mixed (Pickup *et al.* 1993); (a) derivation of indices from band combination, (b) development of continuous layers by using statistical regression techniques and a set of training data, and (c) resolving linear (spectral) mixture models (Smith *et al.* 1990) to map the fraction of classes per pixels for which a spectral signature is available. The first approach (a) is most often used, and we discuss it with more detail below. The second approach (b) was introduced above, and one example of continuous field calibration is presented for a study area in the European Alps below. The third approach (c) is conceptually similar to approach (b). The difference is that in the last approach, pure classes are calibrated spectrally, and then mixtures are derived from spectral gradients between the spectra of pure classes.

The combination of bands to derive indices is often preferred over other approaches, because it tends to give satisfactory results without rigorous atmospheric correction, especially when single scenes from one time step only are used. On the contrary, most other approaches are very sensitive to rigorous atmospheric correction schemes, because small differences in the reflectance signal can significantly affect the results. An array of approaches exist for representing continuous gradients of vegetation, snow, soil or geology characteristics using indices from band combinations. For instance, vegetation indices (VI) have their origin in the late 1960s, when Birth and McVey (1968) and later Jordan (1969) developed ratios of NIR to Red, later called "simple ratio" (see Cohen and Goward [2004] for a discussion of the history of some indices). VIs are empirically-derived linear and non-linear models of the biophysical relationship of vegetation to solar radiation, e.g., leaf physiology, vegetation cover and biomass (Cohen and Goward 2004; Tucker 1979; Sellers 1985). Vegetation indices are generally separated into two groups: perpendicular or linear combinations (orthogonal), and ratios (Baret and Guyot, 1991; Huete *et al.* 1985). Kauth and Thomas (1976) developed a three band "tasseled cap" orthogonal transformation that uses the four MSS bands to create three indices called: (1) soil brightness index (SBI), (2) green vegetation index (GVI), and (3) yellowness or non-such index (NSI). Yellowness is sensitive to haze. The perpendicular vegetation index (PVI) of Richardson and Wiegand (1977) used two Landsat bands (NIR and red) to develop a measure of brightness and greenness. Soil background reflectance usually disturbs VIs; thus, attempts were made to understand and eliminate soil effects. The soil line is a two-dimensional variation of the Kauth-Thomas SBI and a plot of NIR and red (MSS7 and MSS5). Richardson and Wiegand (1977) successfully demonstrated that soils fall on this straight line. Presence of vegetation causes the red radiance to decrease because of strong absorption in the visible spectral region by chlorophylls (Knipling 1970). NIR radiance increases with the presence of vegetation due to internal leaf scattering and no absorption by chlorophylls in the NIR spectral region (Knipling 1970). The orthogonal distance between the soil line and the vegetation spot is a measure of the amount of vegetation (Richardson and Wiegand 1977). This principle remains the same for n-space indices greater than two. The third dimension (yellowness) is measured orthogonally to both brightness and greenness, and the fourth dimension (non- such) is mutually orthogonal to brightness, greenness, and yellowness (Jackson 1983).

These n-space indices were thought to be particularly useful for discriminating vegetation from soil background (Jackson 1983), but Huete *et al.* (1985) showed that normalization of the soil background to a one-dimensional soil line only removes the bare soil spectral influences (differences due to soils of different types in the same brightness range), and not the greater soil brightness influences. Thus GVI and PVI are sensitive to background soil effects that can seriously alter vegetation readings (Huete *et al.* 1985; Huete and Jackson, 1987; Huete 1988). Ratio indices such as the Ratio Vegetation Index (RVI), the Normalized Difference Vegetation Index (NDVI) of Rouse *et al.* (1973), or the Transformed Normalized Difference Vegetation Index (TNDVI) of Deering *et al.* (1975) use various ratios of red and near-infrared bands to characterize vegetation. By this, they are more robust to changes in the background soil effects as well as to variations in atmospheric conditions. They are thus better suited for longer term monitoring and change detection (Pickup *et al.* 1993).

The NDVI is still to date – despite many disadvantages – the most commonly used vegetation index. NDVI has been used as predictor for LAI, vegetation cover, biomass, or even NPP (Ustin *et al.* 2004). Other indices were later derived thereafter to eliminate some of the disadvantages, most recently the Enhanced Vegetation Index (EVI), developed by Huete *et al.* (1994).

Tradeoffs in Resolution

Three types of resolution are relevant to remote sensing data: spatial resolution (grain), temporal resolution (frequency), and spectral resolution (detail). Additionally, sensors differ in the sensitivity in recording electromagnetic energy. Any sensor available combines characteristics of these three domains, but no existing sensor is optimal in more than two of the three domains (Fig. 2). Despite ever increasing computational power, these general tradeoffs in resolution remain, likely due to the limited downlink capacity. For model development, similar tradeoffs have been proposed (Levins 1966), and the consequences seem largely similar. In model building, general theory says there is no single model possible that optimizes precision, generality and reality simultaneously. Thus any model has advantages or disadvantages with respect to the 3 possible characteristics. With respect to sensors, this means that the goal of a study largely determines the selection (or combination) of suitable sensors for a project.

High spatial resolution is one aspect, which traditionally has been important. It is helpful primarily for pattern recognition through image interpretation, classification and segmentation. Usually, high resolution sensors show low temporal resolution, but in the case of airborne systems, they may also provide high spectral resolution (AVIRIS, HYMAP, ROSIS, see e.g. Kötz *et al.* 2004). Modern high resolution images (e.g. IKONOS) are thus well suited to be linked with historical air photography for change detection (Herold *et al.* 2003).

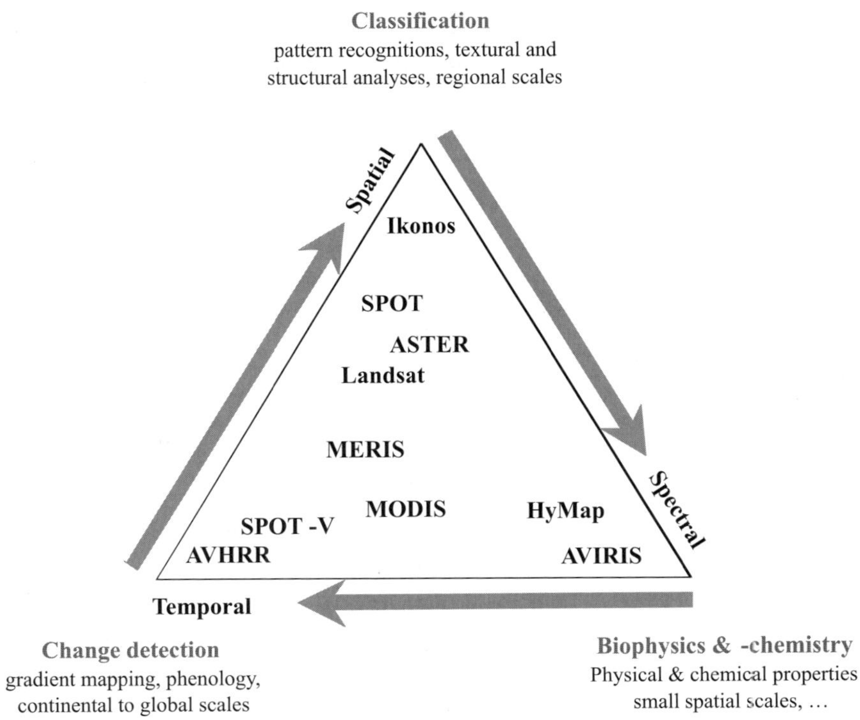

Fig. 2. A classification based on resolutions. No sensor simultaneously optimizes high resolution of more than one property among spatial (grain), temporal (frequency) and spectral (detail) resolution.

High spectral resolution allows for detecting a range of bio-physical and bio-chemical properties, such as foliar nitrogen and/or lignin content (Serrano *et al.* 2002), leaf or soil water content (Baret and Fourty 1997; Danson *et al.* 1992; Kötz *et al.* 2004; Zarco-Tejada *et al.* 2003), leaf area index (Chen *et al.* 1999; Kötz *et al.* 2004), photosynthetic pigments (Daughtry *et al.* 2000; Dawson *et al.* 2003; Haboudane *et al.* 2002; Jacquemoud *et al.* 1995; Zarco-Tejada *et al.* 2000), dry plant matter (Fourty and Baret 1997), or soil properties (Shepherd and Walsh 2002). Additionally, such imagery allows for mapping vegetation properties such as individual plant species, which is not easily feasible from low spectral resolution imagery (e.g. invasive weeds mapping, Aspinall 2002). Careful image processing is a pre-requisite to successfully retrieve such properties. Such data are a great promise for scaling eco-physiological properties from stands to larger landscape. This has significant power to fill the scale gaps between field (plot-level) data and coarse spatial resolution imagery such as MODIS or AVHRR for model calibration and initialization.

High temporal resolution is usually available only for coarse grain space-borne imagery such as AVHRR, MODIS or – to a lesser degree – MERIS. The AVHRR and the MODIS sensors are key platforms to enable the analysis and processing at continental and global scales, such as up-scaling of physiological and biological processes, to evaluate changes in ecosystem response (Myneni *et al.* 1998, 2001; Nemani *et al.* 2003), to detect land cover change (Langevin and Stow 2004; Stow *et al.* 2004), or to simply characterize basic land cover pattern and mixtures (IGBP MODIS, VCF) needed for assessment of global resources and initialization for global process modeling (Running *et al.* 2004). The high temporal resolution allows for evaluation of landscape change at short time intervals (e.g. for analysis of phenology, Duchemin *et al.* 2002; Moulin *et al.* 1997, Myneni *et al.* 1997; Zhou *et al.* 2001), or for rapid assessment of natural hazards such as fires, floodings, or drought (e.g. San-Miguel *et al.* 2000), and compensates for the low spatial resolution to a certain degree. Compositing of 8–10 day sequences by pixel-wise evaluation of the best atmospheric conditions generate images that are global in extent and nearly cloud free. Integrating successive cloud free 8–10 day composites through time allows the analysis of annual phenology pattern, and provides a means for vegetation types to be distinguished that are hard to separate spectrally using a single date. Such high temporal frequency is one of the key factors for the successful separation of fractions of needleleaf and broadleaf tree cover fractions in vegetation continuous fields (Schwarz *et al.* 2004).

Modern Environmental Remote Sensing

Below we present a few examples of modern environmental remote sensing applications focusing on several of the topics discussed above. The material of these efforts is currently under development or has recently been published.

The Southwestern Gap Analysis Project (GAP)

Traditional image classification and feature extraction from remotely sensed imagery has consistently fused imagery with ancillary data such as elevation, slope, aspect, and moisture indices, surface geology, soils, or climate. This is particularly true when large landscapes are being mapped. The tradeoffs in resolution described in this paper are consistent for remotely sensed imagery. However, new techniques and improved sensor engineering are pushing these limits. Another common tradeoff in the application of remote sensing to landscape level mapping is the concept that increasing extent requires decreasing spatial and thematic resolution. Commonly, continental and global mapping efforts focus on low resolution (250 m – 1 km) imagery and a generalized thematic legend while on the other extreme, local area mapping commonly use sensors of higher spatial resolution (1–5 m) and a expanded thematic legend. The challenge to remote sensing is to therefore expand extent and increase resolution. The Gap Analysis Program administered through the Biological Resources Division of the U.S. Geological Survey (USGS-BRD) seeks to map land cover and wildlife habitat at scales appropriate to wildlife studies and usable to land managers and researchers.

The Southwestern Gap Analysis Project that covers the states of Arizona, Colorado, Nevada, New Mexico, and Utah is considered a regional effort covering an area roughly 1/5 the size of the conterminous U.S. It has mapped land cover at relatively fine resolution (30 m) and relatively fine thematic scale (>110 thematic classes). This project utilized three seasonal dates of Landsat Enhanced Thematic Mapper imagery (spring, summer, and fall) over the entire region (240 images) combined with topographic data and landform to map land cover (Fig. 3). These spatial data were trained with approximately 90 000 field data

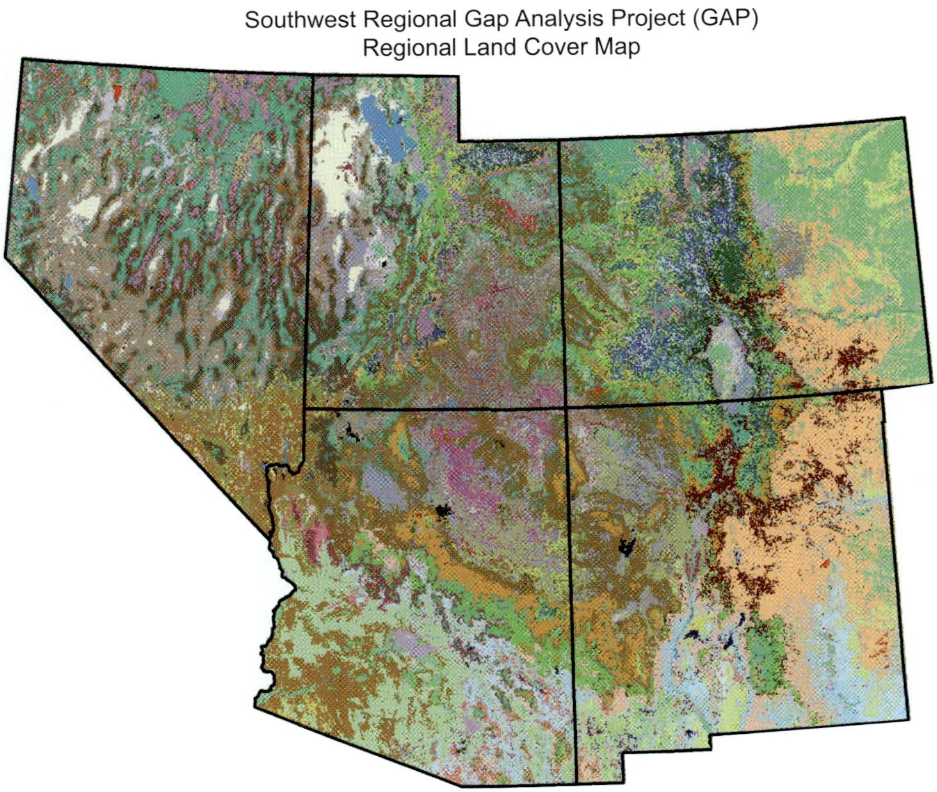

Fig. 3. Soutwestern GAP regional land cover map. The map distinguishes >110 land cover types, classified from 240 multi-temporal Landsat TM images (3 seasons) across the states of Nevada, Utah, Colorado, New Mexico and Arizona. Colors represent individual cover types.

points collected by the mapping team and collaborators. This effort underscored the need for the fusion of imagery and ancillary data, and the importance of field training to drive statistical classification tree models that developed the mapping rules.

We have mentioned the importance of ecological theory to help drive classification models. This effort utilized ancillary data known to predominantly drive land cover distribution in this region (topography and landform). Other ancillary data such as surface geology and soils were also explored, but were found to be heavily correlated to the derived topography and landform layers. Therefore, while classification tree models allow for the inclusion of multiple predictor variables, this study found that a minimal set of quantitatively derived landscape descriptors coupled with spectral information effectively mapped land cover. In this case knowledge of the driving variables in the local ecosystem, coupled with field measures, effectively developed the ecological associations at a quantitative level rather than utilizing existing ecological theory which was more generalized and developed over time by observation and study.

Disentangling climate and grazing effects upon productivity: a spatio-temporal analysis

In a multi-temporal study using 27 years of dry- and wet-season Landsat TM data conducted at 88800 ha of the Deseret Land and Livestock Company ranch (northeastern Utah), Washington and co-workers (Washington-Allen 2003; Washington-Allen *et al.* 2003, 2004) analyzed the combined effects of land use (here grazing intensity) and climate (here

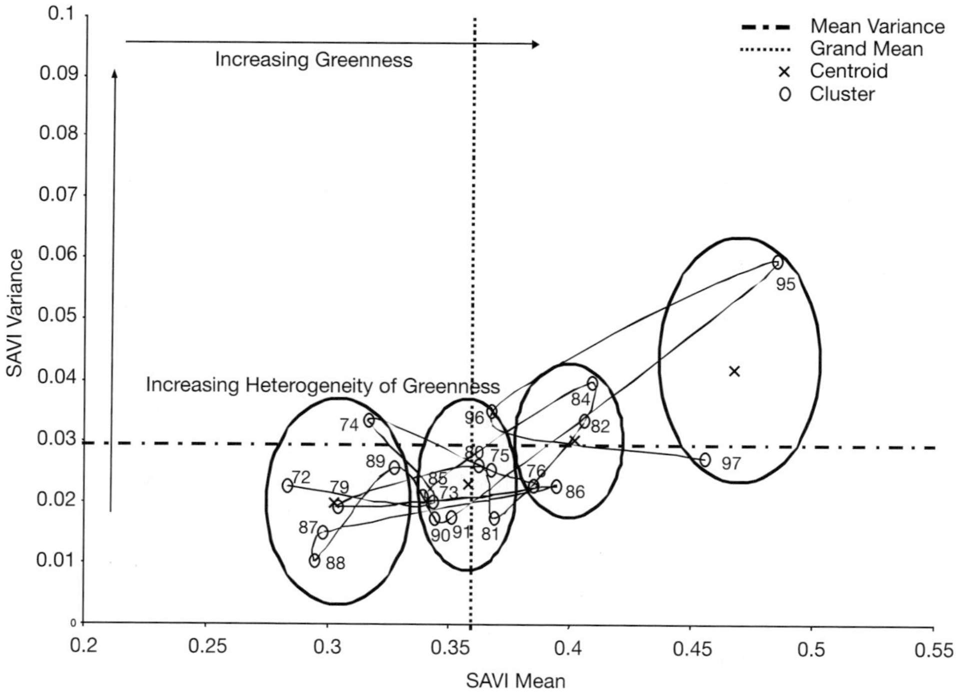

Fig. 4. The SAVI mean-variance analysis from 1972 to 1997 (numbers indicate years, e.g. "95" = 1995) for the sagebrush-grassland community on Deseret Land and Livestock Company ranch, Utah.

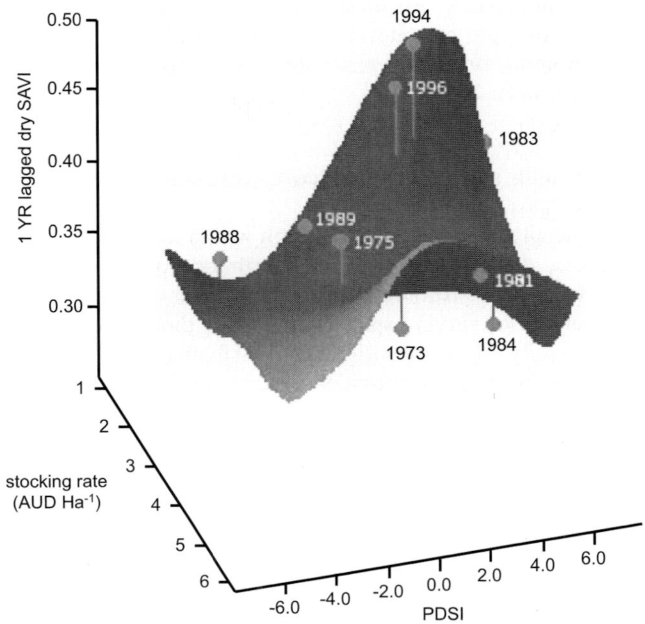

Fig. 5. The one-year lagged soil-adjusted vegetation index (SAVI) response surface with respect to grazing (AUD Ha^{-1} = Animal unit days per area; expressing stocking rate) and climate (PDSI = Palmer Drought Severity Index; a regional-scale measure of soil water availability).

rainfall) on land productivity (expressed as the soil adjusted vegetation index, SAVI, Huete 1988). Time series of TM data were processed using a relative atmospheric correction based on temporally invariant features (Jensen 2004) prior to converting the images to SAVI. In water-limited systems, such as the Great Basin cold desert in Utah, the expectations are low vegetation index (VI) mean and variance during drought periods and high VI mean and variance during wet periods (Hanley 1979). Such system cycling between two states is labeled a limit cycle attractor (Hanley 1979). Higher variance during wet periods is hypothesized to be a function of landscape heterogeneity manifest as differences in soil texture and its effects on soil moisture availability for vegetation productivity (Fernandez-Illecs *et al.* 2001). Hanley (1979) hypothesized that sagebrush steppe vegetation response is primarily constrained by grazing and wet and drought climatic episodes.

Mean-variance analysis (Pickup and Foran 1987) confirmed Hanley's hypothesis by indicating that sagebrush steppe is a limit cycle attractor, with two main states of SAVI (vegetation abundance) during wet and dry periods (Fig. 4). The wet periods are signified by the presence of two very strong El Niño events (in 1983 and 1997) and a 1988 La Niña event: the Great North American Drought (Trenberth *et al.* 1988), respectively. Secondly, the sagebrush steppe landscape's overall SAVI trajectory was increasing non-linearly from 1972 to 1997, due primarily to the very strong El Niño events (Washington-Allen *et al.* 2003). One year lagged SAVI was found to be linearly correlated ($r = 0.62$, $p = 0.006$) with the Palmer Drought Severity Index (PDSI, a regional-scale measure of soil water availability) and non-linearly correlated ($r = 0.63$, $p = 0.04$) with grazing using a first order difference ordinary least-squares regression of climatic factors (Washington-Allen *et al.* 2003; Fig. 5). The surface generated in Figure 5 is a combination of the linear and non-linear responses. Finally, after landscape level impacts induced by repeated droughts in 1977, 1986 to 1991, vegetation recovery from these catastrophic episodes appears to be aided by very strong El Niño events, a finding that presents land managers with an opportunity to exploit this natural subsidy (Holmgren and Scheffer 2001).

Calibrating continuous fields across scales of spatial resolution

One way of conserving small scale spatial information within images of coarse spatial resolution (see also Lischke *et al.* 2007) is to map the distribution of the small scale entities as proportions per pixel of the coarse resolution image. The cover of trees, grass, snow, or open water on the ground may be of varying spatial dimension, though certain objects such as individual tree crowns usually cover areas significantly smaller than 1 hectare. While high resolution imagery allows for mapping broader cover types covering the full pixels, this becomes non-trivial at coarser spatial resolution, since naturally such pixels consist of a mixture of several cover types (Schowengerdt 1996). One alternative is to map fractions of cover types, also termed continuous fields (De Fries *et al.* 1999; Hansen *et al.* 2002).

A key problem for the derivation of such continuous fields is the calibration of images of coarse spatial resolution. Spectral mixture analysis allows resolving the spectral signal of

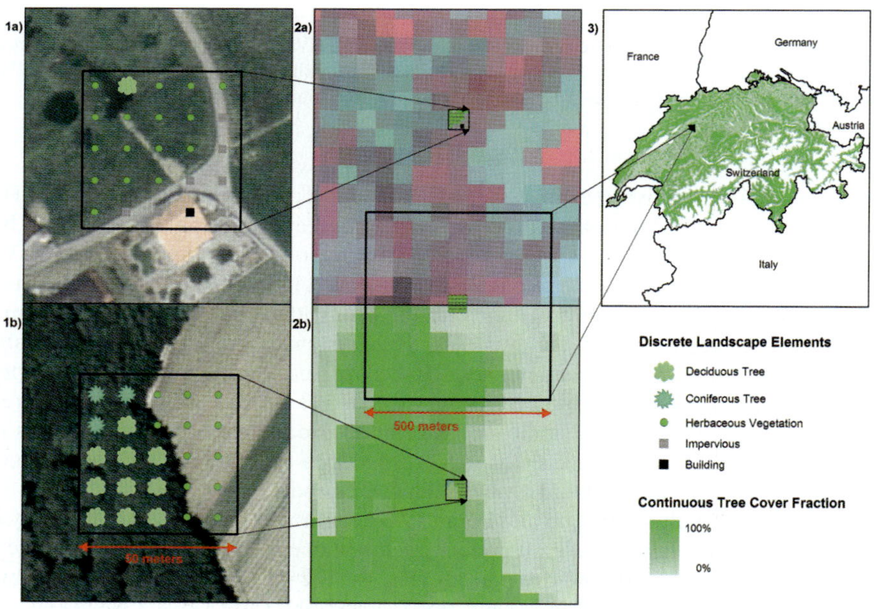

Fig. 6. Predictive up-scaling of land cover information from high spatial resolution (50 cm) aerial photos in (1a,b) to medium resolution (50 m) Landsat TM data (2a,b), and finally to moderate resolution (500 m) MODIS data in (3). Discrete landscape elements are mapped at local scale (Swiss National Forest Inventory sampling sites, 1a,b). Fraction of Tree Cover as derived from these sampling sites is then applied to medium resolution multi-temporal TM data (2a; Landsat composite) using a GLM model to produce spatially continuous fraction of tree cover (2b) at Landsat scale. The model allows map fractional cover across the whole TM scene. When aggregating the TM scene to moderate (500 m) resolution, it can serve as training data for high temporal MODIS reflectance data using GLMs. Direct calibration of 1 to 3 – however – is not possible.

each pixel as a linear mixture of known end members. We propose here another solution by hierarchically calibrating multi-temporal Landsat TM data from field plot data, then calibrating MODIS MOD09 (500 m) data using the up-scaled TM data. The calibration is based on Generalized Linear Models (Schwarz *et al.* 2004; Schwarz and Zimmermann 2005), which are an ideal statistical tool to cope with linear and non-linear mixtures and distributions (Fig. 6). Besides the statistical calibration, the method is highly similar to the derivation of vegetation continuous fields available as global data sets based on MODIS (500 m) and AVHRR (1.1 km) reflectance data (De Fries *et al.* 1999; Hansen *et al.* 2002, 2003).

Imaging spectroscopy for mapping biophysical and -chemical ecosystem properties

Imaging spectroscopy deals with the simultaneous and continuous acquisition of a large number of narrow spectral bands over the solar-reflected range of the electromagnetic spectrum (Goetz *et al.* 1985). The subsequent interpretation of this information for the characterization of earth surface properties is called imaging spectroscopy (Green *et al.* 1998). In contrast to multispectral sensors, imaging spectroscopy exploits quantitative estimates of spectral features caused by the complex absorption and scattering processes at the Earth surface, which can be analyzed to improve the scientific understanding of ecosystem functioning and properties (Asner *et al.* 2000; Ustin *et al.* 2004).

The spectral reflectance of a vegetation canopy, provided by air or spaceborne imaging spectrometers, is known to be primarily a function of the foliage optical properties, the canopy structure, the background reflectance of understory and soil, as well as the illumination conditions and viewing geometry (Asner 1998; Chen *et al.* 2000; Goel 1988; Rast 2001). The complex radiative transfer within a canopy governing the signal recorded by imaging spectrometers can be described by physically based radiative transfer models (RTM), which take into account the above mentioned factors (Goel and Thompson 2000; Kimes *et al.* 2000). The use of such a RTM for a comprehensive retrieval of the biophysical and -chemical canopy properties from imaging spectrometer data was demonstrated in a regional that is discussed below (Kötz *et al.* 2004).

During a large field campaign in the Swiss National Park, in summer 2002, DAIS 7915 and ROSIS imaging spectrometer flights were carried out along with intensive ground measurements of the vegetation properties (Fig. 7). The study site represents an inner-alpine valley covered by boreal sub-alpine forest (dominating species: *Pinus montana* ssp. *arborea,* average altitude 1900 m a.s.l.). Canopy structure was described by two canopy analyzers LAI2000 and hemispherical photographs following well known methods for the characterization of heterogeneous canopies such as coniferous forest (Smolander and Stenberg 1996; Chen *et al.* 1997; Weiss *et al.* 2004). Standard wet-laboratory procedures were used for determination of foliage water content and dry matter. The inversion of the coupled radiative transfer models GeoSAIL (Huemmrich 2001) and PROSPECT (Jacquemoud *et al.* 1996) allowed for a quantitative retrieval of vegetation variables such as fractional cover, LAI, water content and dry matter from imaging spectrometer data. The derived variables represented the actual spatial distribution of the biophysical and –chemical properties as they occur in the landscape (Fig. 8). The spatial information of such quantitative information on forest conditions provide a sound data basis relevant to forest management and to the quantification of ecosystem pools (e.g. carbon). Maps of vegetation variables can also serve as initial starting and boundary conditions of ecological models, improving and validating model predictions on a finer spatial scale (Running *et al.* 1999). If time series of imaging spectrometer data are available, ecological models could also be driven by remote sensing following assimilation or forcing approaches (Delecolle *et al.* 1992; Verhoef and Bach 2003).

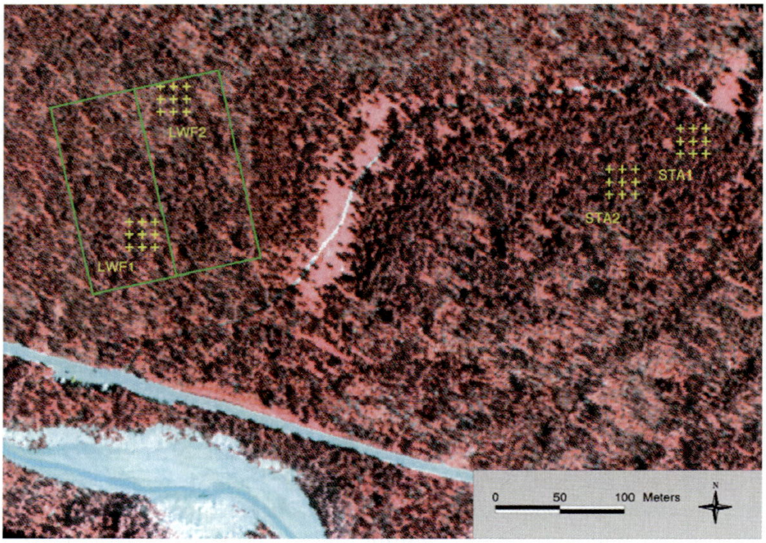

Fig. 7. Airborne imaging spectrometer data over four core test sites in the Swiss National Park (crosses indicate the sampling points). The image composite represents geo-coded and atmospherically corrected data of the spectrometer ROSIS in spatial resolution of one meter resolving the heterogeneity of the observed forest. The Long-term Forest Ecosystem Research site of WSL where forest stand characteristics are acquired is indicated by the green rectangle.

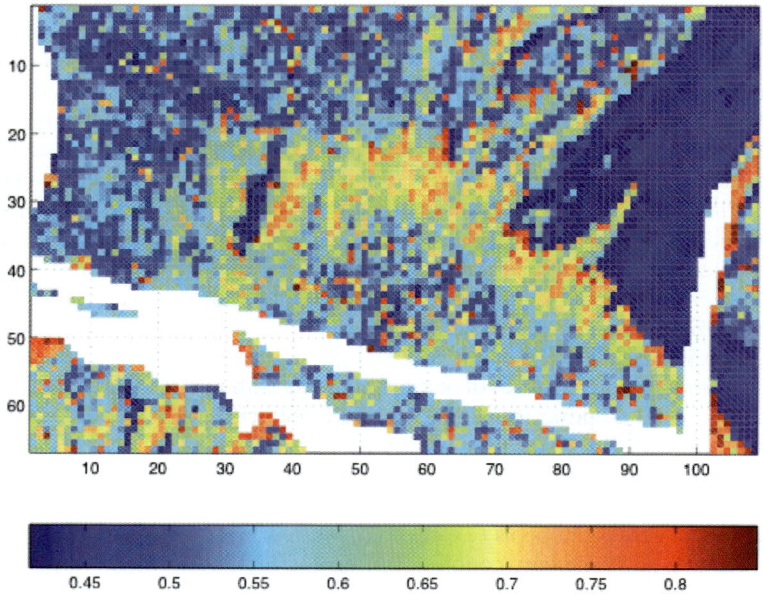

Fig. 8. Map of fractional tree cover, as example of the derived biophysical and -chemical properties in the Swiss National Park. The employed retrieval algorithm follows the approach proposed by Kötz *et al.* (2004). Non-vegetated surfaces are masked.

Guidelines for Linking Field Sampling with Remotely Sensed Data

Issues of positional accuracy and scale

One of the key issues for linking field and other data with remotely sensed imagery is accuracy and scale. This linkage is a prerequisite to calibrate any image for further processing. Any geo-registered image has a certain amount of positional error, which usually is tested against grain. An image is considered well geo-registered if the error is below the side length of one pixel. Field data is nowadays often geo-located by GPS, which on average is of high spatial accuracy. However, we may see positional uncertainties that are significant with respect to pixel size in mountainous terrain, within forests, or in comparison with imagery of small grain (e.g. IKONOS, airborne). If, for example, field data and image both have a positional accuracy of +/– 0.5 pixel size, then in most cases the point may not be within the pixel it is sampling. The second problem associated with linking field data with remote sensing imagery is the spatial extent of the field plots and the grain of the image cover on the ground. Are they of similar extent? Both uncertainty and scale have important implications for sampling and analysis.

Field data collection needs to take the spatial resolution of images into account: a plot size well below one pixel seriously reduces the statistical power of any analysis. If spatial heterogeneity of the spectral response of the surface is high, small plots may represent the reflectance and derived indices of the sampled pixels. This leads to the recommendation that sample plots should be close to one pixel in size, or larger than one pixel. If sampled plots are smaller than one pixel, averaging multiple smaller scale measures per pixel may lead to significantly increased power when relating to image pixels (White *et al.* 1997). Another consideration is that of the high spatial autocorrelation present in these images. In this case adjacent pixels influence the spectral response of neighboring pixels. Therefore samples should be located in areas consisting of small pixel groups (e.g. 3x3 pixels) that are relatively uniform in spectral response. This will ensure an accurate sampling of the spectral response of the target and provide a buffer for spatial errors associated with rectification errors or GPS precision. If heterogeneity is to be incorporated for spectral mixture analysis or subpixel classification, several plots within one pixel provides for a statistical estimate of the mixture of the cover (or spectral signature) classes (Adams *et al.* 1986, Boardman 1989, Boardman 1990).

If positional uncertainty is comparably high, then we recommend considering: (1) averaging neighboring pixels (e.g. Mäkelä and Pekkarinen 2004; Wang *et al.* 2004), and (2) increasing the spatial dimension covered by the field sampling by either sampling larger plots relative to pixel size, or by allocating the plots spatially nested to sample the heterogeneity within the spatial domain larger than one pixel. As an alternative to (2), one could calibrate a finer resolution image first, and use this as an input to calibrate the coarser resolution image (DeFries *et al.* 1999; Hansen *et al.* 2002; Schwarz *et al.* 2004).

Figure 9 illustrates the problem of linking field data with remotely sensed imagery where both have considerable positional uncertainty: the bold circle in the middle represents the true center of a sampling plot, while the grey shaded rectangle represents a pixel of perfect geo-registration. Small dots represent true centers of plots, the open circles illustrate a positional accuracy of ca. 0.5 pixel size around the plots, while the open rectangles represent the center pixel with similar registration error. It is obvious that not all of the points fall within any of the rectangle, and with respect to plot area around these centers (see upper left in figure) it is even worse. Increasing the sample size by sampling plots along radii, as illustrated by the four transects, is one possibility to cope with this problem on the ground. Averaging pixels in a neighborhood helps to cope with the positional uncertainty of the image as well.

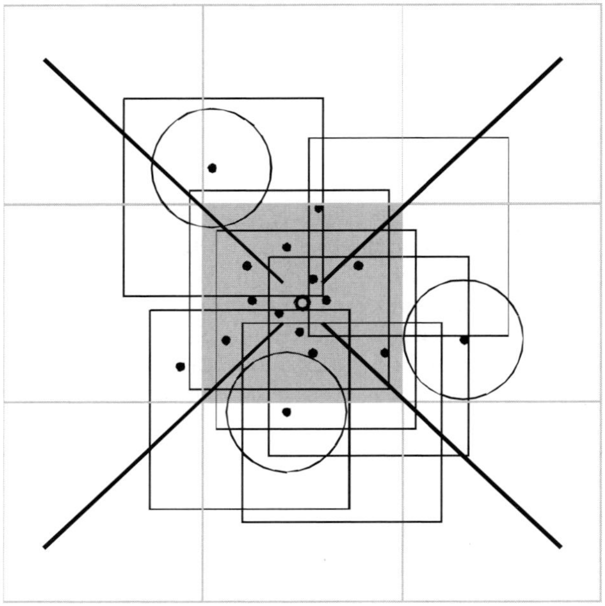

Fig. 9. Issues of positional accuracy and scale. The grey rectangle represents a perfectly geo-located pixel, the bold circle is the true positions of one field plot. Open rectangles represent various possible pixel positions that on average have an error of below 0.5 pixel dimension and the small dots represent true positions of additional field plots. The open circles illustrate plot-related positional uncertainties of similar dimension as for the pixels for three plots. Sampling larger spatial domains (along e.g. transects) and averaging neighboring pixels may help to improve linking field data and imagery that have considerable positional error.

If imaging spectroscopy is used, averaging among spectral bands may additionally help resolve uncertainty in the spectral image processing (e.g. Aspinall 2002).

Such issues not only arise when linking field data with digital imagery, but with any kind of linking spatial data, be it GIS, air- or space-borne imagery, field-surveyed maps, or field data collection. Rigorously testing (and thus knowing) the positional accuracy is a prerequisite for the successful linking of spatial data of similar or different scale. Identification and removal of outliers is another helpful technique to improve such links and scaling. Yet, outliers should only be removed if checked individually and if proper reasons are found that justify removal (e.g. obvious misclassifications, coordinates or labeling, or locations in domains not covered by the study).

Sampling design and analysis

There are two additional domains where linking field data with remotely sensed imagery can significantly improve the analyses. First, calibration of images (for classification or gradient mapping) both can profit from an optimized design of sampling field data. Too often, the design is restricted to what's available, or to simple "checking" of spectral clusters in the field. However, when it comes to large landscapes, a meaningful stratification is critical. While random sampling is a sound statistical approach, it might not generate enough calibration data if complex gradients need to be sampled. Here, a sound theory of the links between spectral and ecological features to be calibrated is a key (see also conclusion). If vegetation has to be classified, or if vegetation cover fraction has to be predicted from an image, then the knowledge of the environmental drivers of these patterns provides a means to optimize sampling designs. A random stratified sampling technique is one alternative that can both sample the full gradient, and maintain statistical independence. Guisan and Zimmermann (2000) provide an overview of some sampling design templates useful in landscape research.

Another potential for improved analyses is the use of modern statistical techniques. While classification approaches are rather well advanced in remote sensing applications, the use of modern regression techniques for calibrating continuous gradients is often restricted to either spectral mixture or ordinary least-square (OLS) regression (often only including linear terms) modeling. This may be suitable in some instances, but often there are non-linear relationships between landscape-level response and spectral response (which causes problems in spectral unmixing approaches; e.g. Hansen *et al.* 2002), or the response to be calibrated shows clearly saturation effects or is bounded between a minimum and a maximum, which impacts the underlying assumptions for OLS regressions (Schwarz and Zimmermann 2005). Neural networks (Atkinson *et al.* 1997; Braswell *et al.* 2003) and regression trees (Hansen *et al.* 2002) are viable alternatives and their application is clearly on the rise. On the other hand, Generalized Linear Models (GLMs) and Generalized Additive models (GAMs) are designed to cope exactly with these types of limitations. While they are often used in various aspects of environmental science (see Guisan and Zimmermann [2000] for a review), there are surprisingly few applications with remote sensing data (Schwarz and Zimmermann 2005). These types of models allow us to cope with many different aspects of limitations by selecting the appropriate link function for specific distributions (GLMs, Hoshmer and Lemeshow 1989; Venables and Ripley 1999), or to fit non-parametric smoothers (GAMs, Hastie and Tibshirani 1987). Additional alternatives include multivariate adaptive regression splines (MARS), or Generalized Boosted Regression Models, also termed gradient boosting (GBMs, Freund and Schapire 1997; Friedman *et al.* 2000, Friedman 2002).

Validation and error analysis

In 1993 Ustin *et al.* stated that validation is perhaps the most overlooked aspect in remote sensing (Ustin *et al.* 1993). While this may have been true back in 1993, this is certainly no longer the case. Still, it is a major challenge to clearly test pattern or gradients derived from remotely sensed imagery thoroughly. Sound statistical tests require the data to be independent and not used for calibration. In the case of remotely sensed imagery, this is not trivial. There are three different ways to comply with this prerequisite: (1) use a spatially random sample of calibration/test data within the domain of an image, split it into two subsamples, one for calibration, and one for validation; (2) take a random sample of calibration in one area of the image and another random sample in a different area of the image for validation; (3) take a random sample of calibration data from one source, then take a second random sample for validation using a different source.

A random sample is usually without replacement and may include any form of stratification. Also, 2 and 3 may be combined. 1 and 2 are applied rather often (e.g. Baret *et al.* 1995; Schwarz and Zimmermann 2005). Where the calibration and test data are not directly measurable but come from models as well, it is important to test for pseudo-errors as a result of uncertainty in the validation data set. A second, independent source for validation is a more secure approach (e.g. Schwarz *et al.* 2004). This is often a problem when models are derived or enhanced using coarse resolution imagery. For example, it is usually not feasible to directly assess tree cover fraction on the ground on a 1 km^2 basis. Thus, some method of scaling is involved to generate calibration and test data sets (often using higher resolution imagery, or DEM-based GIS modeling). This is certainly a viable approach, but requires attention with respect to statistical inference.

Besides the analysis of accuracy, the assessment of errors or uncertainty, and its spatial distribution is an important task (Bastin *et al.* 2002; Tian *et al.* 2002). Often, errors are not randomly distributed across a processed image, as for instance in an image classification exercise (Congalton 1988; Foody 2002), which may have a variety of reasons (see Foody 2002 for a discussion). Users may benefit from a more detailed analysis of the errors and its statistical and geographical characteristics and distributions. Various approaches have been proposed for such error analyses (Foody 2002), including extrapolations from the training data set (Steele *et al.* 1998), analyzing magnitude within partitioned classes (Foody 2000), using geostatistical (Kyriakidis and Dungan 2001) or regression modeling approaches (Leyk and Zimmermann 2004).

Conclusions

Remote sensing is a powerful tool for a wide array of applications. In landscape research, it is very useful for mapping pattern and gradients, though in most cases, remotely sensed imagery is combined with other sources in order to solve scientific and management problems (GIS, field data, etc.). In this latter case, it is important to realize that many aspects of traditional ground-based studies can be adapted, while remotely sensed imagery cannot. Susan Ustin and co-workers (1993) conclude that in order to make the best use of remotely sensing, a new ecological paradigm seems necessary which is consistent with the spectral data. There is still insufficient – though growing – knowledge on how to link spectral response with biophysical and bio-chemical pattern, dynamical processes and small scale mixtures of ecological and landscape features on the ground.

We believe that such a paradigm has to additionally respect grain and frequency at which pattern and processes can be observed. For example, when it comes to calibration or testing landscape level models where remotely sensed imagery is involved, then we are bound to the grain of such images. Careful planning is necessary to collect field and other data at a resolution that makes optimal use of reflectance data. Otherwise, uncertainty and scale mismatch inhibit successful analyses (e.g. Turner *et al.* [2004] for testing ecological models; Woo [2004] for hydrological applications).

On the other hand, using ecological theory and sound ecological understanding is necessary to optimally use, calibrate, or process imagery for environmental research at a landscape scale. Otherwise, models or calibrated features might show unexpectedly high or spatially structured residuals, which clearly have an ecological explanation. In many cases, this also means the incorporation of modern regression techniques, consistent with the quest to calibrate bounded, skewed, or saturated response shapes.

Finally – although not touched in this chapter – the best processing of remote sensing imagery is needed to enable the detection of pattern and gradients at repeatedly high precision. In practice, this means that remote sensing experts and landscape researchers need to collaborate closely, in order to make the best progress in landscape and environmental research.

References

Adams J.B., Smith M.O. and Johnson P.E. 1986. Spectral Mixture Modelling: A New Analysis of Rock and Soil Types at the Viking Lander Site. Journal of Geophysical Research 91: 8098–8112.

Albertz J. 2001. Einführung in die Fernerkundung. Wissenschaftliche Buchgesellschaft, Darmstadt, Germany, 2nd edition.

Asner G.P. 1998. Biophysical and biochemical sources of variability in canopy reflectance. Remote Sensing of Environment 65, 2: 225–226.

Asner G.P., Wessman C.A., Bateson C.A. and Privette J.L. 2000. Impact of tissue, canopy, and landscape factors on the hyperspectral reflectance variability of arid ecosystems. Remote Sensing of Environment 74, 1: 69–84.

Aspinall R.J. 2002. Use of logistic regression for validation of maps of the spatial distribution of vegetation species derived from high spatial resolution hyperspectral remotely sensed data. Ecological Modelling 157: 301–312.

Atkinson P.M., Cutler M.E.J. and Lewis H. 1997. Mapping sub-pixel proportional land cover with AVHRR imagery. International Journal of Remote Sensing 18: 917–935.

Austin M.P. 2002. Spatial prediction of species distribution: an interface between ecological theory and statistical modelling. Ecological Modelling 157: 101–118.

Austin M.P. and Smith T.M. 1989. A new model for the continuum concept. Vegetatio 83: 35–47.

Baret F., Clevers J. and Steven M.D. 1995. The Robustness of Canopy Gap Fraction Estimates from Red and near-Infrared Reflectances – a Comparison of Approaches. Remote Sensing of Environment 54: 141–151.

Baret F. and Fourty T. 1997. Estimation of leaf water content and specific leaf weight from reflectance and transmittance measurements. Agronomie 17, 9–10: 455–464.

Baret F. and Guyot G. 1991. Potentials and limits of vegetation indices for LAI and APAR assessment. Remote Sensing of Environment 35: 161–173.

Bastin L., Fisher P.F. and Wood J. 2002. Visualizing uncertainty in multi-spectral remotely sensed imagery. Computers and Geosciences 28: 337–350.

Belward A.S., Estes J.E. and Kline K.D. 1999. The IGBP-DIS Global 1-Km Land-Cover Dataset DISCover: A Project Overview. Photogrammetric Engineering and Remote Sensing 65: 1013–1020.

Birth G.S. and McVey G.R. 1968. Measuring the color of growing turf with a reflectance spectrophotometer. Agronomy Journal 60: 640–643.

Boardman J.W. 1989. Inversion of Imaging Spectrometry Data Using Singular Value Decomposition. Proceedings of the 1989 International Geoscience and Remote Sensing Symposium (IGARSS'89), and the 12th Canadian Symposium on Remote Sensing, Vancouver, British Columbia, Vol. 4, pp. 2069–2072.

Boardman J.W. 1990. Inversion of High Spectral Resolution Data. Proceedings of SPIE Conference on Imaging Spectrometry of the Terrestrial Environment, Orlando, Florida, Vol. 1298: 222–233.

Braswell B.H., Hage S.C., Frolking S.E. and Salas W.A. 2003. A multivariable approach for mapping sub-pixel land cover distributions using MISR and MODIS: Application in the Brazilian Amazon region. Remote Sensing of Environment 87: 243–256.

Burley F.W. 1988. Monitoring biological diversity for setting conservation priorities. In Wilson, E.O. (ed.), Biodiversity, pp. 227–230. National Academy Press, Washington, D.C., U.S.A.

Chapin F.S. III., Matson P.A. and Mooney H.A. 2002. Principles of terrestrial ecosystem ecology. Springer, New York, NY, U.S.A.

Chen J.M., Leblanc S.G., Miller J.R., Freemantle J., Loechel S.E., Walthall C.L., Innanen K.A. and White H.P. 1999. Compact Airborne Spectrographic Imager (CASI) used for mapping biophysical parameters of boreal forests. Journal of Geophysical Research-Atmospheres 104(D22): 27945–27958.

Chen J.M., Li X., Nilson T. and Strahler A. 2000. Recent Advances in Geometrical Optical Modelling and its Applications. Remote Sensing Reviews 18: 227–2262.

Chen J.M., Rich P.M., Gower S.T., Norman J.M. and Plummer S. 1997. Leaf area index of boreal forests: Theory, techniques, and measurements. Journal of Geophysical Research-Atmospheres 102(D24): 29429–29443.

Cohen W.B. and Goward S.N. 2004. Landsat's role in ecological applications of remote sensing. Bioscience 54, 6: 535–545.

Collins S.L., Glenn S.M. and Roberts D.W. 1993. The hierarchical continuum concept. Journal of Vegetation Science 4: 149–156.

Congalton R.G. 1988. Using Spatial Auto-Correlation Analysis to Explore the Errors in Maps Generated from Remotely Sensed Data. Photogrammetric Engineering and Remote Sensing 54: 587–592.

Danson F.M., Steven M.D., Malthus T.J. and Clark J.A. 1992. High-Spectral Resolution Data for Determining Leaf Water-Content. International Journal of Remote Sensing 13, 3: 461–470.

Daughtry C.S.T., Walthall C.L., Kim M.S., de Colstoun E.B. and McMurtrey J.E. 2000. Estimating corn leaf chlorophyll concentration from leaf and canopy reflectance. Remote Sensing of Environment 74, 2: 229–239.

Dawson T.P., North P.R.J., Plummer S.E. and Curran P.J. 2003. Forest ecosystem chlorophyll content: implications for remotely sensed estimates of net primary productivity. International Journal of Remote Sensing 24, 3: 611–617.

De Bruin S. and Gorte B.G.H. 2000. Probabilistic image classification using geological map units applied to land-cover change detection. International Journal of Remote Sensing 21: 2389–2402.

Deering D.W., Rous J.W., Haas R.H. and Schell J.A. 1975. Measuring forage production of grazing units from Landsat MSS data. Proceedings of the 10th International Symposium on Remote Sensing of Environment, pp. 1169–1178.

DeFries R.S., Townshend J.R.G. and Hansen M.C. 1999. Continuous fields of vegetation characteristics at the global scale at 1-km resolution. Journal of Geophysical Research – Atmospheres 104: 911–916, 925.

Delecolle R., Maas S.J., Guerif M. and Baret F. 1992. Remote-Sensing and Crop Production Models – Present Trends. ISPRS Journal of Photogrammetry and Remote Sensing 47, 2–3: 145–161.

Duchemin B., Goubier J. and Courrier G. 2002. Monitoring phenological key stages and cycle duration of temperate deciduous forest ecosystems with NOAA/AVHRR data. Remote Sensing of Environment 67: 68–82.

Edwards T.C., Deshler E.T., Foster D. and Moisen G.G. 1996. Adequacy of wildlife habitat relation models for estimating spatial distributions of terrestrial vertebrates. Conservation Biology 10: 263–270.

Fernandez-Illecs C.P., Porporato A., Laio F. and Rodriguez-Iturbe I. 2001. The ecohydrological role of soil texture in a water-limited ecosystem. Water Resources Research 37: 2863–2872.

Foody G.M. 2000. Mapping land cover from remotely sensed data with a softened feedforward neural network classification. Journal of Intelligent and Robotic Systems 29: 433–449.

Foody G.M. 2002. Status of land cover classification accuracy assessment. Remote Sensing of Environment 80: 185–201.

Foody G.M. and Hill R.A. 1996. Classification of tropical forest classes from Landsat TM data. International Journal of Remote Sensing 17: 2353–2367.

Fourty T. and Baret F. 1997. Vegetation water and dry matter contents estimated from top-of-the-atmosphere reflectance data: A simulation study. Remote Sensing of Environment 61, 1: 34–45.

Freund Y. and Schapire R.E. 1997. A decision-theoretic generalization of on-line learning and an application to boosting. Journal of Computer and System Sciences 55, 1: 119–139.

Friedl M.A., McIver D.K., Hodges J.C.F., Zhang X.Y., Muchoney D., Strahler A.H., Woodcock C.E., Gopal S., Schneider A., Coope A., Baccini A., Gao F. and Schaaf C. 2002. Global land cover mapping from MODIS: algorithms and early results. Remote Sensing of Environment 83: 287–302.

Friedman J.H., Hastie T. and Tibshirani R. 2000. Additive Logistic Regression: a Statistical View of Boosting. Annals of Statistics 28, 2: 337–374.

Friedman J.H. 2002. Stochastic Gradient Boosting. Computational Statistics and Data Analysis 38, 4: 367–378.

Goel N. and Thompson R. 2000. A snapshot of Canopy Reflectance Models and a Universal Model for the Radiation Regime. Remote Sensing Reviews 18: 197–225.

Goel N.S. 1988. Models of vegetation canopy reflectance and their use in estimation of biophysical parameters from reflectance data. Remote Sensing Reviews 4: 1–212.

Goetz A.F.H., Vane G., Solomon J.E. and Rock B.N. 1985. Imaging Spectrometry for Earth Remote-Sensing. Science 228, 4704: 1147–1153.

Green R.O., Eastwood M.L., Sarture C.M., Chrien T.G., Aronsson M., Chippendale B.J., Faust J.A., Pavri B.E., Chovit, C.J., Solis, M.S., Olah, M.R. and Williams O. 1998. Imaging spectroscopy and the Airborne Visible Infrared Imaging Spectrometer (AVIRIS). Remote Sensing of Environment 65, 3: 227–248.

Guisan A. and Zimmermann N.E. 2000. Predictive habitat distribution models in ecology. Ecological Modelling 135: 147–186.

Haboudane D., Miller J.R., Tremblay N., Zarco-Tejada P.J. and Dextraze L. 2002. Integrated narrow-band vegetation indices for prediction of crop chlorophyll content for application to precision agriculture. Remote Sensing of Environment 81, 2–3: 416–426.

Hall F.G., Strebel D.E., Nickeson J.E. and Goetz J. 1991. Radiometric rectification: Towards a common radiometric response among multidate, multisensor images. Remote Sensing of Environment 35: 11–27.

Hanley T.A. 1979. Application of an herbivore-plant model to rest rotation grazing management on shrub-steppe rangeland. Journal of Range Management 32: 115–118.

Hansen M.C., DeFries R.S., Townshend J.R.G. and Sohlberg R. 2000. Global land cover classification at 1 km spatial resolution using a classification tree approach. International Journal of Remote Sensing 21: 1331–1364.

Hansen M.C., DeFries R.S., Townshend J.R.G., Carroll M., Dimiceli C. and Sohlberg R.A. 2003. Global percent tree cover at a spatial resolution of 500 meters: first results of the MODIS vegetation continuous fields algorithm. Earth Interactions 7, 10: 1–15.

Hansen M.C., DeFries R.S., Townshend J.R.G., Sohlberg R., Dimiceli C. and Caroll M. 2002. Towards an operational MODIS continuous field of percent tree cover algorithm: examples using AVHRR and MODIS data. Remote Sensing of Environment 83: 303–319.

Hastie T. and Tibshirani R. 1987. Generalized Additive Models: Some Applications. Journal of the American Statistical Association 82: 371–386.

He H.S., Mladenoff D.J., Radeloff V.C. and Crow T.R. 1998. Integration of GIS data and classified satellite imagery for regional forest assessment. Ecological Applications 8: 1072–1083.

Herold M., Goldstein N.C. and Clarke K.C. 2003. The spatiotemporal form of urban growth: measurement, analysis and modeling. Remote Sensing of Environment 86: 286–302.

Holmgren M. and Scheffer M. 2001. El Niño as a window of opportunity for the restoration of degraded arid ecosystems. Ecosystems 4: 151–159.

Homer C.G., Ramsey R.D., Edwards T.C. and Falconer A. 1997. Landscape cover-type modeling using a multi-scene thematic mapper mosaic. Photogrammetric Engineering and Remote Sensing 63: 59–67.

Hoshmer D.W. and Lemeshow S. 1989. Applied logistic regression. John Wiley & Sons, New York, NY, U.S.A.

Huemmrich K.F. 2001. The GeoSail model: a simple addition to the SAIL model to describe discontinuous canopy reflectance. Remote Sensing of Environment 75, 3: 423–431.

Huete A.R. 1988. A soil-adjusted vegetation index (SAVI). Remote Sensing of Environment 25: 295–309.

Huete A.R. and Jackson R.D. 1987. Suitability of spectral indices for evaluating vegetation characteristics on arid rangelands. Remote Sensing of Environment 23: 213–232.

Huete A.R., Jackson R.D. and Post D.F. 1985. Spectral response of a plant canopy with different soil backgrounds. Remote Sensing of Environment 17: 37–53.

Huete A.R., Justice C. and Liu H. 1994. Development of Vegetation and Soil Indexes for Modis-Eos. Remote Sensing of Environment 49: 224–234.

Huete A.R., Liu H.Q., Batchily K. and vanLeeuwen W. 1997. A comparison of vegetation indices global set of TM images for EOS-MODIS. Remote Sensing of Environment 59: 440–451.

Jackson R.D. 1983. Spectral indices in n-space. Remote Sensing of Environment 13: 409–421.

Jacquemoud S., Baret F., Andrieu B., Danson F.M. and Jaggard K. 1995. Extraction of Vegetation Biophysical Parameters by Inversion of the Prospect Plus Sail Models on Sugar-Beet Canopy Reflectance Data – Application to Tm and Aviris Sensors. Remote Sensing of Environment, 52, 3: 163–172.

Jacquemoud S., Ustin S.L., Verdebout J., Schmuck G., Andreoli G. and Hosgood B. 1996. Estimating leaf biochemistry using the PROSPECT leaf optical properties model. Remote Sensing of Environment 56, 3: 194–202.

Jensen J.R. 2004. Introductory digital image processing. 3rd ed., Prentice Hall, Upper Saddle River, NJ, U.S.A.

Jordan C.F. 1969. Derivation of leaf-area index from quality of light on the forest floor. Ecology 50: 663–666.

Kauth R.J. and Thomas G.S. 1976. The tasselled cap – a graphic description of the spectral-temporal development of agricultural crops as seen by Landsat. In: Swain P.H. and Morrison D.B. (eds.), Proceedings of the Second Annual Symposium on Machine Processing of Remote Sensing, West Lafayette, IN, 6th June – 2nd July, 1976, New York, U.S.A., pp. 4B–41 – 4B–51.

Kimes D., Knyazikhin Y., Privette J.L., Abuelgasim A.A. and Gao F. 2000. Inversion Methods for Physically-based Models. Remote Sensing Reviews 18: 381–439.

Klostermann R.E. 1999. The what if? collaborative planning support system. – Environment and Planning. B, Planning and Design 26: 393–408.

Knipling E.B. 1970. Physical and physiological basis for the reflectance of visible and near infrared radiation from vegetation. Remote Sensing of Environment 1: 155–159.

Kötz B., Schaepman M., Morsdorf F., Bowyer P., Itten K. and Allgöwer B. 2004. Radiative Transfer Modeling within a Heterogeneous Canopy for Estimation of Forest Fire Fuel Properties. Remote Sensing of Environment 92, 3: 332–344.

Kyriakidis P.C. and Dungan J.L. 2001. A geostatistical approach for mapping thematic classification accuracy and evaluating the impact of inaccurate spatial data on ecological model predictions. Environmental and Ecological Statistics 8: 311–330.

Langevin C. and Stow D.A. 2004. Identifying change in a dynamic urban landscape: a neural network approach to map-updating. Progress in Planning 61: 327–348.

Levins R. 1966. The strategy of model building in population ecology. American Scientist 54: 421–431.

Leyk S. and Zimmermann N.E. 2004. A predictive uncertainty model for field-based survey maps using Generalized Linear Models. Lecture Notes in Computer Science: 3234: 191–205.

Likens G.E., Bormann F.H., Pierce R.S., Eaton J.S., Johnson N.M. 1977. Biogeochemistry of a Forested Ecosystem. Springer, New York, NY, U.S.A.

Lischke H., Loeffler T., Thornton P.E., Zimmemann N.E. 2007. Model up-scaling in landscape research. In: Kienast F., Wildi O. and Ghosh S. (eds.). A Changing World. Challenges for Landscape Research, Vol. 8: 249–272. Springer Landscape Series. Dordrecht.

Loveland T.R. and Belward A.S. 1997. The International Geosphere Biosphere Programme Data and Information System global land cover dataset (DISCover). Acta Astronautica 41: 681–689.

Lunetta R.S., Alvarez R., Edmonds C.M., Lyon J.G., Elvidge C.D., Bonifaz R. and Garcia C. 2002. NALC/Mexico land-cover mapping results: implications for assessing landscape condition. International Journal of Remote Sensing 23: 3129–3148.

Mäkelä H. and Pekkarinen A. 2004. Estimation of forest stand volumes by Landsat TM imagery and stand-level field-inventory data. Forest Ecology and Management 196: 245–255.

Moulin S., Kergoat L., Viovy N. and Dedieu G. 1997. Global-scale assessment of vegetation phenology using NOAA-AVHRR satellite measurements, Journal of Climate 10: 1154–1170.

Mucher C.A., Steinnocher K.T., Kressler F.P. and Heunks C. 2000. Land cover characterization and change detection for environmental monitoring of pan-Europe. International Journal of Remote Sensing 21, 6–7: 1159–1181.

Myneni R.B., Dong J., Tucker C.J., Kaufmann R.K., Kauppi P.E., Liski J., Zhou L., Alexeyev V. and Hughes M.K. 2001. A large carbon sink in the woody biomass of Northern forests. Proceedings of the National Academy of Sciences of the United States of America 98: 14784–14789.

Myneni R.B., Keeling C.D., Tucker C.J., Asrar G. and Nemani R.R. 1997. Increased plant growth in the northern high latitudes from 1981 to 1991. Nature 386: 698–702.

Myneni R.B., Tucker C.J., Asrar G. and Keeling C.D. 1998. Interannual variations in satellite-sensed vegetation index data from 1981 to 1991. Journal of Geophysical Research-Atmospheres 103: 6145–6160.

Nemani R.R., Keeling C.D., Hashimoto H., Jolly W.M., Piper S.C., Tucker C.J., Myneni R.B. and Running S.W. 2003. Climate-driven increases in global terrestrial net primary production from 1982 to 1999. Science 300: 1560–1563.

Payero J.O., Neale C.M.U. and Wright J.L. 2004. Comparison of eleven vegetation indices for estimating plant height of alfalfa and grass. Applied Engineering in Agriculture 20: 385–393.

Pickup G., Chewings V.H. and Nelson D.J. 1993. Estimating changes in vegetation cover over time in arid rangelands using Landsat MSS data. Remote Sensing of Environment 43: 243–263, 1993.

Pickup G. and Foran B.D. 1987. The use of spectral and spatial variability to monitor cover change on inert landscapes. Remote Sensing of Environment 23: 351–363.

Pomeroy L.R. and Alberts J.J. (eds.) 1988. Concepts of ecosystem ecology. Springer, New York, NY, U.S.A.

Rast M. 2001. SPECTRA – Surface Processes and Ecosystem Changes Through Response Analysis. Nordwijk, ESA Publication Division.

Richardson A.J. and Wiegand C.L. 1977. Distinguishing vegetation from soil background information: Photogrammetric Engineering and Remote Sensing 43, 12: 1541–1552, 1977.

Richter R. 1990. A Fast Atmospheric Correction Algorithm Applied to Landsat TM Images. International Journal of Remote Sensing 11, 1: 159–166.

Richter R. 1998. Correction of satellite imagery over mountainous terrain. Applied Optics 37, 18: 4004–4015.

Rouse J.W. Jr., Hass R.H., Deering D.W., Schell J.A. and Harlan J.C. 1973. Monitoring the vernal advancement and retrogression (green wave effect) of natural vegetation. NASA/GSFC Type III Final report, Geenbelt, Maryland, MD, U.S.A.

Running S.W., Baldocchi D.D., Turner D.P., Gower S.T., Bakwin P.S. and Hibbard K.A. 1999. A global terrestrial monitoring network integrating tower fluxes, flask sampling, ecosystem modeling and EOS satellite data. Remote Sensing of Environment, 70, 1: 108–127.

Running S.W., Nemani R.R., Heinsch F.A., Zhao M.S., Reeves M. and Hashimoto H. 2004. A continuous satellite-derived measure of global terrestrial primary production. Bioscience 54: 547–560.

San-Miguel-Ayanz J., Vogt J., de Roo A. and Schmuck G. 2000. Natural Hazards Monitoring: Forest Fires, Droughts, and Floods: the example of European pilot projects. Surveys in Geophysics 21: 291–305.

Schläpfer D. and Richter R. 2002: Geo-atmospheric processing of airborne imaging spectrometry data: Part 1. Parametric orthorectification. Int. J. Remote Sensing, 23, 13: 2609–2630.

Schowengerdt R.A. 1996. On the estimation of spatial-spectral mixing with classifier likelihood functions. Pattern Recognition Letters 17: 1379–1387.

Schwarz M. and Zimmermann N.E. 2005. A new GLM-based method for mapping tree cover continuous fields using MODIS reflectance data. Remote Sensing of Environment 95: 428–443.

Schwarz M., Zimmermann N.E., Wildi O. and Kienast F. 2004. Erfassen kontinuierlicher Bodenbedeckungsdaten anhand von MODIS Daten in der Schweiz. Botanica Helvetica 114, 2: 151–167.

Scott J.M., Davis F., Csuti B., Noss R., Butterfield B., Groves C., Anderson H., Caicco S., Derchia F., Edwards T.C., Ulliman J. and Wright R.G. 1993. Gap Analysis – a Geographic Approach to Protection of Biological Diversity. Wildlife Monographs 123: 1–41.

Scott J.M. and Jennings M.D. 1998. Large-area mapping of biodiversity. Annals of the Missouri Botanical Garden 85, 1: 34–47.

Sellers P.J. 1985. Canopy reflectance, photosynthesis and transpiration, International Journal of Remote Sensing 6: 1335–1372.

Sellers P.J., Dickinson R.E., Randall D.A., Betts A.K., Hall F.G., Mooney H.A., Nobre C.A., Sato N., Field C.B. and Henderson-Sellers A. 1997. Modelling the exchanges of energy, water and carbon between continents and the atmosphere. Science 275: 502–509.

Serrano L., Penuelas J. and Ustin S.L. 2002. Remote sensing of nitrogen and lignin in Mediterranean vegetation from AVIRIS data: Decomposing biochemical from structural signals. Remote Sensing of Environment 81, 2–3: 355–364.

Shepherd K.D. and Walsh M.G. 2002. Development of reflectance spectral libraries for characterization of soil properties. Soil Science Society of America Journal 66: 988–998.

Smith M.O., Ustin S.L., Adams J.R. and Gillespie A.R. 1990. Vegetation in deserts. I: A regional measure of abundance from m ultispectral images. Remote Sensing of Environment 31: 1–26.

Smolander H. and Stenberg P. 1996. Response of LAI-2000 estimates to changes in plant surface area index in a Scots pine stand. Tree Physiology 16, 3: 345–349.

Steele B.M., Winne J.C. and Redmond R.L. 1998. Estimation and mapping of misclassification probabilities for thematic land cover maps. Remote Sensing of Environment 66: 192–202.

Stow D.A., Hope A., McGuire D., Verbyla D., Gamon J., Huemmrich F., Houston S., Racine C., Sturm M., Tape K., Hinzman L., Yoshikawa K., Tweedie C., Noyle B., Silapaswan C., Douglas D., Griffith B., Jia G., Epstein H., Walker D., Daeschner S., Petersen A., Zhou L.M. and Myneni R.B. 2004. Remote sensing of vegetation and land-cover change in Arctic Tundra Ecosystems. Remote Sensing of Environment 89: 281–308.

Tian Y.H., Woodcoc C.E., Wang Y.J., Privette J.L., Shabanov N.V., Zhou L.M., Zhang Y., Buermann W., Dong J.R., Veikkanen B., Hame T., Andersson K., Ozdogan M., Knyazikhin Y. and Myneni R.B. 2002. Multiscale analysis and validation of the MODIS LAI product – I. Uncertainty assessment. Remote Sensing of Environment 83: 414–430.

Townshend J.R.G., Justice C.O., Skole D., Malingreau J.-P., Cihlar J., Teillet P., Sadowski F. and Ruttenberg S. 1994. The 1 km resolution global dataset: needs of the International Geosphere Biosphere Program. International Journal of Remote Sensing 15: 3417–3442.

Trenberth K.E., Branstator G. and Arkin P.A. 1988. Origins of the 1988 North American drought. Science 242: 1640–1645.

Tucker C.J. 1979. Red and photographic infrared linear combinations for monitoring vegetation. Remote Sensing of Environment 8: 127–150.

Tucker C.J., Townshend J.R.G. and Goff T.E. 1985. African landcover classification using satellite data. Science 227: 369–374.

Turner M.G., Gardner R.H. and O'Neill R.V. 2001. Landscape Ecology in Theory and Practice: Pattern and Process. Springer, New York, NY, U.S.A.

Turner D.P., Ollinger S., Smith M.L., Krankina O. and Gregory M. 2004. Scaling net primary production to a MODIS footprint in support of Earth observing system product validation. International Journal of Remote Sensing 25: 1961–1979.

Ustin S.L., Roberts D.A., Gamon J.A., Asner G.P. and Green R.O. 2004. Using imaging spectroscopy to study ecosystem processes and properties. Bioscience 54: 523–534.

Ustin S.L., Smith M.O. and Adams J.B. 1993. Remote Sensing of Ecological Processes: A strategy for developing and Testing Ecological Models Using Spectral Mixture Analysis. In: Ehleringer J.R. and Field C.B. (eds). Scaling Physiological Processes: Leaf to Globe. Academic Press, San Diego, CA, U.S.A., pp. 339–357.

Venables W.N. and Ripley B.D. 1999. Modern Applied Statistics with S-PLUS. -Springer, New York, NY, U.S.A.

Verhoef W. and Bach H. 2003. Remote sensing data assimilation using coupled radiative transfer models. Physics and Chemistry of the Earth 28, 1–3: 3–13.

Wang Y.P., Xia H., Fu J.M. and Sheng G.Y. 2004. Water quality change in reservoirs of Shenzhen, China: detection using LANDSAT/TM data. Science of the Total Environment 328: 195–206.

Washington-Allen R.A. 2003. Retrospective ecological risk assessment of commercially-grazed rangelands using multitemporal satellite Imagery. PhD Thesis, Utah State Univ., Logan, UT, U.S.A.

Washington-Allen R.A., Ramsey R.D. and West N.E. 2004. Spatiotemporal mapping of the dry season vegetation response of sagebrush steppe. Community Ecology 5, 1: 69–79.

Washington-Allen R.A., West N.E. and Ramsey R.D. 2003. Remote sensing-based dynamical systems analysis of sagebrush steppe vegetation in rangelands. pp. 416–418. In: Allsopp N., Palmer A.R., Milton S.J., Kirkman K.P., Kerley G.I.H., Hurt C.R. and Brown C.J. (eds.). Proc. VII International Rangelands Congress in Durban, South Africa July 26 – August 1, 2003.

Weiss M., Baret F., Smith G.J., Jonckheere I. and Coppin P. 2004. Review of methods for in situ leaf area index (LAI) determination Part II. Estimation of LAI, errors and sampling. Agricultural and Forest Meteorology 121, 1–2: 37–53.

White J.D., Running S.W., Nemani R., Keane R.E. and Ryan K.C. 1997. Measurement and remote sensing of LAI in Rocky Mountain montane ecosystems. Canadian Journal of Forest Research 27: 1714–1727.

Woo M.K. 2004. Boundary and border considerations in hydrology. Hydrological Processes 18: 1185–1194.

Zarco-Tejada P.J., Miller J.R., Mohammed G.H., Noland T.L. and Sampson P.H. 2000. Chlorophyll fluorescence effects on vegetation apparent reflectance: II. Laboratory and airborne canopy-level measurements with hyperspectral data. Remote Sensing of Environment 74: 596–608.

Zarco-Tejada P.J., Rueda C.A. and Ustin S.L. 2003. Water content estimation in vegetation with MODIS reflectance data and model inversion methods. Remote Sensing of Environment 85, 1: 109–124.

Zhou L., Tucker C., Kaufmann R., Slayback D., Shabanov N. and Myneni R., 2001. Variations in northern vegetation activity inferred from satellite data of vegetation index during 1981 to 1999. Journal of Geophysical Research 106(D17): 20069–20083.

A Large-scale, Long-term view on Collecting and Sharing Landscape Data

Adrian Lanz*, Martin Brändli and Andri Baltensweiler

Swiss Federal Research Institute WSL, Zürcherstrasse 111, CH-8903 Birmensdorf, Switzerland
* Corresponding author: e-mail: adrian.lanz@wsl.ch

Abstract
Modeling and analysis of temporal and spatial processes in landscapes are based on large amounts of collected data today. These data, however, are often stored at multiple locations and are only accessible from distributed repositories. This chapter discusses techniques for efficient and effective data integration, and presents an approach for accessing distributed data based on open standards. Data from different providers may have different levels of accuracies. Therefore, data interoperability implies comprehensive reports on data accuracy and other data quality elements (metadata). We illustrate the importance of metadata within the scope of long-term monitoring surveys, where technological and scientific progress leads to permanent improvement of the sampling procedures and estimation techniques, and discuss resampling techniques and model building with existing data aiming at the optimization of data collection and the enhancement of comparability, reliability and accuracy of estimates of population parameters over a long period.

Keywords: data interoperability, data integration, inventory planning, optimization of data collection, metadata

F. Kienast, O. Wildi & S. Ghosh (eds.), A Changing World. Challenges for Landscape Research, 93–111.

Data Interoperability: A New Focus for Management of Ecological Data

Data and information are basic resources of scientific research. Information can be defined as raw data combined with the knowledge about the context of the data (Worboys and Duckham 2004, p. 5) including facts about interrelationship between existing data, about how data is collected, processed, used, and understood within an application. Both data and information form the indispensable foundation for thorough analysis of landscape patterns and comprehensive exploration and modeling of temporal and spatial processes in landscapes. Effective handling of data leading to successful data modeling, analysis, and derivation of meaningful information requires comprehensive data management. Brunt (2000) presents an idealized view of managing ecological data as a process starting with the conception and design of a research project, continuing with data acquisition and quality control, data manipulation and quality assurance, analysis and interpretation, and finally leading to archiving and publication of gathered, processed and derived data as well as making them directly accessible (see Fig. 1). Implementing and integrating these individual steps as interrelated software components contribute to a comprehensive data management system for ecological data.

In ecology and other fields, this scheme of data management is traditionally applied by individuals and small groups of investigators. The tasks presented in Figure 1 are usually performed on small plots over relatively short time periods (Michener 2000). However, new research challenges, progress in information technology, and increasing user needs for always up-to-date and accurate information ask for a substantial modification of this traditional view of the data management scheme:

- Ecological phenomena are increasingly analyzed in long-term monitoring projects. Data gathering may vary significantly in frequency and duration. For instance, in an ongoing Swiss long-term Forest Ecosystem Research project ("LWF", Cherubini and Innes 2000)

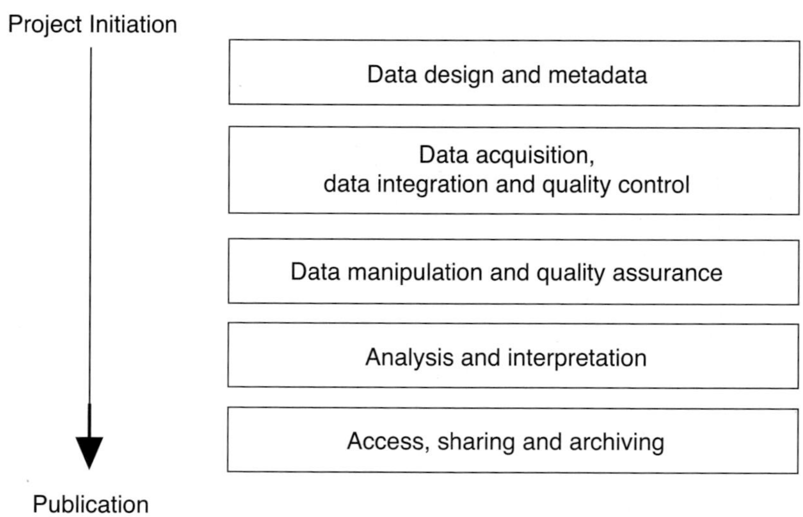

Fig. 1. Components of the management of ecological data and their relationship to research (adapted from Brunt (2000)). In addition to Brunt (2000), data integration and sharing are added as major challenges of current environmental data management tasks.

meteorological data have been collected continuously at an interval of 10 minutes for more than 7 years. In contrast, in the Swiss National Forest Inventory ("Swiss NFI", Brassel and Lischke 2001) the terrestrial data are collected every ten years.
- The volume of the datasets containing ecological and related data required by environmental scientists substantially increased over the last years. The amount of remote sensing data enabling photogrammetric data interpretation in the NFI-project is in the range of several terabytes, for instance. As a consequence of the huge increase in data volume size, data of different ecological phenomena are not stored within single data repositories any longer but are held in distributed thematic databases.
- Exploration and analysis of ecological phenomena – such as biodiversity for instance – progressively require a more interdisciplinary approach. Consequently, techniques for the integration of data from distributed sources are becoming more and more important.
- The scale of ecological phenomena exploration and analysis is shifting from a rather narrow regional to an enlarged continental and global view, since many of the current ecological challenges such as global change, sustainability, and biodiversity require large-scale investigation (Michener *et al.* 1997). We mention here two initiatives at the European level, which aim at standardization and harmonization of data collection and data sharing. The INSPIRE initiative (Infrastructure for Spatial Information in Europe, http://www.ec-gis.org/inspire/) promotes the availability of relevant, harmonized and high quality geographic information. The ENFIN initiative (European National Forest Inventory Network) aims in a new research action (http://www.metla.fi/eu/cost/e43/) at the harmonization of definitions, data collection and estimation techniques applied in national forest inventories.

As a consequence of these trends, data management has to be improved in the following fields:

- *Data integration:* There is a clear need for techniques and methods enabling the combination of data from distributed data repositories. Besides technical solutions for acquiring data from distributed computing platforms, concepts and methods for dealing with heterogeneous, erroneous data, and data with different semantics and degrees of uncertainty must be provided (Vckovski *et al.* 1999; Lutz *et al.* 2003; Brodeur *et al.* 2003; Samal *et al.* 2004; Wilson 2004). Data integration techniques are also needed in order to extend or harmonize time series data because data for similar locations may be available from different data sources and/or methodological changes may have occurred.
- *Data sharing:* While data integration concentrates on data combination, data sharing consists of the task of making data available to involved partners, peers and the interested public. Data must be prepared and communicated in a standardized and uniform way in order to make them easily accessible by other computing platforms.
- *Quality control and quality assurance:* The efforts to integrate quality control and quality assurance components into long-term monitoring programs have to be strengthened in the future. While already present in certain ecological programs monitoring especially sensitive data like air pollutions or radiation, standards for quality control and quality assessment are not systematically known in landscape monitoring programs. However, quality control and quality assurance tools are needed to guarantee methodologically consistent data collection and data processing over long periods and adequate data quality for efficient data integration and data sharing.

In the next section we take a look at quality assurance and quality control techniques applied in two long-term, national forest monitoring programs, and illustrate the use of quality control data generated during the data collection stage of the programs for the continuous development of sampling techniques and estimation procedures in this programs. The chapter continues by discussing the importance of metadata and corresponding quality elements for effective data assessment. We present a new method based on a process-oriented data model which enables tracking of data history for the handling of the particular data quality element of lineage. In the third section we address the two strongly related tasks of data integration and data sharing by discussing problems related to providing and accessing distributed data repositories. A particular solution – the Virtual Database – which is based on the use of open standards working on the Internet as the exchange platform is introduced.

Data Collection: Implementation and Use of Quality Control Data

Implementation of quality control

Institutions collecting and evaluating data must assure their quality. Quality assurance is defined as all systematic actions necessary to provide confidence that a service or product will satisfy given requirements (Shampine 1993; Lanea 1997). In practice, successful quality assurance means that the data are of a known quality that is adequate to meet the program's objectives (Caughlan and Oakley 2001). Quality controls are important elements of quality assurance and comprise the operational techniques and activities that are used to control a process. In long-term landscape monitoring programs, quality assurance activities should be implemented at all stages of the survey. Taking the FIA (Forest Inventory and Analysis) program of the USDA Forest Service and the Swiss National Forest Inventory (Swiss NFI) program as examples, quality assurance includes the following components (Pollard 2002; Brassel and Lischke 2001):

- *Planning:* The FIA program, for instance, prepares a formal management system and field implementation plan compliant with the American National Standards for quality assurance systems documents (ANSI/ASQC E4, 1994). Quality assurance planning is reviewed and approved by regional and national managers.
- *Documentation of methods:* All phases of the programs produce documentations. Documentation should not only include field manuals, but also documentation of data management, data analysis and data processing. Today, most documents of these two programs are available on the Internet (http://fia.fs.fed.us and http://www.lfi.ch).
- Training: Production crews must be trained, tested and certified for their ability to generate data that conform to the measurement quality objectives and tolerances established for the program.
- *Checks for data quality:* This includes mainly control surveys, performed either as hot checks (in presence of an inspector during data collection), cold checks (control of completed data collection by an inspector) or blind checks (complete remeasurement by a qualified inspection crew or by an other production team without access to the production crew's data). The uncertainty in collected data is assessed by analysis of remeasured data (blind checks).
- *Information management, analysis and reporting:* Nationwide data provision requires consistent and standardized data analysis, definitions (like forest types or volume models) and reporting. The list of accepted codes for a variable is enforced by the database system.
- *Continuous feedback:* The programs include a variety of internal and external feedback and review procedures. For instance, following the completion of the second Swiss NFI and in the run-up to the third NFI, NFI services (outputs) and NFI utilization (impacts) were evaluated and recommendations for the third NFI were formulated (Bättig *et al.* 2002).

The following criteria have been used to judge data quality in data collection: precision, unbiasedness, completeness, comparability, plausibility, homogeneity, representativeness and reproducibility (Stierlin 2001, Traub 2001). Most of these criteria can be judged using statistical tests and resampling techniques on production, training and control survey data (Kaufmann and Schwyzer 2001; Paschedag and Keller 2001). In the Swiss NFI, approximately 10% of data are collected on control surveys (blind checks), and nearly 10% of the overall work time for terrestrial data collection was spent on training (Zinggeler 2001).

Using quality control data for design optimization

The Swiss NFI example: Methods and techniques for data collection in long-term, large-scale landscape monitoring programs are invariably subject to constant changes. Not only technological and scientific progress but also ever changing user needs require a periodic review and adaptation of sampling techniques and estimation procedures. Experiences made and data collected in earlier surveys (metadata) are the key to adopt efficient sample designs at all times. In this section, the Swiss NFI serves as an example to illustrate the role of metadata in landscape monitoring programs.

Design changes in the Swiss NFI: The sampling points of the Swiss NFI are distributed on a regular, quadratic grid over the whole country. The Swiss NFI is a combined, permanent inventory: ground data collection and interpretation of aerial photographs are repeated at the same locations every 10 years. In the first NFI, both samples had the same size. The resulting combined sample included more than 10,000 sampling points systematically distributed on a quadratic grid (one sampling point per square kilometer). Photogrammetric data have mainly been used to prepare the terrestrial inventory (forest or non-forest decision, reference points for field crews). In the second NFI, the number of sampling points on aerial photographs (first phase) were increased by a factor of 4 (grid width of 0,500 km) and the number of remeasured terrestrial sampling points (second phase, a sub-sample of first phase sampling points) was reduced by a factor of 2 (grid width of 1,412 km). The first phase sample provides estimates for strata sizes. The classification of the first phase sampling points into strata is essentially based on tree height measurements in a 50 m by 50 m interpretation area around the sampling point. For the most important parameters of the Swiss NFI (total and mean timber volume and number of stems for five large production regions of Switzerland), the new design was implemented without loss in the precision of the estimates, while reducing substantially the overall cost of data collection (Köhl 1994).

A second important change affected the selection of sampling trees on terrestrial sampling points. Several measurements are taken on these trees; most important are measurements of stem diameter and tree height to predict the tree volume. In both inventories the sampling trees were selected on two concentric circles. Smaller trees on a circle of 200 m^2 surface area, and larger trees on a circle of 500 m^2 surface area. While the stem diameter at breast height (DBH, 1,3 m above ground) is measured on every tree in the sample, the time consuming measurements of an upper stem diameter (D7, 7 m above ground) and the tree height (H) are only taken on a sub-set of trees, the so called tariff trees. In the first NFI nearly every second tree in the sample was a tariff tree. In the second NFI, the sample of tariff trees was reduced to around every fifth tree. Again, no loss in the precision of the overall volume estimates had to be accepted. Furthermore, the new estimation technique also corrects possible bias in the volume functions (Kaufmann 2001).

Like in other terrestrial forest inventories, the mean and total volume estimates in the Swiss NFI are extrapolated from volume predictions on individual trees, the sampling trees.

Two models are used to predict individual tree volumes. An accurate, site and stand independent model, called the volume function, predicting the individual tree volume as a function of DBH, D7, H for main tree species groups, and a less accurate, site and stand dependent model, called the tariff function, using DBH, tree species and several variables describing the site and stand condition of the tree.

The volume functions could not have been established in such a short time and high accuracy without reliable data from growth and yield studies established during the last decades and going back to the beginning of the last century. Approximately 38,000 trees from such studies, with exact measurements of tree volume in 2-meter sections of the stem, have been used to derive nine volume functions for the main tree species of Switzerland. These functions are unbiased at the 95% level for all diameter classes and for all species, and the standard deviation of the residuals is in most cases below 10%. The tariff trees in the NFI sample have been used to derive the tariff functions. These functions are for most species and diameter classes unbiased at a high level, but the relative standard deviation of the residuals is with 26% much higher (Kaufmann 2001).

Once the volume and tariff functions are established, at least three different methods for overall timber volume estimation are available: (a) utilize all trees in the sample with individual volumes predicted with the tariff function, (b) utilize only the tariff trees with individual tree volumes predicted with the volume function, or (c) utilize all trees in the sample with individual tree volumes predicted with the tariff function and correct the overall estimate with residuals between tariff and volume function observed on tariff trees. Method (a) was applied in the first NFI and method (c) on later NFIs. The main reason for the choice of this new method was that it corrects potential bias in tariff functions.

The role of quality control data for design optimization: Control survey data can be used to draw up an error-budget of the estimates and to identify error sources that have a substantial influence on the reliability of the results (Köhl and Gertner 1992). For the volume estimation in the Swiss NFI – with a percent root mean square error (RMSE) of 1,28% – they found that the sampling error is by far the most important source of error (98,42% of RMSE), and that the tariff function is the second most important source of error (1,12% of RMSE). Even if doubled compared to random measurement errors observed in control surveys, random measurement and classification errors were of minor importance.

However, systematic measurement errors would have had dramatic consequences in the first NFI. Assuming systematic deviations of only 1% for DBH, D7 and H measurements would result in an increase of the RMSE of the overall volume estimate to 3,9% (compared to 1,28% under random measurement errors only). This is because the assumed bias – also only 1% under this assumption – does not decrease with increasing sampling size, while the sampling error (variance component) decreases substantially for such a large sample (Köhl and Gertner 1992). Control surveys have shown that the probability for systematic measurement errors in the Swiss NFI is very low. Nevertheless, these findings lead to a new, two stage volume estimation procedure that protects not only against possible bias in the tariff functions, but also leads to an optimized selection of tariff trees (Kaufmann 2001; Mandallaz and Ye 1999).

Cost and variance parameters from the first NFI were also indispensable for the advancement of the point sampling design as proposed by Köhl (1994) for the second NFI. In the meantime, a new theoretical framework for forest inventory optimization was developed (Mandallaz and Ye 1999; Mandallaz and Lanz 2001; Mandallaz 2002), which allows the planner to optimize the tree and point selection scheme at the same time. For these tools, control survey data and other metadata from large-area forest inventories are an important information source.

Data Assessment: The Importance of Metadata

Metadata and data quality

The last section illustrated the use of metadata for planning and optimization of a forest survey. In this section we focus the discussion of data quality and metadata on data interoperability. Data interoperability allows users of information systems to share their information in a distributed and heterogeneous environment (Bishr 1999). More formally, data interoperability is defined as the "capability to communicate, execute programs, or transfer data among various functional units in a manner that requires the user to have little or no knowledge of the unique characteristics of those units" (International Organization for Standardization 1993). Data interoperability for comprehensive data exploration, modeling and analysis requires a thorough understanding and assessment of involved data sets. The data, in particular the quality of the data, must be assessed for the fitness of use for a given modeling or analysis task on the one hand, and – if data heterogeneities exist – corresponding methods must be developed and applied in order to make the data interoperable on the other hand. This section gives an overview of existing standards for metadata provision and data quality descriptions. Subsequently, an example and solution for the handling of a particular data quality element is presented.

Comprehensive metadata essentially support broad and long-term use and interpretation of scientific data (Michener 2000). For metadata to be useful it is crucial that they comply with widely and internationally accepted and agreed standards (Longley *et al.* 2001). Until recently, metadata used in the geographic information community mainly obeyed the US Federal Geographic Data Committee's Content Standard for Digital Geospatial Metadata (CSDGM, Federal Geographic Data Committee, 1998). The standard contains information on identification, data quality, spatial data organization, spatial reference, entity types and attributes among many other features. The CSDGM also served as the basis for the recently ratified ISO International Standard 19115 for metadata concerning geographic information (International Organization for Standardization 2003). The ISO standard is currently the main standard for geographic metadata and implemented in several geographic information systems, such as ArcGIS, for instance. A major characteristic of the ISO standard as well as its predecessors is that they have a strong geospatial component. Michener (2000) argues that for most ecological studies spatially-oriented metadata standards only incompletely satisfy the needs in ecology. A generic set of non-geospatial metadata descriptors particularly designed for the management of ecological data was therefore introduced (Michener *et al.* 1997; Michener 2000). Michener (2000) admits, though, that the distinction between spatial and non-spatial metadata is rather artificial, and indeed, the ISO standard enables the definition of particular application profiles which offer the opportunity to extend the standard to accommodate the inclusion of detailed ecological non-geospatial metadata.

We thus proceed by discussing the ISO metadata standard for a detailed look at spatial data quality elements which are in our opinion generic enough to include non-spatial ecological data as well. The quality of spatial data was first identified by five elements (Moellering 1987) – lineage, positional accuracy, attribute accuracy, completeness, and logical consistency – and extended later by the elements of semantic accuracy and temporal information (Morrison 1995). The characteristics of these elements are defined and described as follows:

- *Lineage:* Provides information on the history of a dataset. Is defined as "the recounting of the life cycle of a data set, from its collection or acquisition, through the many stages of compilations, corrections, conversions and transformation to the generation of new interpreted products" (Clarke and Clark 1995, p. 13).

- *Positional accuracy:* The locations of real world entities are described by values in a given spatial reference system. According to Drummond (1995, p. 32), positional accuracy may be defined as "the representation of the nearness of those values to the entity's "true" position in that system".
- *Attribute accuracy:* Characteristics or the semantics of real world entities are described by means of attributes. The accuracy of attributes is defined as "the difference between a measurement and some comparable measurement known to be of higher accuracy" (Goodchild 1995, p. 66). The inherently relative definition of accuracy requires an agreement on rules that can identify a source of being of higher accuracy (Goodchild 1995).
- *Completeness:* This element is defined as the description of the relationship between the objects in a data set and the abstract universe of all objects (Morrison 1988, p. 135). A detailed discussion, clarification and application of the term "abstract universe" is provided by Brassel *et al.* (1995).
- *Logical consistency:* This element "deals with the logical rules of structure and attribute rules for spatial data and describes the compatibility of a datum with other data in a data set" (Kainz 1995, p. 109). It is mainly a way to express the correctness of data and relationship encoding in accordance with given integrity constraints (Morrison 1995, p. 10).
- *Semantic accuracy and temporal information:* These two elements may be considered as particular extensions to the first five elements: Semantic accuracy extends completeness and consistency with a specific focus on semantic issues, and temporal information relates to all quality elements by explicitly considering temporal aspects.

The ISO metadata standard models data quality using these categories by providing more specialized and concise sub-elements. Furthermore, for all elements, except lineage, measures and methods of applied data quality checks may be provided. Exploring these measures and the results of applied data quality evaluation methods should allow for a comprehensive assessment of an existing data set.

An example of a process-oriented data model

Lineage of data: From the list of the five data quality elements, lineage is selected to demonstrate the application and use of quality characteristics for comprehensive data assessment in landscape ecology. Lineage describes the history of a data set. It is recognized that in particular lineage tracking and thus the comprehensive description of lineage is one of the key requirements for the operational use of geographic information within a multi-user and multi-organization application (Sargent 1999). Lineage plays an important role in the research project "Data Center for Nature and Landscape (DNL)" which focuses on the design and implementation of a database for the storage and management of data and metadata of endangered and protected biotopes in Switzerland. The goal of the database design is to provide a recording and archiving tool for actions and decisions of environmental protection measures particularly originating from the Swiss Federal Office for the Environment. The database should essentially enable to follow the complete genesis of the existing data records.

Endangered biotopes, such as mires, flood plains and semi-arid meadows are usually managed by inventories including biotope objects known by name, number, spatial extent and additional characteristics such as type of vegetation, etc. Many of the endangered biotopes are subject to a specific set of protection goals and measures. The selection of environmental protection areas normally consists of several steps, triggered for instance by a governmental decision which initiates the selection process. In a next step, scientific experts determine criteria for choosing appropriate areas. Subsequently, base documents (images, maps) are used for digitizing potential protection areas. During succeeding field work, potential areas and their extents are verified, followed by a refinement of digital objects. Finally, a legal document published by the government determines the exact position and extent of the protection areas as well as protection measures.

Storing and communicating lineage shall deliver comprehensive knowledge about this realization procedure. It is essential to provide and ensure valid information necessary as a legal basis for potential lawsuits or appeals, subsidies or compensation payments, to be fully documented and to have all the data readily available.

The ISO metadata standard formalizes lineage by a list of descriptions of process steps and affected data sources. Process step descriptions include a textual description of the event, corresponding parameters and tolerances, as well as a rationale, the processing data and the name of the processor. Source descriptions include the citation of the source, its scale and a short textual description.

Metadata-centered database design: By the very nature of metadata, this information normally comes as a companion of a data set. Due to the importance of lineage information for the DNL project, it was decided in the database design phase to set the lineage element as the center of the design approach. The result was a so-called process-oriented data model which structures the above mentioned selection procedure for protection areas into clearly identifiable and distinguishable steps. Every step is considered an individual process which is related to a certain type of data and corresponding metadata. Similar models are used for workflow management which represents repeatedly executed tasks and their interrelationships (Ailamaki *et al.* 1998). Figure 2 shows the data model using the Unified Modeling Language notation (UML, Fowler and Scott 2000). UML is a data modeling language which

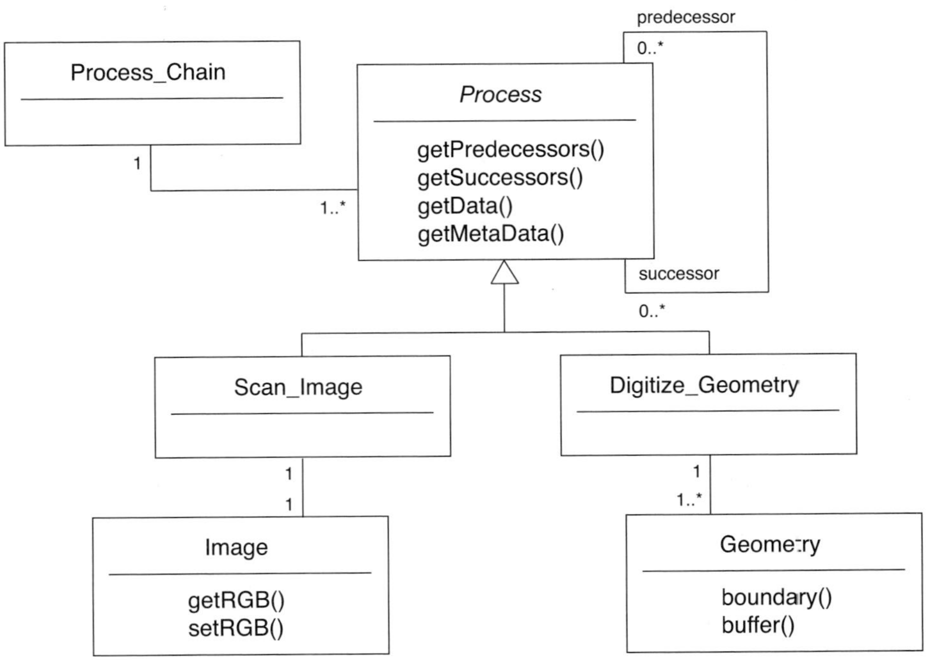

Fig. 2. Process-oriented data model of the Swiss Data Center for Nature and Landscape (DNL) in UML-notation.

is based on object-oriented design methods (Booch, 1994, Rumbaugh *et al.* 1991). The UML-model presented in Figure 2 can be described in short as follows (a more detailed description is available from Baumberger and Haegeli [2000]): The main part of the model is built by individual processes which are grouped (generalized) to a general abstract class *Process*. It comes with a generic set of process-related metadata including the time and date of processing, the processor, used technology and methods among others. Particular types of processes are modeled as subclasses of this generic class and are associated with corresponding data and more specialized metadata. Examples of such subclasses are for instance a class *Digitize_Geometry* which serves for digitizing geometries associated with geometric data, or a class *Scan_Image* responsible for the scanning of images associated with the corresponding image data. The history of data genesis is modeled using the class *Process_Chain* and the role names *predecessor* and *successor* of the *Process* class. For each process there are zero or more predecessor or successor processes. These associations essentially model a directed graph, representing the links between connected processes.

The model allows recapitulation of the history of all entities stored within the database by traversing the graph of processes through the *predecessor* and *successor* roles. Since the model includes final data sets as well as intermediate and source data, lineage tracking also

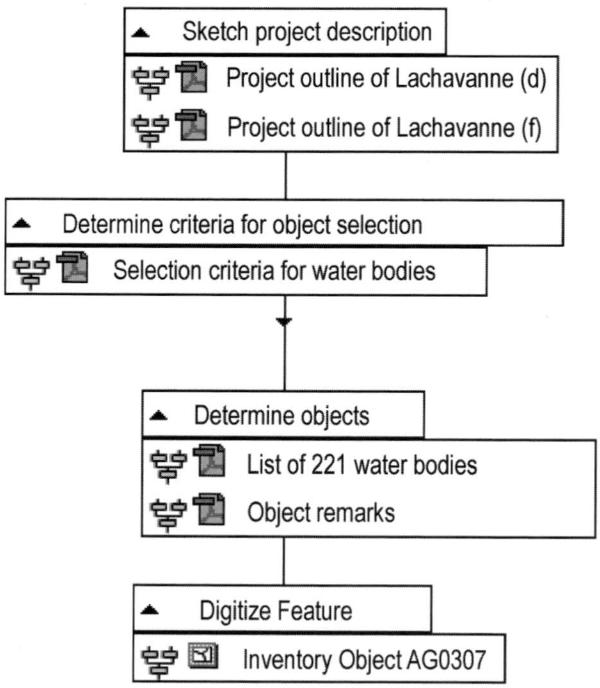

Fig. 3. An example of a simple directed graph which represents the history of creating an inventory object in a Nature and Landscape database. The object has the identification AG0307. It is part of an inventory consisting of ponds and small lakes for the establishment of an ecological typology and the description of their biodiversity. The graph shows the relations between the different process classes (for example "Determine objects") and the corresponding processes. These processes consist of several pdf-documents such as selection criteria and further remarks for object delineation determining the extent of the inventory object.

allows the assessment of impacts of process steps on a given data set. In contrast to alternative lineage-based metadata bases (Lanter 1991), this model is neither restricted to track geometric data uniquely nor tied to an aggregated feature level of a geometric layer or a coverage. Lineage tracking is also a reliable supporting tool for the database administrator or data providers. It allows verification of the database content as concerns its completeness, provided the entered data were accessed visually. This is due to the fact that missing links between successive processes may be identified quickly. Visual access of the data and metadata is provided by a web-interface which is described in Baltensweiler and Brändli (2004). An example for the lineage of a particular inventory object is provided in Figure 3.

Data Integration: Interoperability of Distributed Data Repositories

The Internet as a uniform platform for data sharing and access

Comprehensive metadata as described in the last section are an indispensable prerequisite when data from distributed repositories must be integrated. This section focuses on technical issues related to data access and data combination in order to enable data interoperability. Data integration and data sharing allowing the combined and integrated use of ecological data which are available from distributed data repositories benefit from research efforts in the information technology and particularly the geographic information science community. Traditionally, geographic information systems (GIS) provide spatial data handling capabilities for data input, storage, retrieval, management, manipulation, analysis, and output (Aronoff 1989; Burrough and McDonnell 1998). Geoprocessing functionality is usually supplied by a single and monolithic system, and the data is normally stored in a single database. Due to the popular use of the Internet this closed architecture paradigm of GIS today is gradually shifting towards a distributed geographic information services paradigm (Tsou and Buttenfield 2002; Preston *et al.* 2003). Distribution includes both the storage of data in physically distributed database systems and dispersed geoprocessing capabilities offered as so-called geo-services (Peng and Tsou 2003). The price of this paradigm shift is the requirement for enhanced interoperability, reusability and flexibility of both data and geo-services.

From a technical point of view, the requirement for enhanced data interoperability is currently only partly satisfied, because the majority of spatial data handling applications on the Internet concern web mapping or web cartography offering functionality for the use, distribution and production of maps by means of the Internet (Kraak 2001; Orthofer and Loibl 2004). Additionally, they support viewing operations on spatial data and submitting simple queries. Current standardization efforts such as the initiatives by the Open Geospatial Consortium (OGC, http://www.opengeospatial.org) foster this type of geospatial data handling. The OGC released the Web Map Service (WMS) Implementation Specification which standardizes the way map images, service-level metadata, and information about particular map features contained in a map are requested and presented (OGC 2001). An OGC-compliant WMS does not necessarily include any further tools for spatial analysis and modeling, however. Additionally, as data are returned in an image format, they cannot be accessed for analysis and modeling (later on) anymore. Distributed access to the real data is necessary, though, in order to get into a position to move from simple visualization to detailed analyses, predictions, and projections such as extrapolation of biodiversity estimates, predictions of species invasions and the projections of potential geographic species distributions in the future (Wiley and Peterson 2004).

The need for exchanging and sharing spatial data on the Internet which goes beyond viewing and querying images is well recognized. Software producers such as ESRI are extending their web map server products by including functionality that exceeds the basic WMS requirements. The OGC released the Web Feature Service (WFS, see below [OGC 2002]) and Geography Markup Language (GML) Implementation Specification (OGC 2003) which focus on the exchange of geographic data in a format that allows additional client-side processing. The WFS in particular enables access to vector data which serves as input for a wide variety of advanced geoprocessing functions such as spatial overlays.

An example of a Virtual Database

Goals and architecture of the Virtual Database: The Virtual Database described here is an example for an integrated environmental and landscape database system which follows and implements the trend towards Internet-based data exchange, the use of open standards and open interfaces. It aims at integrating different distributed databases of the Swiss Federal Office for the Environment storing fauna and flora data of nature protection in order to build an integrated data federation enabling combined exploration, modeling and analysis of available ecological data. A concept with a very similar name – Virtual Data Set (VDS) – was developed by Stephan *et al.* (1993) and is described in detail in Vckovski (1998). The VDS concept argues for an interface which hides as many of a single data source's details behind the interface. In contrast to the Virtual Database which offers an integrated view on physically distributed data repositories, the VDS concept concentrates on the particular task of interface-compliant access of data sets.

The Virtual Database is mostly based on development efforts and specifications of current open standardization initiatives such as the program of the OGC and the suite of interoperable technologies proposed and ratified by the W3C Consortium (http://www.w3c.org).

The goals of a Virtual Database are in particular:

- Integration of distributed data repositories (called data components) using possibly different database management systems or other storage software to build a data federation. This federation must not restrict the autonomy of the individual components.
- Implementation of distributed queries. Other database management functionality such as inserts and updates are handled by applications of the individual components.
- Adaptation and implementation of standardized interfaces for data access.
- Query, analysis and presentation of data from distributed database systems by means of Internet browser, map and feature server technology.

The design and implementation of the Virtual Database follow the principle of loose coupling of individual data components (Sheth and Larson 1990; Geppert and Dittrich 2001). Data components and necessary software modules are separated and interoperably arranged as presented in Figure 4:

- *Databases and files:* The bottom layer consists of data repositories that have to be integrated. Data may be either stored in database management systems or as files of different types.
- *Access layers:* Integration of data repositories is addressed by using interfaces which handle data access. The interfaces specify the way data is served on the one hand and the way data can be queried on the other hand. In addition, interfaces to access corresponding metadata must be provided in order to enable assessment of data served. The access layers also control access constraints on the data.

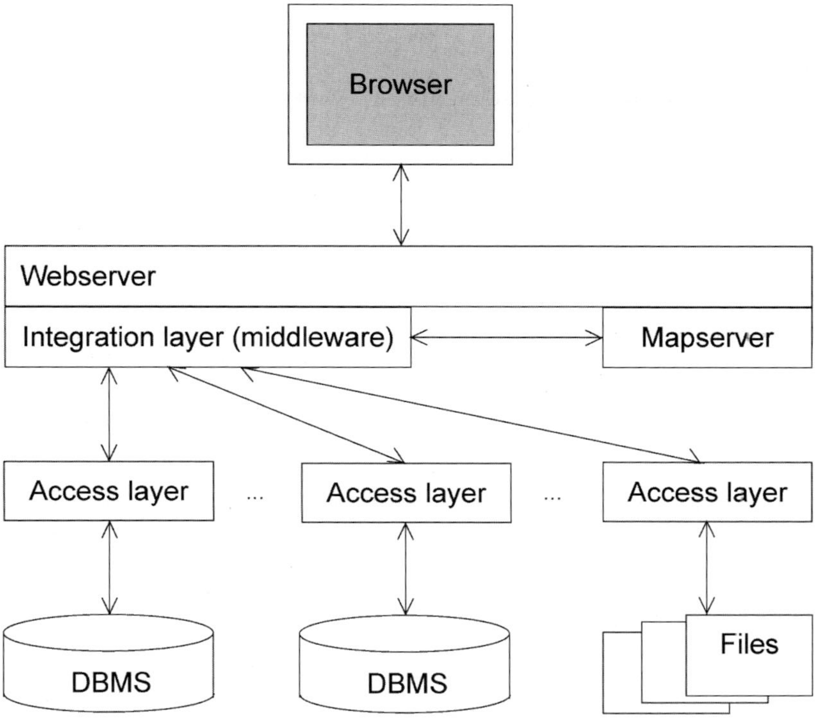

Fig. 4. Architecture of a Virtual Database.

- *Integration layer:* The integration layer mainly controls access to the distributed data repositories and integrates the data retrieved from the access layers. Map server software then creates user-friendly graphical output such as maps, charts and tables by suitable rendering of the spatial data.
- *Browser:* Data retrieved from the map server are displayed by Internet browser software. The browser is designed as a thin client handling basic user interaction and display.

Implementation of the Virtual Database: The interfaces for accessing the data of individual components (access layers, see Fig. 4) are implemented using the Java 2 platform including Java servlet technology for network programming and remote data access (http://java.sun.com/webservices/). The implementation of the interfaces is compliant with the WFS Implementation Specification of the OGC (OGC 2002). Providing data according to internationally agreed and open standards enables uniform access of data which also simplifies data sharing for other evolving data infrastructures such as INSPIRE and ENFIN mentioned above. The specification provides interfaces for the manipulation of spatial features and bases the communication between distributed computing platforms on HTTP. The full functionality of a WFS consists of methods for querying, inserting, updating and deleting data. Due to the defined goals, the Virtual Database only implements the minimum requirements of a WFS which consist of the following subset of interfaces:

- *GetCapabilities:* Reports on the data and operations a WFS offers. Results of a GetCapabilities request are transmitted as an XML document containing the characteristics of offered data.
- *DescribeFeatureType:* Reports on the structure of the spatial objects called features the WFS offers. Querying available feature types results in an XML Schema describing the structure of the data. The XML schema is constrained to XML schema components specified by the Geography Markup Language (GML), a particular XML-based language for the exchange and description of spatial data (OGC 2003).
- *GetFeature:* Enables the access of the data using spatially and non-spatially constraining queries. Results of a data query are sent as GML documents.

Implementing the integration layer consists of two main tasks: integration of data from different repositories and the graphical compilation for combined rendering in a browser window. These two tasks are also implemented using the Java 2 platform. Currently, collecting and integrating data is restricted to a simple merge of data served by the different data repositories. Existing metadata, such as the data quality elements discussed in the last section are indeed reported but not automatically taken into consideration for data integration at the current version of the implementation. The interpretation of metadata would for instance allow detection of possible heterogeneities such as differences in scale or resolution in order to apply methods which try to harmonize these differences. Automatic interpretation and handling of data heterogeneities, data uncertainties and erroneous data are postponed to future developments of the Virtual Database.

The task of rendering the integrated data takes advantage of ESRI's Internet mapping software ArcIMS. ArcIMS is based on a typical three-tier architecture consisting of a presentation, business logic and data tier, respectively. Visualization and query of spatial data and related attribute data is based on combining and extending out-of-the-box components of this software.

Currently, a prototype is running in a test phase integrating three different data components with heterogeneous database management systems: The first database is the above mentioned "Data Center for Nature and Landscape" (DNL) which stores data according to the presented process-oriented data model. It mainly holds inventory data of protected biotopes in Switzerland (Baltensweiler and Brändli 2004). Oracle is used as database management system in combination with ESRI's Spatial Database Engine (SDE) for handling and processing spatial data types. The second database is located at the Centre Suisse de Cartographie de la Faune (CSCF) and stores data on endangered animal species using Oracle as DBMS. In contrast to the DNL database, the geometries (coordinates) of locations of discovered animals are stored as standard columns in regular database tables. The third data component installed at the Institute of Systematic Botany, University of Zurich, contains discovered locations of endangered and rare moss species. Again, attribute data are stored in an Oracle database. Locational data are, however, stored in ESRI's shapefile format and thus represent a show case for the integration of a particular file type (as shown in Fig. 4).

The prototype is currently extended in order to enable integrated data modeling and analysis. Overlay functions, such as intersects and unions will allow the combined analysis of distributed data in the near future (Frehner *et al.* 2004) and will thus provide a solid base for effective data management.

Conclusions

The management of ecological data is faced with several new challenges as mentioned in the introductory section. Long-term monitoring programs, increase of dataset volume, distributed data storage and a scale shift towards a global view require that data management strengthens data interoperability. This paper presented a selection of promising techniques and solutions for data integration and data sharing, and emphasized the importance of quality assurance and quality controls in ecological data collection and management.

Quality controls and metadata are especially important for long-term monitoring programs. Technical innovations and changing user needs lead to continuous adaptations of measurement techniques, sampling strategies and data analysis procedures. The monitoring programs themselves are a most important cause for adaptations in data collection and analysis of environmental data. Today, integration and sharing of distributed and related ecological data for data analysis results in a considerable expansion of our system knowledge. In this context, metadata and quality controls are necessary pre-requisites for efficient and up-to-date long-term monitoring and the production of comparable results between regions and countries, as well as over time.

The metadata section of the paper concentrated on the particular topic of lineage metadata in order to underpin the importance of metadata for effective data assessment. The proposed process-oriented database both supports data users and data providers with a strong tool for data and metadata exploration. Data users are able to relate available data to a temporal, spatial and semantic context. By visually accessing entered data, lineage tracking allows data providers to check the content of the database for completeness and consistency. Additionally, the genesis of the data always remains traceable and comprehensible.

Metadata of different kinds are also involved in building data infrastructures from distributed environmental data repositories with the goal of combined data exploration and analysis. The Virtual Database, as a particular example of an integrated environmental and landscape information system, is based on metadata access related to data structuring and general characteristics of involved data. Using an interface approach for metadata and data access results in high flexibility for the integration of heterogeneous data sources. The use of interfaces based on internationally agreed and open standards also simplifies software development since adequate software components already exist and are readily available for reuse.

Although the proposed solutions substantially contribute to successful data interoperability in environmental data management, major problems are still waiting to be solved. The tools provided for the exploration of data quality elements, metadata and data mainly require visual investigation by expert users. Little support is offered for automatic data and metadata interpretation. However, machine-based interpretation of existing data heterogeneities, data errors and data uncertainties is extremely important in particular for integrating and combining environmental data from heterogeneous data repositories. Current research approaches focus on the specification and use of ontologies for specific application domains in order to better formalize the semantics of available data (for instance Fonseca *et al.* 2002, Kuhn 2002). The availability of ontologies offers the potential for detecting existing heterogeneities and provides a framework for data homogenization.

Besides the implementation of methods for data homogenization concerning data semantics, functions for automatic handling of data errors and data uncertainties must be developed. Reliable interpretation and assessment of output resulting from the integrated use and analysis of distributed environmental data requires corresponding integration and reporting tools.

The use of ontologies and the development of automatic interpretation methods of metadata for data integration, exploration and subsequent analysis will be key research topics in the near future.

References

Ailamaki A., Ioannidis Y.E. and Livny M. 1998. Scientific workflow management by database management. Proceedings of the 10th International Conference on Scientific and Statistical Database Management, Capri, July 1–3.

Aronoff S. 1989. Geographic Information Systems: A Management perspective. WDL Publications, Ottawa, Canada.

Baltensweiler A. and Brändli M. 2004. Web-based exploration of environmental data and corresponding metadata, in particular lineage information. In: Scharl A. (ed.), Environmental Online Communication, Advanced Information and Knowledge Processing Series, pp. 127–132. Springer, London. UK.

Bättig Ch., Bächtiger Ch., Bernasconi A., Brändli U.-B. and Brassel P. 2002. Wirkungsanalyse zu LFI1 und 2 und Bedarfsanalyse für das LFI3. Umweltmaterialien Nr. 143, Bundesamt für Umwelt, Wald und Landschaft BUWAL, Bern, Switzerland. 89 pp.

Baumberger N. and Hägeli M. 2000. Using metadata in multistep preprocessing and longterm monitoring. GIS Work Session UN/ECE, Conference of European Statisticians, Working Paper No. 15, April 10–12, Neuchâtel, Switzerland. 8 pp.

Bishr Y.A. 1999. A global unique persistent object id for geospatial information sharing. In: Vckovski A., Brassel K.E. and Schek H.-J. (eds.). Interoperating geographic information systems, Proceedings of the second international conference, INTEROP'99, Lecture Notes in Computer Science, Vol. 1580, Springer, Berlin, Heidelberg, Germany.

Booch, G. 1994. Object-oriented analysis and design with applications. Second edition, The Benjamin/Cummings Publishing Company, Redwood City, U.S.A.

Brassel K.E., Bucher F., Srephan E.-M., and Vckovski A. 1995. Completeness. In: Guptill S.C and Morrison J.L. (eds.). Elements of spatial data quality, pp. 81–108. Elsevier Science Ltd, Oxford, UK.

Brassel P. and Lischke H. 2001. Swiss national forest inventory: Methods and models of the second assessment. Swiss Federal Research Institute WSL, Birmensdorf, Switzerland. 336 pp.

Brodeur J., Bédard Y., Edwards G. and Moulin B. 2003. Revisiting the concept of geospatial data interoperability within the scope of human communication processes. Transactions in GIS 7, 2: 243–265.

Brunt J.W. 2000. Data management principles, implementation and administration. In: Michener W.K. and Brunt J.W. (eds.). Ecological data: design, management and processing, pp. 25–47. Blackwell Science, Oxford, UK.

Burrough P.A. and McDonnell R.A. 1998. Principles of geographical information systems. Oxford University Press, Oxford, UK. 333 pp.

Caughlan L. and Oakley K.L. 2001. Cost considerations for long-term ecological monitoring. Ecological Indicators 1: 123–134.

Cherubini P. and Innes J.L. 2000. Switzerland: The Swiss long-term forest ecosystem research programme. In: Sprott P. (ed.). The international long term ecological research network 2000: Perspectives from participating networks, pp. 56–59. Albuquerque, N.M., U.S. Long Term Ecological Research Network Office.

Clarke D.G. and Clark D.M. 1995. Lineage. In: Guptill S.C. and Morrison J.L. (eds.). Elements of spatial data quality, pp. 13–30. Elsevier Science Ltd, Oxford, UK.

Drummond J. 1995. Positional accuracy. In: Guptill S.C. and Morrison J.L. (eds.). Elements of spatial data quality, pp. 31–58. Elsevier Science Ltd, Oxford, UK.

Federal Geographic Data Commitee 1998. FGDC-STD-001-1998. Content standard for digital geospatial metadata (revised June 1998). Federal Geographic Data Committee. Washington, D.C., U.S.A. (http://www.fgdc.gov/standards/documents/standards/metadata/v2_0698.pdf).

Fonseca F.T., Egenhofer M.J., Agouris P. and Câmara G. 2002. Using ontologies for integrated geographic information systems. Transactions in GIS 6, 3: 231–257.

Fowler M. and Scott K. 2000. UML Distilled, A brief guide to the standard object modeling language, Second edition. Addison-Wesley, Reading.

Frehner M., Brändli M. and Schenker J. 2004. The Virtual Database – A tool for integrated data processing in a distributed environment. Proceedings EnviroInfo 2004, 18th International

Conference Informatics for Environmental Protection, October 21–23, Geneva, Switzerland, pp. 537–548.

Geppert A. and Dittrich K.R. 2001. Component database systems: Introduction, foundations, and overview. In: Dittrich K.R. and Geppert A. (eds.). Component database systems, pp. 1–28. Morgan Kaufmann Publishers, San Francisco, U.S.A.

Gertner G. and Köhl M. 1992. An assessment of some nonsampling errors in a national survey using an error budget. Forest Science 38, 3: 525–538.

Goodchild M.F. 1995. Attribute accuracy. In: Guptill S.C and Morrison J.L. (eds.). Elements of spatial data quality, pp. 59–79. Elsevier Science Ltd., Oxford, UK.

International Organization for Standardization 1993. ISO/IEC 2382–1: Information technology –Vocabulary – Part 1: Fundamental terms, International Standard ISO/IEC 2382–1:1993.

International Organization for Standardization 2003. ISO 19115 – Geographic information – Metadata, International Standard ISO 19115:2003.

Kainz W. 1995. Logical consistency. In: Guptill S.C and Morrison J.L. (eds.). Elements of spatial data quality, pp. 109–137. Elsevier Science Ltd., Oxford, UK.

Kaufmann E. 2001. Estimation of standing timber, growth and cut. In: Brassel P. and Lischke H. (eds.). National forest inventory: methods and models of the second assessment, pp. 162–196. Swiss Federal Research Institute WSL, Birmensdorf, Switzerland.

Kaufmann E. and Schwyzer A. 2001. Control survey of the terrestrial inventory. In: Brassel P. and Lischke H. (eds.). National forest inventory: methods and models of the second assessment, pp. 88–92. Swiss Federal Research Institute WSL, Birmensdorf, Switzerland.

Köhl M. 1994. Statistisches Design für das Schweizerische Landesforstinventar: Ein Folgeinventurkonzept unter Verwendung von Luftbildern und terrestrischen Aufnahmen. Mitteilungen der Eidgenössischen Forschungsanstalt für Wald, Schnee und Landschaft. Swiss Federal Research Institute WSL, Birmensdorf, Switzerland. 69, 1: 141.

Köhl M. 2001. Error sources and their influence on the NFI inventory results. In: Brassel P. and Lischke H. (eds.). National forest inventory: methods and models of the second assessment, pp. 297–305. Swiss Federal Research Institute WSL, Birmensdorf, Switzerland.

Köhl M., Gertner G. 1992. Geostatische Auswertungsmöglichkeiten für Waldschadeninventuren: Methodische Überlegungen zur Beschreibung räumlicher Verteilungen. Forstwiss. Cent.bl. 111: 320–331.

Kraak M.-J. 2001. Settings and needs for web cartography. In: Kraak M.-J., and Brown A. (eds.). Web cartography. Developments and prospects. Taylor and Francis, London and New York.

Kuhn W. 2002. Modeling the semantics of geographic categories through conceptual integration. Proceedings of Geographic Information Science, Second International Conference, GIScience 2002, September 2002, Boulder, U.S.A., pp. 108–118.

Lanea A.M.J. 1997. The UK environmental change network database: An integrated information resource for long-term monitoring and research. Journal of Environmental Management 51, 1: 87–105.

Lanter D.P. 1991. Design of a lineage-based meta-data base for GIS. Cartography and Geographic Information Systems 18, 4.

Longley P.A., Goodchild M.F., Maguire D.J. and Rhind D.W. 2001. Geographic information systems and science. John Wiley and Sons, Chichester. 454 pp.

Lutz M., Riedmann C. and Probst F. 2003. A classification framework for approaches to achieve semantic interoperability between GI web services. In: Kuhn W., Worboys M.F. and Timpf S. (eds.). COSIT 2003, Lecture Notes in Computer Sciences, 2825: 186–2003. Springer, Heidelberg, Germany.

Mandallaz D. 2002. Optimal sampling schemes based on the anticipated variance with lack of fit. Canadian Journal of Forest Research 32, 12: 2236–2243.

Mandallaz D. and Lanz A. 2001. Forest inventory: further aspects for optimal sampling schemes based on the anticipated variance. Canadian Journal of Forest Research 31, 10: 1845–1853.

Mandallaz D. and Ye R. 1999. Forest inventory with optimal two-phase, two-stage sampling schemes based on the anticipated variance. Canadian Journal of Forest Research 29, 11: 1691–1708.

Michener W.K. 2000. Research design: Translating ideas to data. In: Michener W.K. and Brunt J.W. (eds.). Ecological data: Design, management and processing, pp. 1–24. Blackwell Science, Oxford, UK.

Michener W.K. 2000. Metadata. In: Michener W.K. and Brunt J.W. (eds.). Ecological data: Design, management and processing, pp. 92–116. Blackwell Science, Oxford, UK.

Michener W.K., Brunt J.W., Helly J., Kirchner T.B. and Stafford S.G. 1997. Non-geospatial metadata for the ecological sciences. Ecological Applications 7, 1: 330–342.

Moellering H. 1987. A draft proposed standard for digital cartographic data. National Commitee for Digital Cartographic Standards, American Congress on Surveying and Mapping, Report No. 8.

Morrison J.L. 1988. The proposed standard for digital cartographic data. The American Cartographer 15, 1: 129–135.

Morrison J.L. 1995. Spatial data quality. In: Guptill S.C. and Morrison J.L. (eds.). Elements of spatial data quality. Elsevier Science Ltd., Oxford, UK.

OGC 2001. Web Map Service implementation specification. Version: 1.1.1. Open Geospatial Consortium, Inc. (http://www.opengeospatial.org/docs/01–068r2.pdf).

OGC 2002. Web Feature Service implementation specification. Version: 1.0.0. Open Geospatial Consortium, Inc. (http://www.opengeospatial.org/docs/02–058.pdf).

OGC 2003. OpenGIS Geography Markup Language (GML) implementation specification. Version: 3.00. Open Geospatial Consortium, Inc. (http://www.opengeospatial.org/docs/02–023r4.pdf).

Orthofer R. and Loibl W. 2004. Sharing environmental maps on the Web: The Austrian EnviroMap System. In: Scharl A. (ed.). Environmental Online Communication, Advanced Information and Knowledge Processing Series, pp. 133–144. Springer, London, UK.

Paschedag I. and Keller M. 2001. Control survey of the aerial photo interpretatation. In: Brassel P. and Lischke H. (eds). National forest inventory: Methods and models of the second assessment, pp. 88–92. Swiss Federal Research Institute WSL, Birmensdorf, Switzerland.

Peng Z. and Tsou M. 2003. Internet GIS: Distributed geographic information services for the internet and wireless networks. John Wiley and Sons, New Jersey, U.S.A.

Pollard J. 2002. Quality Assurance. FIA Fact Sheet Series. http://fia.fs.fed.us/library/Factsheets/QA.doc.

Preston M., Clayton P. and Wells G. 2003. Dynamic run-time application development using CORBA objects and XML in the field of distributed GIS. International Journal of Geographical Information Science 18, 5: 459–489.

Rumbaugh J., Blaha M., Premerlani W., Eddy F. and Lorenzen W. 1991. Object-oriented modeling and design. Prentice Hall, Englewood Cliffs.

Samal A., Seth S. and Cueto K. 2004. A feature-based approach to confluation of geospatial sources. International Journal of Geographical Information Science 17, 4: 321–341.

Sargent P. 1999. Feature identities, descriptors and handles. In: Vckovski A., Brassel K.E. and Schek H.-J. (eds.). Interoperating geographic information systems, Proceedings of the second international conference, INTEROP'99, Lecture Notes in Computer Science, Vol. 1580, Springer, Berlin, Heidelberg, Germany.

Shampine W.J. 1993. Quality assurance and quality control in monitoring programs. Environmental Monitoring and Assessment 26: 143–151.

Sheth A.P. and Larson J.A. 1990. Federated database systems for managing distributed, heterogeneous, and autonomous databases. ACM Computing Surveys 22, 3: 183–236.

Stephan E.M., Vckovski A. and Bucher F. 1993. Virtual Data Set: an approach for the integration of incompatible data. Proceedings Auto-Carto 11, Minneapolis, Minnesota, October 1993, pp. 93–102.

Stierlin H.-R. 2001. Criteria and previsions for quality assurance. In: Brassel P. and Lischke H. (eds.). National forest inventory: Methods and models of the second assessment, pp. 88–92. Swiss Federal Research Institute WSL, Birmensdorf, Switzerland.

Traub B. 2001. Representativenes of the sample grid. In: Brassel P. and Lischke H. (eds.). National forest inventory: Methods and models of the second assessment, pp. 132–149. Swiss Federal Research Institute WSL, Birmensdorf, Switzerland.

Tsou M.-H. and Buttenfield B.P. 2002. A dynamic architecture for distributing geographic information services. Transactions in GIS 6, 4: 355–381.

Vckovski A. 1997. Interoperable and distributed processing in GIS. Taylor and Francis, London.

Vckovski A., Brassel K.E. and Schek H.-J. 1999. Interoperating geographic information systems. Proceedings of the second international conference, INTEROP'99, Lecture Notes in Computer Science, Vol. 1580, Springer, Berlin, Heidelberg, Germany. 327 pp.

Wiley E.O. and Peterson A.T. 2004. Biodiversity and the Internet: Building and using the Virtual World Museum. In: Scharl A. (ed.). Environmental Online Communication, Advanced Information and Knowledge Processing Series, pp. 91–99. Springer, London, UK.

Worboys M. and Duckham M. 2004. GIS – a computing perspective. Second Edition, CRC Press, Boca Raton. 426 pp.

Wilson N. 2004. The beginnings of a logical semantics framework for the integration of thematic map data. International Journal of Geographical Information Science 18, 4: 389–416.

Zinggeler J. 2001. Expenditure of the terrestrial inventory. In: Brassel P. and Lischke H. (eds.). National forest inventory: Methods and models of the second assessment, pp. 88–92. Swiss Federal Research Institute WSL, Birmensdorf, Switzerland.

On Selected Issues and Challenges in Dendroclimatology

Jan Esper[1*], David C. Frank[1] and Jürg Luterbacher[2]

[1] Swiss Federal Research Institute WSL, Zürcherstrasse 111, CH-8903 Birmensdorf, Switzerland
[2] NCCR Climate and Institute of Geography, University of Bern, CH-3012 Bern, Switzerland
[*] Corresponding author: e-mail: jan.esper@wsl.ch

Abstract
We discuss selected issues in palaeoclimatic research, with a focus on tree ring based temperature reconstructions. Topics include the difficulty to retain long-term temperature variations in tree ring based reconstructions, the effects of this and other limitations on the estimation of the absolute temperature amplitude over the past millennium, and the potentials and limitations of including precipitation sensitive tree ring data in large-scale temperature reconstructions. To address these issues, we begin with a brief introduction into some principles of proxy data and specific characteristics of tree ring time-series.

Keywords: dendrochronology, tree rings, temperature reconstruchtion, age-trend, proxy, palaeoclimate

F. Kienast, O. Wildi & S. Ghosh (eds.), A Changing World. Challenges for Landscape Research, 113–132.

Introduction

From the range of potential research questions that can be addressed using dendroclimatic methods, the documentation of natural climate variability from periods prior to evident human impact stands out prominently (Watson *et al.* 2001). An essential component of this more recent impact is the emission of greenhouse gases such as CO_2 into the atmosphere since the beginning of widespread industrialization in the middle of the 19th century (Keeling

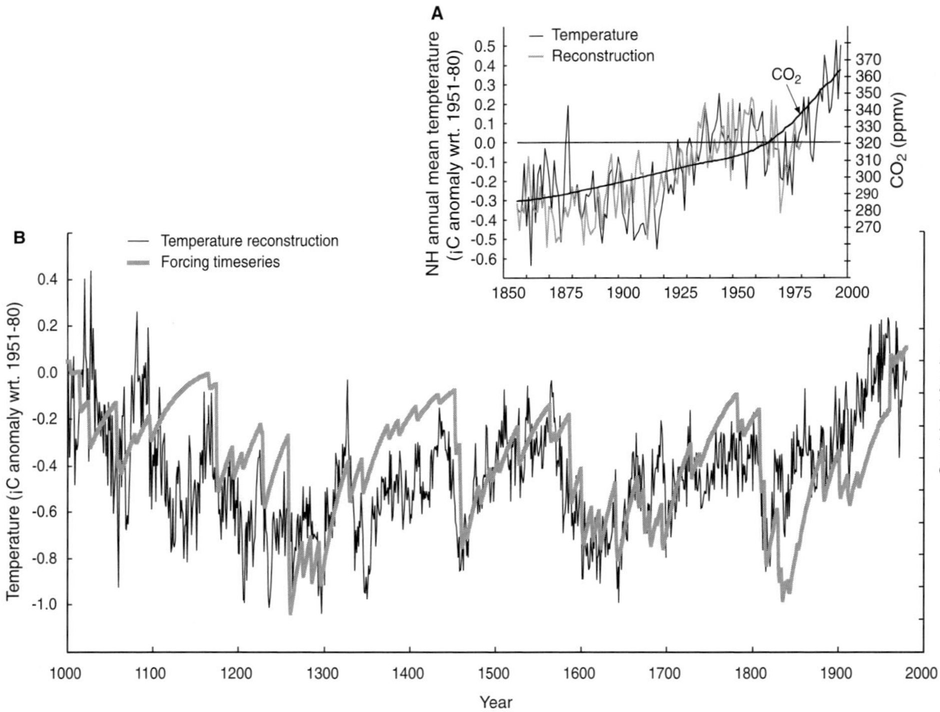

Fig. 1. Recent temperature and CO_2 timeseries, together with a millennial-long temperature reconstruction and a simple forcing timeseries. (A) Annual mean temperatures obtained from averaging Northern Hemisphere meteorological station data indicate a warming trend of about 0,6 °C over the past 140 years (Jones *et al.* 1999). These data represent 90% of the Northern Hemisphere surface area in the 1950s, 50% in the 1900s, and 20% in the 1860s. Numbers are derived utilizing a 1200 km radius for each single met station (Hansen *et al.* 1999). In comparison, the CO_2 content (Robertson *et al.* 2001) increases more steadily from about 285 to 370 ppm, but does not show the significant inter-decadal fluctuations as seen in the temperature data. To understand the forcing of CO_2 (and other greenhouse gases) on temperature, it is necessary to reconstruct natural, pre-industrial, climate variations from periods before the 19th century, and to relate these findings with reconstructions of natural forcing factors, such as the variability of solar irradiance and volcanic eruptions. (B) The large-scale temperature reconstruction over the past millennium averages information from 14 tree ring sites north of 30 °N (Esper *et al.* 2002a). This record is calibrated against annual temperatures, averaged over Northern Hemisphere land and sea surface areas, using the 1856–1980 period as shown in A (Jones *et al.* 1999). Accordingly, the temperature amplitude over the past millennium is in the order of 1 °C. The reconstruction indicates warmth at the beginning of the past millennium, similar to the temperatures recorded during about the middle of the 20th century. Some of the inter-decadal to centennial scale variations in this reconstruction are in line with an average series combining volcanic (Crowley 2000, Robock und Free 1996), solar (Bard *et al.* 2000, Crowley 2000, Lean *et al.* 1995), CO_2 (Etheridge *et al.* 1996), and an estimate of tropospheric aerosol forcing (Crowley 2000, Etheridge *et al.* 1996).

et al. 1996; Robertson *et al.* 2001). During this time, the concentration of CO_2 has risen from about 280 parts per million to about 370 in recent years (Fig. 1A). With this, and projected future emissions, comes the concern that the anthropogenic forcing of temperature (Hansen *et al.* 1999) and change of weather and climate extremes (Katz and Brown 1992; Kharin and Zwiers 2005; Klaus 1993; Stainforth *et al.* 2005; Stott *et al.* 2004) will also become more significant in the future. Such changes would significantly impact natural ecosystems (Nemani *et al.* 2003) and the utilization of these resources by humans (Ahmad *et al.* 2001).

To quantify the influence of greenhouse gases and to be able to develop reliable projections of future climate variations, it is important to understand the forcing from both natural and anthropogenic factors (Fig. 1B). In contrast to the recent period since industrialization, late Holocene greenhouse gas variation prior to the middle of the 19th century was negligible. During this pre-industrial time, changes in solar radiation and stratospheric reaching aerosols from volcanic eruptions were likely the primary forcing factors for climate variations. The quantification of these different forcing factors represents a key objective for model calibration and testing (Gerber *et al.* 2003) and future predictions (Boer *et al.* 2000).

To provide information on longer term regional and large-scale climate history, so-called proxy data are analyzed. Following the discussion of some principles of such data, we address the climatic signals retained in certain tree ring parameters, and stress the challenging issue of preserving low frequency temperature variations in dendroclimatology. This topic is followed by a discussion of the estimation of absolute temperature variations over the past millennium, and the problems and potential to include precipitation sensitive tree ring data in large-scale temperature reconstructions. The review closes by highlighting some challenges of future dendroclimatic research.

Proxy Sources

Proxy data, such as timeseries of the thickness and composition of lake sediments, glacial ice layers, growth increments in corals, speleothems and trees, borehole temperature profiles, and documentary evidence are key to the understanding and quantification of past climatic variations (overview in Bradley and Jones 1992; Jones *et al.* 1996; Jones and Mann 2004; Moberg *et al.* 2005), and, thus, the current global climatic change debate (Watson *et al.* 2001). Data from these archives are generally compared and correlated with instrumental measurements to quantify their climatic sensitivities and signals, and are subsequently used to document climate prior to the period of instrumental data.

The different proxy sources have varying strengths and weaknesses, and comparison of several independent sources, or even their combination, is desirable to develop robust reconstructions of past climate variability (Casty *et al.* 2005a, 2005b; Luterbacher *et al.* 2004, 2005; Mann 2002; Xoplaki *et al.* 2005). Proxy sources are usually confined to certain geographic regions, with ice cores limited to high mountain and polar regions (Watanabe *et al.* 2003), corals to low latitude sea shores (Felis *et al.* 2000), tree rings primarily to extratropical forested ecosystems (Schweingruber 1996; Stahle 1999; Worbes 1999), and documentary evidence to regions from which historic reports are available, such as Europe (Brázdil *et al.* 2005; Bürgi *et al.* 2007; Pfister 1999; Pfister *et al.* 1998), eastern Asia (Ge *et al.* 2005; Qian *et al.* 2003; Wang *et al.* 2001; Yang *et al.* 2002; Zhang and Crowley 1989), and South America (Prieto *et al.* 2004). Also the number and type of measured parameters varies considerably between the proxy sources. In the family of available proxies, ice cores tend to provide the largest variety of commonly measured parameters, including direct measures of CO_2 concentration from trapped air bubbles (Smith *et al.* 1997), the isotopic fractionation of oxygen, and levels of sulfate from volcanic eruptions (Petit *et al.* 1999; EPICA 2004).

Additional differentiation between proxy sources concerns their lengths and resolutions. Whereas tree ring series are of yearly resolution, the resolution and precision of other proxy sources is generally lower (except for documentary evidence which can even resolve daily weather), or decreases with depth, such as for ice core data. In exchange for these resolution tradeoffs, ice core data can extend many hundreds of thousands of years back, whereas continuous tree ring data rarely extend back ten thousand years (Briffa and Matthews 2002; EPICA 2004; Grudd *et al.* 2002; Schaub *et al.* 2003).

A distinctive characteristic of tree ring data, and various other proxies, is that the quality of the chronology is not stable over time, with greater uncertainties typically occurring further back in time. Furthermore the climatic sensitivity also varies through time (Esper *et al.* 2001a). This phenomenon can be illustrated in so-called extreme year analysis in dendroclimatology (Schweingruber *et al.* 1990). Certain years stand out from all others, whereby the majority of trees synchronously produces an exceptionally narrow or wide ring (Neuwirth *et al.* 2003; Schweingruber *et al.* 1991). These years possess greater signal strength than "average" years, in which say seven trees produce wide rings, while 10 produce narrow rings (Esper and Gärtner 2001). Analyses of extreme years (Esper *et al.* 2001b; Neuwirth *et al.* 2003; Schweingruber *et al.* 1991) show that extremes (both positive or negative) of a similar magnitude can be triggered by a wide variety of climatic circumstances. This can complicate dendroclimatic studies, where the influence of a single climatic parameter on tree growth is generally sought. Such complications can be reduced however by careful consideration of site ecology.

Climatic Signals in Tree Ring Parameters, and Trend Problems

The ecology of the most suitable sites for dendroclimatic analyses are well known (overview in Schweingruber 1996). A clear climatic signal can be obtained at e.g. treeline sites, whereas intermediate sites often yield somewhat fuzzy signals. Trees from drought stressed locations, such as the lower forest border in the American southwest (LaMarche 1974), Mongolia (Pederson *et al.* 2001) or the Mediterranean area (Akkemik and Aras 2005; Chbouki *et al.* 1995; D'Arrigo and Cullen 2001; Touchan *et al.* 2005) tend to most clearly show a precipitation signal.

Similarly the temperature signal is maximized at the temperature limits of tree growth at, for example, the northern treeline in Siberia (Briffa *et al.* 1998) or upper treeline in the Tien Shan and Karakorum (Esper *et al.* 2002b, 2003b). Even though the vast majority of dendroclimatic reconstructions are based on trees, shrubs and even dwarf shrubs that develop annual rings (Schweingruber 2001) are generally usable. While ring width is the most easily measured parameter used for dendroclimatic reconstructions, other parameters, such as the maximum latewood density (Schweingruber *et al.* 1978) and even specific anatomical features (Schweingruber 2001), can also have great reconstructive power.

Stable isotope ratios measured in tree rings are another data source that is increasingly used for climate reconstruction (Borella *et al.* 1998a, 1998b; Leavitt and Long 1984; Saurer *et al.* 1997; Treydte *et al.* 2001). Here the stable isotope ratio of, for example, ^{13}C and ^{12}C (or isotopes of oxygen), is compared with an internationally recognized standard (Craig 1957) and is expressed as the deviations of $\delta^{13}C$ in parts per thousand (‰). For tree cellulose the values are negative. It has been shown that such measurements can be linked, depending upon site ecology, with temperature or precipitation variations (Treydte 2003; Treydte *et al.* 2001). Indeed, highly significant relationships between $\delta^{13}C$ series and temperature measurements can be obtained, particularly in the high-frequency domain. These correlations are only valid, however, after removal of an increasingly negative trend in $\delta^{13}C$ measurements in time (Farquhar *et al.* 1982; Leavitt and Long 1989; Marshall and Monserud 1996). The negative trend results from the enrichment of air from fossil fuel carbon (which has a highly negative value) and also from plant physiological effects in response to the increasing partial pressure of CO_2. As the plant physiological mechanisms of $\delta^{13}C$ fixation are not totally understood, the low frequency, centennial scale variations can not yet be confidently estimated due to the greater uncertainties that exist during the anthropogenically influenced calibration period.

Specific wood anatomical features that result from relatively few and/or highly specific influences can also be used for reconstructing past environmental changes. For example, occurrences of late (in spring) and early (in autumn) frosts have been reconstructed (Hantemirov *et al.* 2000). To do so deformed cell walls (callous tissues), that result from frost, are used as the characteristic features (Fig. 2). The location of these callous cells within the tree ring, e.g. can be used to reconstruct frost events even at a sub-annual resolution (Schweingruber 2001).

Raw measurements of tree ring width (in mm) and maximum latewood density (in g/cm^3) are the result of a multitude of physiological drivers. Some of them are closely linked to climate, some are linked to non-climatic sources, e.g. the age trend (Bräker 1981; Cook and Kairiukstis 1990; Fritts 1976). The age trend is responsible for decreasing ring width or

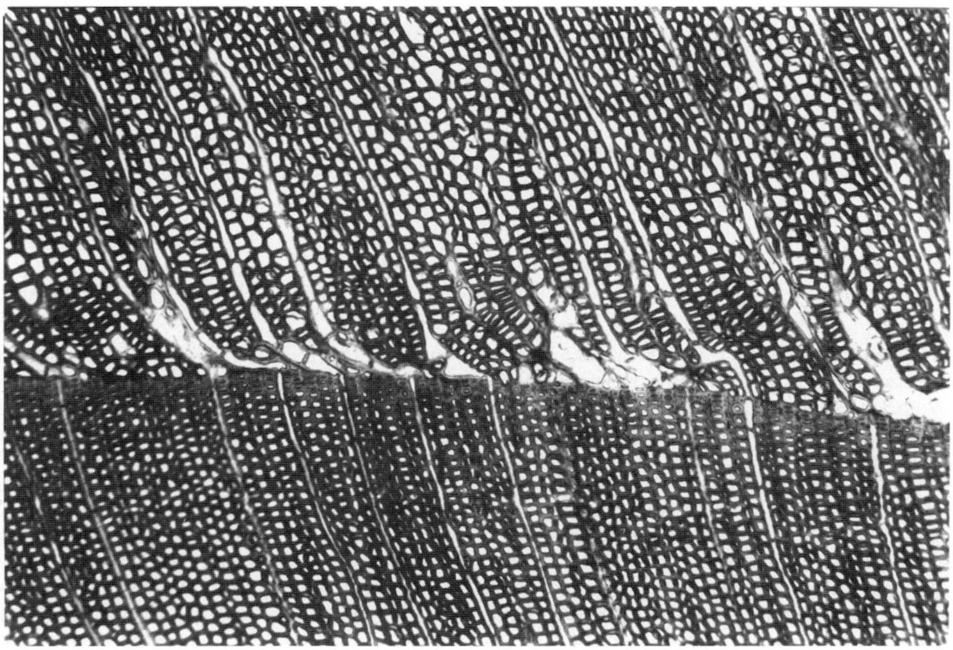

Fig. 2. Callous tissue (frost ring) in the earlywood of a juniper tree *(Juniperus turkestanica)* from the upper timberline in the Tien Shan Mountains, Kirghizia. In the center of the picture a tree ring boundary separating two rings (below and above) is seen. The lower part shows the latewood portion of the (older) ring with smaller cells and thicker cell walls, followed by the earlywood portion of the (younger) ring with larger cells and thinner cell walls. During the beginning of the formation of this younger ring, a frost occurred affecting the cambium. The freezing of the cambium resulted in deformed, callous cells (center of the picture). Typical for such frost rings is also the curvilinear offset of the ray cells cutting through the ring boundary. The frost occurred in spring, immediately after the tree started to build earlywood.

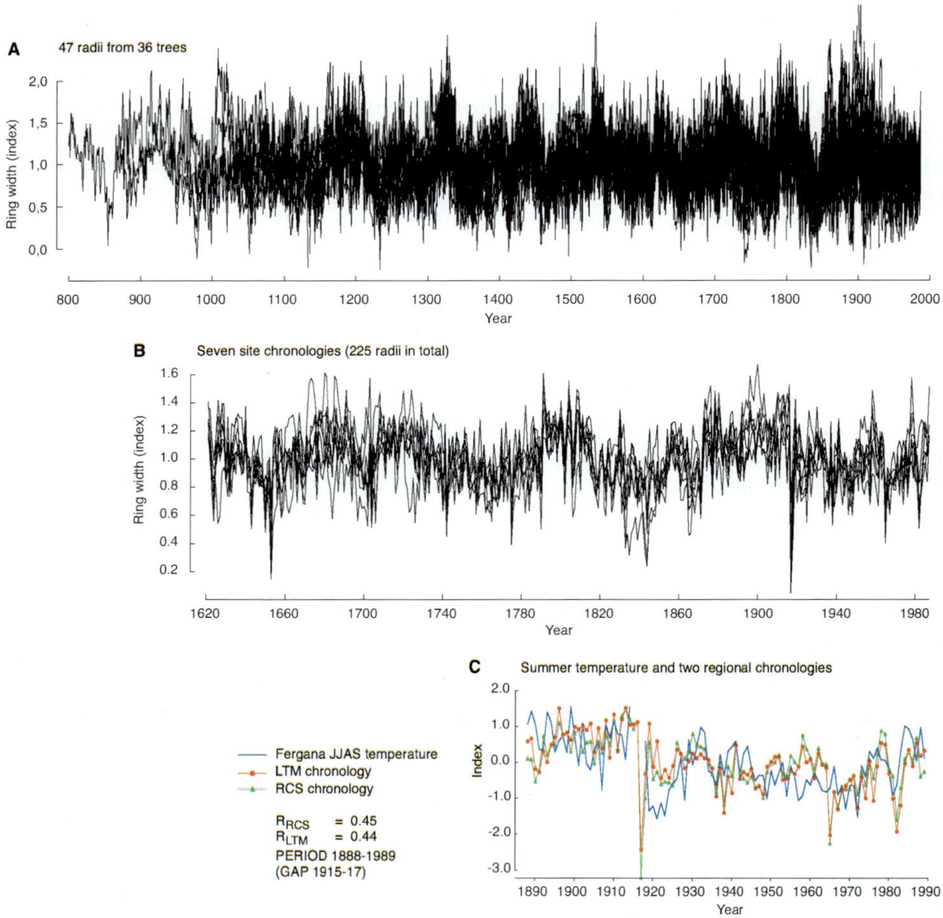

Fig. 3. Synchronous ring width variations in high elevation Juniper trees *(Juniperus turkestanica)* from the Tien Shan Mountains, and correlation with instrumental temperature data. (A) 47 ring width series from 36 trees near the upper treeline in 3,200 m a.s.l. show distinct inter-decadal scale variations. These variations are synchronous throughout most periods of the past millennium, indicative of a strong common signal between the trees. Age trends were removed with a spline standardization technique (Cook 1985) and an adaptive power-transformation (Cook and Peters 1997). Details are given in Esper *et al.* (2003b). (B) Ring width variation between seven high-elevation stands in the Tien Shan Mountains since 1620. The variation here is extremely synchronous suggesting that a spatially common climate signal (likely temperature) is controlling tree growth. (C) This connection can be additionally validated through a comparison between two regional mean chronologies (LTM, RCS) and summer temperature measured at the Fergana instrumental station. Of particular note is the absence of a 20th century warming trend in western central Asia that is, for example, evident in Europe and large-scale temperature timeseries. The portions of the RCS and LTM chronologies shown here were standardized using methods to specifically preserve multi-centennial trends (Esper *et al.* 2003b). These trends are not preserved using the more traditional spline standardization shown in A.

density with increasing tree age. It results from the geometric property of adding more or less constant biomass to an increasing surface area as the tree grows. This trend must be eliminated prior to climatic analyses, otherwise the ring width or maximum latewood density values are biased and reflect tree age rather than a climate signal. This age trend is commonly removed with detrending procedures, while still preserving the common climatic signal of interest. By common variation, we mean the positive and negative deviations (after standardization) of ring width or density, that are synchronous between trees at a given site, and ultimately between different study sites (Fig. 3) (Esper *et al.* 2001a; Wigley *et al.* 1984). Such synchronous variations within and between sites can only result from common environmental influences over larger areas.

While the synchronous variation between trees and stands represents a clear conceptual strength of dendroclimatology, the age-trend and its necessary removal represent a substantial weakness or limitation of the ring width and density parameters. In general, it is quite difficult to distinguish the biological age-trend from long-term climatic signals, particularly when the long-term climate represents a cooling (e.g. the transition from the Medieval Warm Period into the Little Ice Age), which for temperature sensitive trees, can mimic the shape of the biological age trend. The separation of these trends represents a substantial and critical task in modern dendroclimatology (for further information see Briffa *et al.* 1992, 2001; Cook *et al.* 1995; Cook and Peters 1997; Esper *et al.* 2002a, 2003a).

Temperature Amplitude

To understand the role of different anthropogenic and natural forcings on ecosystems, an assessment of past climatic changes is needed over time periods longer than the instrumental interval. Currently, the temperature variation over the past 1000 years receives considerable attention (Watson *et al.* 2001). The characteristics, timing and regional particularities of the Medieval Warm Period, the subsequent Little Ice Age, and the present warm period are particularly relevant (Broecker 2001, Mann *et al.* 2003b). Related questions are: *(i)* how large was the temperature amplitude (in °C) over the past 1000 years, and *(ii)* how quickly has the temperature varied without anthropogenic impacts such as greenhouse gases (Esper *et al.* 2004, 2005a, 2005b; Moberg *et al.* 2005; Mann *et al.* 2003b; von Storch *et al.* 2004).

To help answer these questions, millennial-long large-scale reconstructions with annual resolution have been developed (Briffa 2000; Esper *et al.* 2002a; Jones *et al.* 1998; Mann *et al.* 1999). Some of these series (e.g. Briffa 2000; Esper *et al.* 2002a; see also Moberg *et al.* 2005) show that in large parts of the Northern Hemisphere a distinct warm period existed about 1000 years prior to present. According to these reconstructions, temperature patterns similar to those of the middle 20th century existed within the past millennium. The warm periods are separated by distinct cooler episodes associated with the Little Ice Age. In contrast, the reconstruction from Mann *et al.* (1999) shows relatively warm conditions about 1000 years ago, with a very gradual cooling for 900 or so years until a fairly abrupt modern increase.

Much discussion about common features in large-scale reconstructions, and the possibility to retain common multi-centennial climatic variations currently exists (Briffa and Osborn 2002; Cook *et al.* 2004a; Esper *et al.* 2002a; 2004a; Mann *et al.* 2002). It was suggested (e.g. by Esper *et al.* 2004a) that differences in the lower frequency domains of currently used long-term tree-ring series (Fig. 4) may be a result of detrending procedures that were sub-optimal to fully preserve multi-centennial wavelength information. This problem is particularly critical, as mentioned above, for the transition from the Medieval Warm Period into the Little Ice Age, where age-trends and long-term climatic evolution both exhibit decreasing trends.

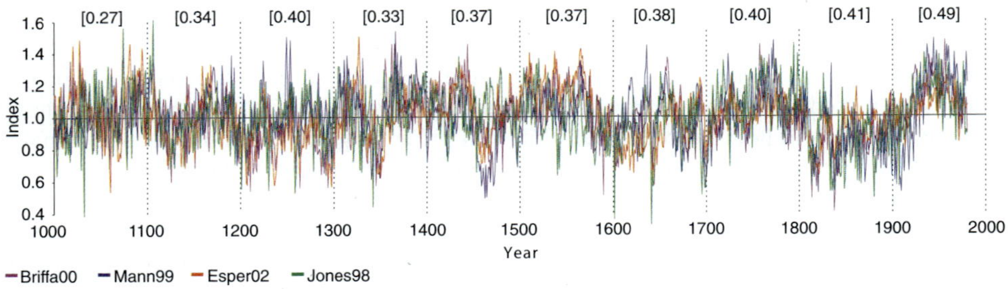

Fig. 4. Annually resolved, large-scale temperature reconstructions (Briffa 2000, Esper *et al.* 2002, Jones *et al.* 1998, Mann *et al.* 1999) showing synchronous multi-decadal variations. The records were detrended using a spline filter to remove low frequency, multi-centennial trends. The significant similarities in the remaining (higher) frequencies suggest that differences in the lower frequencies likely result from differing standardization techniques applied in the original reconstructions (for details see Esper *et al.* 2004a). The average interseries correlation of the records as shown here, is 0.42 over the 1000–1980 period. Average correlations for each century are indicated in the figure. While these reconstructions share some data, tests that minimized this overlap did not reveal substantial differences (Esper *et al.* 2004a).

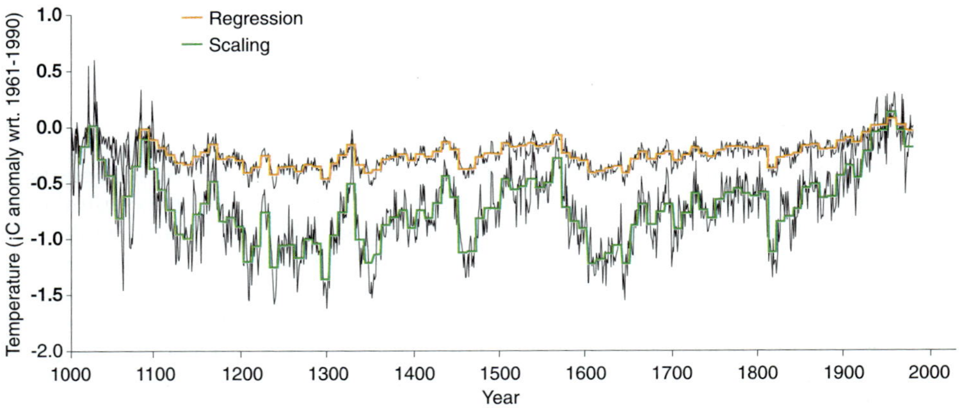

Fig. 5. Varying temperature amplitudes (in °C) obtained after calibrating the same proxy reconstruction (Esper *et al.* 2002a) with different (currently accepted) methods. The record displaying a smaller amplitude was regressed to warm season land and sea surface temperature data over the 1856–1980 period, while that showing greater amplitude, by scaling (mean and variance equalization) to annual land temperature data over the 1900–1977 period. The spatial domain for both instrumental datasets is 20–90 °N latitude (Jones *et al.* 1999). Various estimates of the amplitude of past variability from these calibration choices add additional uncertainty in the magnitude of past temperature variability, and can result in widely diverging views, particularly when instrumental data are spliced to the modern end of such records. See Esper *et al.* 2005b for more details.

In addition to the different "shapes" of these curves, the temperature amplitudes reconstructed from them also tend to differ considerably. The approximate decadal scale amplitude (difference between the warmest to coldest decade) derived from the Mann *et al.* (1999) reconstruction is about 0.5 °C, whereas the amplitude reconstructed by Esper *et al.* (2002a), and then recalibrated by Cook *et al.* (2004a) are both about 1.0 °C (see also Moberg *et al.* 2005). This latter figure is more consistent with large-scale estimates derived from borehole reconstructions over the past 500 years (Beltrami 2002; González-Rouco *et al.* 2003; Huang *et al.* 2000; Pollack and Huang 2000; Pollack and Smerdon 2004), although other analyses using the same data indicate somewhat lower amplitudes (Mann *et al.* 2003a; Rutherford and Mann 2004).

Reconstruction of the temperature amplitude can also be hampered by calibration methods and data used (see e.g. Briffa and Osborn 2002). This issue was recently tested by von Storch *et al.* (2004) using general circulation model (GCM) results as a surrogate for the "true" climate over the past millennia, and "pseudo proxy data" with statistical characteristics similar to real proxy (e.g. long instrumental, tree ring, documentary, coral, etc.) data. Their results suggest that regression based calibration methods – similar to those used by Mann *et al.* (1999) – may consistently underestimate the true temperature amplitude, and in such a way where there tends to be a bias towards greater error, and hence reduced amplitude, in frequencies outside those well captured in the calibration interval (i.e. low frequencies). However, recent work (Mann *et al.* 2005) suggests that the von Storch *et al.* (2004) results depend on the selection of the GCM model, the (varying) radiative forcing applied to these models, and the long-term performance (drift) in climate simulations. It was, for example, shown that the GKSS simulation (as used by von Storch *et al.* 2004) is biased by a 'spin-up' artifact, i.e. the simulation was initialized from a warm 20th century state at AD 1000, prior to the application of pre-anthropogenic radiative forcing, leading to a long-term drift in mean temperature (Goosse *et al.* 2005). The Mann *et al.* (2005) results seem to contradict the suggestion that empirical proxy-based temperature reconstructions suffer from systematic underestimations of low-frequency variability (von Storch *et al.* 2004). Further tests are necessary to solve this issue, and particularly to determine the impact of employing different climate models and forcing series (and their weighting) on reconstructed temperature amplitudes.

Esper *et al.* (2005a) addressed the same issue by systematically surveying the effects on large-scale reconstructions' amplitude that result from the calibration to a variety of instrumental targets with a variety of methods – all of which are used in current literature. The results indicate that both the selection of various "reasonable" instrumental data and calibration periods, as well as various fitting procedures, adds a methodological uncertainty to the reconstructions that easily approaches 0.5 °C (Fig. 5). It is evident that the exact assessment of this variation, and particularly the amplitude, has significant consequences for the quantitative estimations of greenhouse gas forcing in the past 150 years and hence for future predictions (Esper *et al.* 2005b).

Separating Temperature and Precipitation Signals in Tree-ring Measurements

It is important for dendroclimatic reconstructions that in many regions temperature and precipitation variations are significantly negatively correlated. This is, for example, the case in the Alps during summer when precipitation occurs with generally cooler atmospheric conditions or more local convective systems (Böhm *et al.* 2001; Wanner *et al.* 2000). This covariance between temperature and precipitation can have the effect that it is difficult to demonstrate the dominance of a single factor's influence, such as summer temperature, on

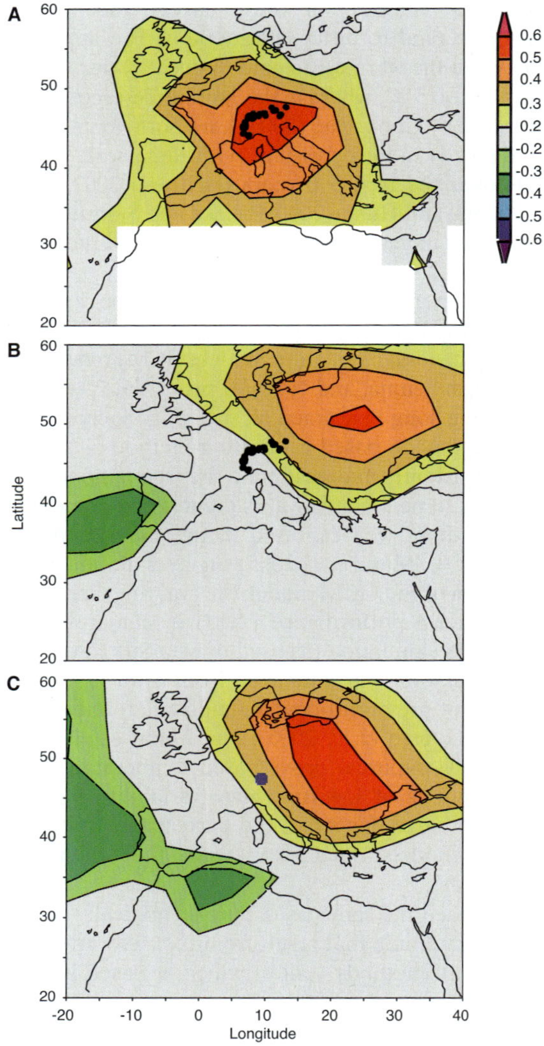

Fig. 6. Correlation fields with the first principal component of 45 high elevation (above 1,500 m a.s.l.) ring width sites (black circles) from the western and central Alps with (A) average June-August temperature, and (B) average June-August sea-level pressure, computed over the 1900–1973 period (white region in A represents missing data). This principle component explains 20% of the network variance over the 1850–1973 period, capturing the dominant mode of ring width variability in this network that is closely related to summer temperature variability. The trees provide temperature information approximately centered over their average geographical locations, but still yield significant correlations over much of Europe. Highest correlations between Alpine ring width and pressure are located further east, likely reflecting continental synoptic influences on temperature. (C) For comparison, average June-August temperatures from the Säntis meteorological station (blue square) in Switzerland correlated with the same pressure field as in B. Similar patterns are evident including the eastward shift in the center of highest positive correlations, and areas of negative correlation west of Spain. The 90% significance level for correlations corresponds approximately to the mapped colored regions. Maps generated using the KNMI Climate Explorer. Säntis temperature data are from GHCNv2 (Peterson and Vose 1997); surface temperature data are from HadCRUT2v (Jones and Moberg 2003; Rayner *et al.* 2003); and SLP data (Trenberth and Paulino 1980).

tree growth. In such cases, with only a positive temperature influence on tree-growth, a negative (secondary) correlation with precipitation is also obtained.

The difficulties in extracting a particular single signal can, however be advantageous for some reconstructions. If, for example, a precipitation sensitive chronology provides a strong correlation with temperature, it might well be incorporated into large-scale temperature reconstructions. For example, 1000-year ring width series from cedar trees growing in Morocco (Stockton 1985; Verstege *et al.* 2004) correlate significantly with precipitation (Chbouki *et al.* 1995) and through this, their growth is also linked with the North Atlantic Oscillation (Glueck and Stockton 2001; Hurrell 1995, Wanner *et al.* 2001). Such Moroccan cedar timeseries were used by Mann *et al.* (1999) for their Northern Hemisphere temperature reconstruction, for example, indicating that it is feasible to use some of the cross correlation and synoptic interaction between temperature and precipitation for temperature reconstructions. Similar considerations would apply to other primarily precipitation sensitive tree ring chronologies from the Mediterranean region (e.g. Akkemik and Aras 2005; Touchan *et al.* 2005).

When including such precipitation sensitive data in large-scale temperature reconstructions, however, an analysis of the frequency spectra of precipitation (and temperature) sensitive proxy timeseries seems useful. This is because measured precipitation data generally possess "whiter" spectra than those for temperature data. If precipitation sensitive tree ring material is included in large-scale temperature records, this limitation could be addressed by allowing only the inter-annual to inter-decadal precipitation information to enter the composite record at decadal and higher frequencies, and let the "true" temperature sensitive proxy data determine the lower frequency trends. Recent work by Cook *et al.* (2004b), however, shows indication of longer term variability in area aridity indices derived using Palmer Drought Severity Index (PDSI; Palmer 1965) reconstructions, thus initiating a discussion on the lower frequency behavior of precipitation related parameters.

Statistical methods, such as Principal Component Analysis (Peters *et al.* 1981; Preisendorfer 1988; von Storch and Zwiers 1999), can be effectively used to isolate a certain fraction of variance from tree ring timeseries, with these fractions subsequently used to explain climatic parameters, such as summer temperature (Frank and Esper 2005; Cook *et al.* 2003). These methods work particularly well if a larger network of tree ring sites is used, and can provide climatic information on regional to continental scales (Fig. 6). Though methods, such as PCA, perhaps perform superiorly for certain applications, it should be noted that highly regarded results based on correlation analysis for huge networks exist and have their own advantages (e.g. Briffa *et al.* 1998). In any case, it is recommendable in dendroclimatic studies to first study and understand connections with simple correlations, prior to using other methods where some of the basic relationships that exist can be more easily obscured. Such correlation approaches strengthen the basic foundations for which climate parameters truly have an influence on growth or isotopic values.

Challenges in Dendroclimatology

From the range of challenges this discipline currently faces, the task to produce robust estimates of low-frequency (multi-centennial) climate variations is particularly notable (Briffa *et al.* 2001; Cook *et al.* 1995, 2000; Esper *et al.* 2002a, 2003b). This assumes, however, that the chosen climate parameter actually varies in the lower frequency domain. This seems to be the case with temperature, and from various examples with precipitation as well, depending upon the time-scale of interest (Cook *et al.* 2004b; Dai *et al.* 1997). In this context, due to the greater challenge, it is particularly important to preserve long-term cooling trends, such as from the Medieval Warm Period into the Little Ice Age with an equal fidelity as warming trends, such as those since the Little Ice Age. Composite detrending methods, such as Regional Curve Standardization (RCS; Briffa *et al.* 1992, 1996; Becker *et al.* 1995; Mitchell 1967) and Age-Banding (Briffa *et al.* 2001) will increasingly be applied to preserve and study multi-centennial trends. These methods depend upon extensive datasets (Esper *et al.* 2003a), whereby it is likely that shifts in sampling strategies will need to occur to meet these requirements. More trees per stand and all age-classes (young through old) should be collected. At the same time, it will be necessary to conduct network analyses to tie together new and existing timeseries for the above-mentioned applications. This will also allow comparison of long-term trends in different, independent datasets, which is necessary to help overcome the limited statistical tests that can be conducted to calibrate centennial scale proxy variations against instrumental data. The aggregation of local tree ring series, where standardization methods were applied that were not designed to preserve long-term variability, should be avoided to study lower frequency climatic changes (Esper *et al.* 2004a).

Multi-proxy comparisons will perhaps play a greater role in the future. It is, however, rather ambitious to merge different archives, developed with discipline specific methods, with different temporal responses, and different climatic and seasonal sensitivities (Moberg *et al.* 2005). However, in principle, the merging should only serve to strengthen the picture of past climate variability by using the strength from the individual archives, rather than transporting unexplained portions of variance from individual records. Projects that seek to study and compare various proxy archives within a defined region, such as the project VITA (Varves, Ice cores and Tree ring Archives with annual resolution) within the Swiss NCCR-Climate program, should prove valuable towards this objective. Within this project, ice core, lake sediment (and organisms deposited therein), and tree ring data are collected and compared from a geographically focused region in the central Alps (Bigler 2002).

Additionally, opportunities exist in dendroclimatology to help approach important questions from related disciplines. For example, quantifying carbon sequestration and fluxes in terrestrial ecosystems (Janssens *et al.* 2003) is a challenging task where tree ring data can be used to provide insight. So far, this topic has received only marginal efforts by dendrochronologists, and instead has been driven by shorter term estimations through eddy flux measurements (Ehman *et al.* 2002), and verified by forest inventories (Goodale *et al.* 2002) and model calculations (Gurney *et al.* 2002). At the same time, current estimations for large-scale terrestrial carbon fluxes are highly variable and inconsistent (Houghton 2003; Körner 2003). In our opinion, dendrochronology can provide a substantial contribution to these efforts by quantifying biomass dynamics in forests over long timescales. Furthermore, dendrochronology can be used to study whether mid- to long-term fluctuations in biomass (as a surrogate for carbon) have been stimulated by climatic or other (CO_2, nitrogen) factors (Graumlich *et al.* 1989).

A rather long-term challenge in dendroclimatology is to potentially question and at the same time iteratively confirm early instrumental measurements. Currently, instrumental measurements are used almost exclusively to calibrate tree ring timeseries and other proxy data. However, even the temperature and precipitation measurements themselves contain uncertainties, and are fundamentally changed during necessary homogenization (Auer *et al.* 2005; Barriendos *et al.* 2002; Begert *et al.* 2005; Bergström and Moberg 2002; Böhm *et al.* 2001; Brunetti *et al.* 2004; Camuffo 2002a, 2002b; Cocheo and Camuffo 2002; Demarée *et al.* 2002; Klein Tank *et al.* 2005; Klingbjer and Moberg 2003; Maugeri *et al.* 2002a, 2002b; Moberg *et al.* 2002; Peterson *et al.* 1998; Slonosky 2002). These homogenization methods are particularly relevant, yet limited, in the early instrumental time period (e.g. around 1800 in Switzerland) during which few station records exist for comparison and verification, and also during more recent times through trying to understand the so-called urbanization effects (warming in cities through construction and changes near climate stations; Arnfield 2003, Kalnay and Cai 2003, Landsberg 1981, Peterson 2003, Parker 2004). In particular, there is high potential for dendroclimatic studies to validate the instrumental homogenization at locations where hundreds of tree ring sites (Briffa *et al.* 2002) and only a few long instrumental timeseries (e.g. Siberia) exist.

Acknowledgements

Supported by the Swiss National Science Foundation, project Millennium (# 2100-066628), and the Swiss National Science Foundation, NCCR-climate program. We thank Ulf Büntgen, Holger Gärtner, and Kerstin Treydte and two anonymous reviewers for helpful comments.

References

Ahmad Q.K., *et al.* (Eds.) 2001. Climate change 2001 – impacts, adaptation, and vulnerability. Summary for policy makers. Cambridge Univ. Press, Cambridge, 17 pp.

Akkemik Ü. and Aras A. 2005. Reconstruction (1689–1994 AD) of April–August precipitation in the southern part of central Turkey, International Journal of Climatology 25: 537–548.

Arnfield A.J. 2003. Two decades of urban climate research – a review of turbulence, exchanges of energy and water, and the urban heat island. International Journal of Climatology 23: 1–26.

Auer I., Böhm R., Jurkovic A., Orlik A., Potzmann R. and 20 co-authors 2005. A new instrumental precipitation dataset for the greater alpine region for the period 1800–2002. International Journal of Climatology 25: 139–166.

Bard E., Raisbeck G., Yiou F. and Jouzel J. 2000. Solar irradiance during the last 1200 years based on cosmogenic nuclides. Tellus 52: 985–992.

Barriendos M., Martin-Vide J., Pena J.C. and Rodriguez R. 2002. Daily meteorological observations in Cadiz – San Fernando. Analysis of the documentary sources and the instrumental data content (1786–1996). Climatic Change 53: 151–170.

Becker M., Bert G.D., Bouchon J., Dupouey J.L., Picard J.F. and Ulrich E. 1995. Long-term changes in forest productivity in Northeastern France – the dendroecological approach. In: Landmann G., Bonneau, M. (eds.). Forest decline and atmospheric deposition effects in the French mountains. Springer, Berlin, 143–156.

Begert M., Schlegel T. and Kirchhofer W. 2005. Homogeneous temperature and precipitation series of Switzerland from 1864 to 2000. International Journal of Climatology 25: 65–80.

Beltrami H. 2002. Climate from borehole data – energy fluxes and temperatures since 1500. Geophysical Research Letters 29, doi: 10.1029/2002GL015702.

Bergström H. and Moberg A. 2002. Daily air temperature and pressure series for Uppsala (1722–1998). Climatic Change 53: 213–252.

Bigler C. 2002. VITA – Varves, ice cores, and tree rings – archives with annual resolution. ProClim-Flash 24: 8–9.

Böhm R., Auer I., Brunetti M., Maugeri M., Nanni T. and Schöner W. 2001. Regional temperature variability in the European Alps 1760–1998 from homogenized instrumental time series. International Journal of Climatology 21: 1779–1801.

Boer G.J., Flato G. and Ramsden D. 2000. A transient climate change simulation with greenhouse gas and aerosol forcing: projected climate for the 21st century. Climate Dynamics 16: 427–450.

Borella S., Leuenberger M. and Saurer M. 1998a. Analysis of $\delta 18O$ analysis in tree rings – wood-cellulose comparison and method dependent sensitivity. Journal of Geophysical Research 106: 19267–19273.

Borella S., Leuenberger M., Saurer M. and Siegwolf R. 1998b. Reducing uncertainties in $\delta 13C$ analysis of tree rings – pooling, milling and cellulose extraction. Journal of Geophysical Research 103: 19519–19526.

Bradley R.S. and Jones P.D. (eds.) 1992. Climate since A.D.1500. Routledge, London, 679 pp.

Bräker O.U. 1981. Der Altersttrend bei Jahrringdichten und Jahrringbreiten von Nadelhölzern und sein Ausgleich. Mitteilungen der Forstlichen Bundesversuchsanstalt Wien 142: 75–102.

Brázdil R., Pfister C., Wanner H., von Storch H. and Luterbacher J. 2005. Historical climatology in Europe – The state of the art. Climatic Change 70: 363–430.

Briffa K.R. 2000. Annual climate variability in the Holocene – interpreting the message from ancient trees. Quaternary Science Reviews 19: 87–105.

Briffa K.R., Jones P.D., Bartholin T.S., Eckstein D., Schweingruber F.H., Karlén W., Zetterberg P. and Eronen M. 1992. Fennoscandian summers from AD 500: temperature changes on short and long timescales. Climate Dynamics 7: 111–119.

Briffa K.R., Jones P.D., Schweingruber F.H., Karlén W. and Shiyatov S.G. 1996. Tree-ring variables as proxy-indicators – problems with low-frequency signals. In: Jones P.D. *et al.* (eds.) Climatic variations and forcing mechanisms of the last 2000 years. Springer, Berlin, 9–41.

Briffa K.R., Schweingruber F.H., Jones P.D., Osborn T.J., Shiyatov S.G. and Vaganov E.A. 1998. Reduced sensitivity of recent tree-growth to temperature at high northern latitudes. Nature 391: 678–682.

Briffa K.R., Osborn T.J., Schweingruber F.H., Harris I.C., Jones P.D., Shiyatov S.G. and Vaganov E.A. 2001. Low-frequency temperature variations from a northern tree ring density network. Journal of Geophysical Research 106: 2929–2941.

Briffa K.R. and Matthews J.A. 2002. ADVANCE-10K – a European contribution towards a hemispheric dendroclimatology for the Holocene. The Holocene 12: 639–642.

Briffa K.R. and Osborn R.J. 2002. Blowing hot and cold. Science 295: 2227–2228.

Briffa K.R., Osborn T.J., Schweingruber F.H., Jones P.D., Shiyatov S.G. and Vaganov E.A. 2002. Tree-ring width and density data around the Northern Hemisphere – part 2, spatio-temporal variability and associated climate patterns. The Holocene 12: 759–789.

Broecker W.C. 2001. Was the Medieval Warm Period Global? Science 291: 1497–1499.

Brunetti M., Maugeri M., Monti F. and Nanni T. 2004. Changes in daily precipitation frequency and distribution in Italy over the last 120 years. Journal of Geophysical Research 109, doi: 10.1029/2003JD004296.

Bürgi M., Hersperger A.M., Hall M., (Russell) Southgate E.W.B. and Schneeberger N., 2007. Using the past to understand the present land use and land cover. In: Kienast F., Wildi O. and Ghosh S. (eds.). A Changing World. Challenges for Landscape Research, Vol. 8: 133–144. Springer Landscape Series, Dordrecht.

Camuffo D. 2002a. Calibration and instrumental errors in early measurements of air temperature. Climatic Change 53: 297–330.

Camuffo D. 2002b. Errors in early temperature series arising from changes in style of measuring time, sampling schedule and number of observations. Climatic Change 53: 331–354.

Casty C., Handorf D. and Sempf M. 2005a. Combined winter climate regimes over the North Atlantic/European sector 1766–2000. Geophysical Research Letters 32, doi: 10.1029/2005GL022431.

Casty C., Luterbacher J., Wanner H., Esper J. and Böhm R. 2005b. Temperature and precipitation variability in the European Alps since 1500. International Journal of Climatology 25: 1855–1880.

Chbouki N., Stockton C.W. and Myers D.E. 1995. Spatio-temporal patterns of drought in Morocco. International Journal of Climatology 15: 187–205.

Cocheo C. and Camuffo D. 2002. Corrections of systematic errors and data homogenisation in the Padova series (1725 - today). Climatic Change 53: 77–100.

Cook E.R. 1985. A time series analysis approach to tree-ring standardization. Lamont-Doherty Geological Observatory, New York, 171 pp.

Cook E.R. and Kairiukstis L.A. 1990. Methods of dendrochronology – applications in the environmental science. Kluwer, Dordrecht, 394 pp.

Cook E.R., Briffa K.R., Meko D.M., Graybill D.A. and Funkhouser G. 1995. The 'segment length curse' in long tree-ring chronology development for palaeoclimatic studies. The Holocene 5: 229–237.

Cook E.R. and Peters K. 1997. Calculating unbiased tree-ring indices for the study of climatic and environmental change. The Holocene 7: 361–370.

Cook E.R., Buckley B.M., D'Arrigo R.D. and Peterson M.J. 2000. Warm-season temperatures since 1600 BC reconstructed from Tasmanian tree rings and their relationship to large-scale sea surface temperature anomalies. Climate Dynamics 16: 79–91.

Cook E.R., Krusic P.J. and Jones P.D. 2003. Dendroclimatic signals in long tree-ring chronologies from the Himalayas of Nepal. International Journal of Climatology 23: 707–732.

Cook E.R., Esper J. and D'Arrigo R. 2004a. Extra-tropical Northern Hemisphere temperature variability over the past 1000 years. Quaternary Science Reviews 23: 2063–2074.

Cook E.R., Woodhouse C., Eakin C.M., Meko D.M. and Stahle D.W. 2004b. Long-term aridity changes in the Western United States. Science 306: 1015–1018.

Craig H. 1957. Isotopic standards for carbon and oxygen and correction factors for mass-spectrometric analysis of carbon dioxide. Geochimica et Cosmochimica Acta 12: 133–149.

Crowley T.J. 2000. Causes of climate change over the past 1000 years. Science 289: 270–277.

Dai A., Fung I.Y. and Del Genio A.D. 1997. Surface observed global land precipitation variations during 1900–1988. Journal of Climate 10: 2943–2962.

D'Arrigo R.D. and Cullen H.M. 2001. A 350-year (AD 1628–1980) reconstruction of Turkish precipitation. Dendrochronologia 19: 167–177.

Demarée G.R., Lachaert P.J., Verhoeve T. and Thoen E. 2002. The long-term daily central Belgium temperature (CBT) series (1767–1998) and early instrumental meteorological observations in Belgium. Climatic Change 53: 269–293.

Ehman J.L., Schmid H.P., Grimmond C.S.B., Randolph J.C., Hanson P.J., Wayson C.A. and Cropley F.D. 2002. An initial intercomparison of micrometeorological and ecological inventory estimates of carbon exchange in a mid-latitude deciduous forest. Global Change Biology 8: 575–589.

EPICA Community Members 2004. Eight glacial cycles from an Antarctic ice core. Nature 429: 623–628.

Esper J. and Gärtner H. 2001. Interpretation of tree-ring chronologies. Erdkunde 55: 277–287.

Esper J., Neuwirth B. and Treydte K. 2001a. A new parameter to evaluate temporal signal strength of tree ring chronologies. Dendrochronologia 19: 93–102.

Esper J., Treydte K., Gärtner H. and Neuwirth B. 2001b. A tree ring reconstruction of climatic extreme years since AD1427 for Western Central Asia. Palaeobotanist 50: 141–152.

Esper J., Cook E.R. and Schweingruber F.H. 2002a. Low-frequency signals in long tree-ring chronologies for reconstructing past temperature variability. Science 295: 2250–2253.

Esper J., Schweingruber F.H. and Winiger M. 2002b. 1,300 years of climate history for Western Central Asia inferred from tree-rings. The Holocene 12: 267–277.

Esper J., Cook E.R., Peters K. and Schweingruber F.H. 2003a. Tests of the RCS method for preserving low-frequency variability in long tree-ring chronologies. Tree-Ring Research 59: 81–98.

Esper J., Shiyatov S.G., Mazepa V.S., Wilson R.J.S., Graybill D.A. and Funkhouser G. 2003b. Temperature-sensitive Tien Shan tree ring chronologies show multi-centennial growth trends. Climate Dynamics 8: 699–706.

Esper J., Frank D.C. and Wilson R.J.S. 2004. Temperature reconstructions – low frequency ambition and high frequency ratification. EOS 85: 113, 120.

Esper J., Frank, D.C., Wilson, R.J.S., Briffa and K.R. 2005b. Effect of scaling and regression on reconstructed temperature amplitude for the past millennium. Geophysical Research Letters 32, doi:10.1029/2004GL021236.

Esper J., Wilson R.J.S., Frank D.C., Moberg A., Wanner H. and Luterbacher J. 2005b. Climate – past ranges and future changes. Quaternary Science Reviews 24: 2164–2166.

Etheridge D.M., Steele L.P., Langenfelds R.L., Francey R.J., Barnola J.M. and Morgan V.I. 1996. Natural and anthropogenic changes in atmospheric CO_2 over the last 1000 years from air in Antarctic ice and firn. Journal of Geophysical Research 101: 4115–4128.

Farquhar G.D., O'Leary M.H. and Berry J.A. 1982. On the relationship between carbon isotope discrimination and the intercellular carbon dioxide concentration in leaves. Australian Journal of Plant Physiology 9: 2599–2609.

Felis T., Pätzold J., Loya Y., Fine M., Nawar A.H. and Wefer G. 2000. A coral oxygen isotope record from the northern Red Sea documenting NAO, ENSO, and North Pacific teleconnections on Middle East climate variability since the year 1750. Paleoceanography 15: 679–694.

Frank D. and Esper J. 2005. Temperature reconstructions and comparisons with instrumental data from a tree-ring network for the European Alps. International Journal of Climatology 25: 1437–1454.

Fritts H.C. 1976. Tree rings and climate. Academic Press, New York, 567 pp.

Ge Q.S., Zheng J.Y., Hao Z.X., Zhang P.Y. and Wang W.C. 2005. Reconstruction of historical climate in China, high-resolution precipitation data from Qing Dynasty archives. Bulletin of the American Meteorological Society 86: 671–674.

Gerber S., Joos F., Brügger P.P., Stocker T.F., Mann M.E., Sitch S. and Scholze M. 2003. Constraining temperature variations over the last millennium by comparing simulated and observed atmospheric CO_2. Climate Dynamics 20: 281–299.

Glueck M.F. and Stockton C.W. 2001. Reconstruction of the North Atlantic Oscillation, 1429–1983. International Journal of Climatology 21: 1453–1465.

González-Rouco J.F., von Storch H. and Zorita E. 2003. Deep soil temperature as a proxy for surface air-temperature in a coupled model simulation of the last thousand years. Geophysical Research Letters 30, doi: 10.1029/2003GL018264.

Goodale C.L., Apps M.J., Birdsey R.A., Field C.B., Heath L.S. and 8 co-authors 2002. Forest carbon sinks in the northern hemisphere. Ecological Applications 12: 891–899.

Goosse H., Crowley T.J., Zorita E., Ammann C.M., Renssen H. and Driessschaert E. 2005. Modeling the climate of the last millennium: what causes the differences between simulations? Geophysical Research Letters 32, doi: 10.1029/2005GL022368.

Graumlich L.J., Brubaker L.B. and Grier C.C. 1989. Long-term trends in forest net primary productivity – Cascade Mountains, Washington. Ecology 70: 405–410.

Grudd H., Briffa K.R., Karlen W., Bartholin T.S., Jones P.D. and Kromer B. 2002. A 7400-year tree-ring chronology in northern Swedish Lapland – natural climatic variability expressed on annual to millennial timescales. The Holocene 12: 657–665.

Gurney K.R. *et al.* 2002. Towards robust regional estimates of CO_2 sources and sinks using atmospheric transport models. Nature 415: 626–630.

Hansen J., Ruedy R., Glascoe J. and Sato M. 1999. GISS analysis of surface temperature change. Journal of Geophysical Research 104: 30997–31022.

Hantemirov R.M., Gorlanova L.A. and Shiyatov S.G. 2000. Pathological tree-ring structures in Siberian juniper (Juniperus sibirica Burgsd.) and their use for reconstructing extreme climatic events. Russian Journal of Ecology 31: 167–173.

Houghton R.A. 2003. Why are estimates of the terrestrial carbon balance so different? Global Change Biology 9: 500–509.

Huang S., Pollack H.N. and Shen P.Y. 2000. Temperature trends over the past five centuries reconstructed from borehole temperature. Nature 403: 756–758.

Hurrell J.W. 1995. Decadal trends in the North Atlantic Oscillation – regional temperatures and precipitation. Science 269: 676–679.

Janssens I. *et al.* 2003. Europe's terrestrial biosphere absorbs 7 to 12% of European anthropogenic CO_2 emissions. Science 300: 1538–1542.

Jones P.D., Bradley S.R. and Jouzel J. (eds.) 1996. Climatic variations and forcing mechanisms of the last 2000 years. Springer, Berlin, 649 pp.

Jones P.D., Briffa K.R., Barnett T.P. and Tett S.F.B. 1998. High-resolution palaeoclimatic records for the past millennium – interpretation, integration and comparison with general circulation model control-run temperatures. Holocene 8: 455–471.

Jones P.D. and Mann M.E. 2004. Climate over past millennia. Review of Geophysics 42, doi: 10.1029/2003RG000143.

Jones P.D. and Moberg A. 2003. Hemispheric and large-scale surface air temperature variations – an extensive revision and an update to 2001. Journal of Climate 16: 206–223.

Jones P.D., New M., Parker D.E., Martin S. and Rigor I.G. 1999. Surface air temperature and its changes over the past 150 years. Review of Geophysics 37: 173–199.

Kalnay E. and Cai M. 2003. Impact of urbanization and land-use change on climate. Nature 423: 528–531.

Katz R.W. and Brown B.G. 1992. Extreme events in changing climate: variability is more important than averages, Climatic Change 21: 289–300.

Keeling C.D., Chin J.F.S. and Whorf T.P. 1996. Increased activity of northern vegetation inferred from atmospheric CO_2 measurements. Nature 382: 146–149.
Kharin V.V. and Zwiers F.W. 2005. Estimating extremes in transient climate change simulations. Journal of Climate 18: 1156–1173.
Klaus D. 1993. Zirkulations- und Persistenzänderungen des Europäischen Wettergeschehens im Spiegel der Großwetterlagenstatistik. Erdkunde 47: 85–104.
Klein Tank A.M.G., Können G.P. and Selten F.M. 2005. Signals of anthropogenic influence on European warming as seen in the trend patterns of daily temperature variance. International Journal of Climatology 25: 1–16.
Klingbjer P. and Moberg A. 2003. A composite monthly temperature record from Tornedalen in northern Sweden. International Journal of Climatology 23: 1465–1494.
Körner C. 2003. Slow in, rapid out – carbon flux studies and Kyoto targets. Science 300: 1242–1243.
LaMarche Jr. V.C. 1974. Paleoclimatic inferences from long tree-ring records. Science 183: 1043–1048.
Landsberg H.E. 1981. The urban climate. Academic Press, New York, 275 pp.
Lean J., Beer J. and Bradley R. 1995. Reconstruction of solar irradiance since 1610 – implications for climate change. Geophysical Research Letters 22: 3195–3198.
Leavitt S.W. and Long A. 1984. Sampling strategy for stable carbon isotope analysis of tree rings in pine. Nature 311: 145–147.
Leavitt S.W. and Long A. 1989. The atmospheric δ13C record as derived from 56 pinyon trees at 14 sites in the southwestern United States. Radiocarbon 31: 469–474.
Luterbacher J., Dietrich D., Xoplaki E., Grosjean M. and Wanner H. 2004. European seasonal and annual temperature variability, trends, and extremes since 1500. Science 303: 1499–1503.
Luterbacher J., Xoplaki E., Casty C., Wanner H., Pauling A., Küttel M. and 38 co-authors 2005. Mediterranean climate variability over the last centuries; a review. In: Lionello P., Malanotte-Rizzoli P. and Boscolo R. (eds.). The Mediterranean climate variability. Elsevier pp. 27–148.
Mann M.E. 2002. The value of multiple proxies. Science 297: 1481–1482.
Mann M.E., Bradley R.S. and Hughes M.K. 1999. Northern Hemisphere temperatures during the past millennium – inferences, uncertainties, and limitations. Geophysical Research Letters 26: 759–762.
Mann M.E., Hughes M.K., Cook E.R. and Esper J. 2002. Tree-ring chronologies and climate variability. Science 296: 848–849.
Mann M.E., Rutherford S., Bradley R.S., Hughes M.K. and Keimig F.T. 2003a. Optimized surface temperature reconstructions using terrestrial borehole data. Journal of Geophysical Research 108, doi: 10.1029/2002JD002532.
Mann M.E., Ammann C., Bradley R., Briffa K., Jones P., Osborn T. and others 2003. On past temperatures and anomalous late-20th century warmth. EOS 84: 256–258.
Mann M.E., Rutherford S., Wahl E.R. and Ammann C.M. 2005. Testing the fidelity of methodologies used in proxy-based reconstructions of past climate. Journal of Climate 18: 4097–4107.
Marshall J.D. and Monserud R.A. 1996. Homeostatic gas-exchange parameters inferred from 13C/12C in tree rings of conifers. Oecologia 105: 13–21.
Maugeri M., Buffoni L. and Chlistovsky F. 2002a. Daily Milan temperature and pressure series (1763–1998) – history of the observations and data and metadata recovery. Climatic Change 53: 101–117.
Maugeri M., Buffoni L., Delmonte B. and Fassina A. 2002b. Daily Milan temperature and pressure series (1763–1998) – completing and homogenising the data. Climatic Change 53: 119–149.
Mitchell V.L. 1967. An investigation of certain aspects of tree growth rates in relation to climate in the central Canadian boreal forest. Technical Report 33. University of Wisconsin, Department of Meteorology, Wisconsin.
Moberg A., Bergström H., Ruiz Krigsman J. and Svanered O. 2002. Daily air temperature and pressure series for Stockholm (1756–1998). Climatic Change 53: 171–212.
Moberg A.D., Sonechkin M., Holmgren K., Datsenko N.M. and Karlen W. 2005. Highly variable Northern Hemisphere temperatures reconstructed from low- and high-resolution proxy data. Nature 433: 613–617.

Nemani R.R., Keeling C.D., Hashimoto H., Jolly W.M., Piper S.C., Tucker C.J., Myneni R.B. and Running S.W. 2003. Climate-driven increases in global terrestrial net primary production from 1982 to 1999. Science 300: 1560–1563.

Neuwirth B., Esper J., Schweingruber F.H. and Winiger M. 2003. Site ecological differences to the climatic forcing of spruce pointer years from the Lötschental, Switzerland. Dendrochronologia 21: 69–78.

Palmer W.C. 1965. Meteorological drought. Research Paper No. 45, U.S. Department of Commerce (Weather Bureau, Washington, D.C.).

Parker D. 2004. Climate – large-scale warming is not urban. Nature 432: 290.

Pederson N., Jacoby G.C., D'Arrigo R.D., Cook E.R. and Buckley B.M. 2001. Hydrometeorological reconstructions for northeastern Mongolia derived from tree rings: 1651–1995. Journal of Climate 14: 872–881.

Peters K., Jacoby G.C. and Cook E.R. 1981. Principal component analysis of tree-ring sites, Tree-Ring Bulletin 41: 1–19.

Peterson T.C. and Vose R.S. 1997. An overview of the Global Historical Climatology Network temperature data base. Bulletin of the American Meteorological Society 78: 2837–2849.

Peterson T.C., Easterling D.R., Karl T.R., Groisman P., Nicholls N., Plummer N. and 15 co-authors 1998. Homogeneity adjustments of in situ climate data – a review. International Journal of Climatology 18: 1493–1517.

Peterson T.C. 2003. Assessment of urban versus rural in situ surface temperatures in the contiguous United States: no difference found. Journal of Climate 16: 2941–2959.

Petit J.R., Jouzel J., Raynaud D., Barkov N.I., Barnola J.-M., Basile I., Bender M. and 12 co-authors 1999. Climate and atmospheric history of the past 420000 years from the Vostok ice core, Antarctica. Nature 399: 429–436.

Pfister C. 1999. Wetternachhersage – 500 Jahre Klimavariationen und Naturkatastrophen (1496–1995). Haupt, Bern, 304 pp.

Pfister C., Luterbacher J., Schwarz-Zanetti G. and Wegmann M. 1998. Winter air temperature variations in western Europe during the Early and High Middle Ages (AD 750–1300). The Holocene 8: 535–552.

Pollack H.N. and Huang S. 2000. Climate reconstruction from subsurface temperatures. Annual Reviews of Earth Planetary Sciences 28: 339–365.

Pollack H.N. and Smerdon J.E. 2004. Borehole climate reconstructions – spatial structure and hemispheric averages, Journal of Geophysical Research 109, doi: 10.1029/2003JD004163.

Preisendorfer R.W. 1988. Principal component analysis in meteorology and oceanography. Developments in Atmospheric Sciences 17: 293–318.

Prieto M.R., García-Herrera R. and Hernandez Martin E. 2004. Early records of icebergs in the south Atlantic ocean from Spanish documentary sources. Climatic Change 66: 29–48.

Qian W., Hu Q., Zhu Y. and Lee D.K. 2003. Centennial-scale dry-wet variations in east Asia. Climate Dynamics 21: 77–89.

Rayner N.A., Parker D.E., Horton E.B., Folland C.K., Alexander L.V., Rowell D.P., Kent E.C. and Kaplan A. 2003. Globally complete analyses of sea surface temperature, sea ice and night marine air temperature, 1871–2000. Journal of Geophysical Research 108, doi: 10.1029/2002JD002670.

Robertson A., Overpeck J., Rind D., Mosley-Thompson E., Zielinski G., Lean J., Koch D., Penner J., Tegen I. and Healy R. 2001. Hypothesized climate forcing time series for the last 500 years. Journal of Geophysical Research 106: 14783–14803.

Robock A. and Free M.P. 1996. The volcanic record in ice cores for the past 2000 years. In: Jones P.D. *et al.* (Eds.) Climatic variations and forcing mechanisms of the last 2000 years. Springer, Berlin, 533–546.

Rutherford S. and Mann M.E. 2004. Correction to "Optimal surface temperature reconstructions using terrestrial borehole data". Journal of Geophysical Research 109, doi: 10.1029/2003JD004290.

Saurer M., Borella S., Schweingruber F.H. and Siegwolf R. 1997. Stable carbon isotopes in tree rings of beech – climatic versus site-related influences. Trees 11: 291–297.

Schaub M., Kaiser K.F. and Kromer B. 2003. Subfossil pines from loamy slope sediments reflect late glacial climatic variations. In: Schleser *et al.* (eds.) Trace – Tree rings in archaeology, climatology and ecology, Proceedings of the Dendrosymposium 2002, Bonn, 40–46.

Schweingruber F.H. 1996. Tree rings and environment – dendroecology. Haupt, Bern, 609 pp.

Schweingruber F.H. 2001. Dendroökologische Holzanatomie. Haupt, Bern, 472 pp.

Schweingruber F.H., Fritts H.C., Bräker O.U., Drew L.G. and Schär E. 1978. Dendroclimatic studies on conifers from central Europe and Great Britain. Tree-Ring Bulleting 38: 61–91.

Schweingruber F.H., Eckstein D., Serre-Bachet F. and Bräker O.U. 1990. Identification, presentation and interpretation of event years and pointer years in dendrochronology. Dendrochronologia 8: 9–38.

Schweingruber F.H., Wehrli U., Aellen-Rumo K. and Aellen M. 1991. Weiserjahre als Zeiger extremer Standortseinflüsse. Schweizerische Zeitschrift für Forstwesen 142: 33–52.

Slonosky V. C. 2002. Wet winters, dry summers? Three centuries of precipitation data from Paris. Geophysical Research Letters 29, doi: 10.1029/2001GL014302.

Smith H.J., Wahlen M. and Mastroianni D. 1997. The CO_2 concentration of air trapped in GISP2 ice from the last glacial Maximum-Holocene transition. Geophysical Research Letters 24: 1–4.

Stahle D.W. 1999. Effective strategies for the development of tropical tree-ring chronologies. IAWA Journal 20: 249–253.

Stainforth D.A., Aina T., Christensen C., Collins M., Faull N., Frame D.J. and 10 co-authors 2005. Uncertainty in predictions of the climate response to rising levels of greenhouse gases. Nature 433: 403–406.

Stockton C.W. 1985. Current research progress toward understanding drought. In: Proceedings, Drought, Water Management and Food Production. Imprimerie de Fedala, Mohammedia, 21–35.

Stott P.A., Stone D.A. and Allen M.R. 2004. Human contribution to the European heatwave of 2003. Nature 432: 610–614.

Touchan R., Xoplaki E., Funkhouser G., Luterbacher J., Hughes M.K., Erkan N., Akkemik Ü. and Stephan J. 2005. Reconstructions of spring/summer precipitation for the Eastern Mediterranean from tree-ring widths and its connection to large-scale atmospheric circulation, Climate Dynamics, doi: 10.1007/s00382–005–0016–5.

Trenberth K. and Paolino D. 1980. Northern Hemisphere sea-level pressure data set – trends, errors and discontinuities. Monthly Weather Reviews 108: 855–872.

Treydte K. 2003. Dendro-Isotope und Jahrringbreiten als Klimaproxis der letzten 1200 Jahre im Karakorumgebirge/Pakistan. Schriften des Forschungszentrums Jülich, Reihe Umwelt/ Environment 38, 190 pp.

Treydte K., Schleser G.H., Schweingruber F.H. and Winiger M. 2001. The climatic significance of $\delta^{13}C$ in subalpine spruce (Lötschental, Swiss Alps) – a case study with respect to altitude, exposure and soil moisture. Tellus 53: 593–611.

Verstege A., Esper J., Neuwirth B., Alifriqui M. and Frank D. 2004. On the potential of cedar forests in the Middle Atlas (Morocco) for climate reconstructions. In: Jansma E. *et al.* (eds.) Proceedings International Conference TRACE (Utrecht, May 1–3, 2003) 78–84.

von Storch H. and Zwiers F.W. 1999. Statistical analysis in climate research. Cambridge University Press, UK.

von Storch H., Zorita E., Jones J.M., Dimitriev Y., González-Rouco F. and Tett S.F.B. 2004. Reconstructing past climate from noisy data. Science 306: 679–682.

Wang S.W., Gong D. and Zhu J. 2001. Twentieth-century climatic warming in China in the context of the Holocene. The Holocene 11: 313–321.

Wanner H., Luterbacher J., Gyalistras D., Rickli R., Salvisberg E. and Schmutz C. 2000. Klimawandel im Schweizer Alpenraum. ETH Zürich, Zürich, 285 pp.

Wanner H., Brönnimann S., Casty C., Gyalistras D., Luterbacher J., Schmutz C., Stephenson D.B. and Xoplaki E. 2001. North Atlantic Oscillation – concepts and studies. Surveys in Geophysics 22: 321–381.

Watanabe O., Jouzel J., Johnsen S., Parrenin F., Shoji H. and Yoshida N. 2003. Homogeneous climate variability across east Antarctica over the past three glacial cycles. Nature 422: 509–512.

Watson R.T. (eds.) 2001. Climate change 2001 – the physical basis of climate change. Synthesis Report. Cambridge Univ. Press, Cambridge, 34 pp.

Wigley T.M.L., Briffa K.R. and Jones P.D. 1984. On the average of correlated time series, with applications in dendroclimatology and hydrometeorology. Journal of Climate and Applied Meteorology 23: 201–213.

Worbes M. 1999. Annual growth rings, rainfall dependent growth and long-term growth patterns of tropical trees from the Forest Reserve Caparo in Venezuela. Journal of Ecology 87: 391–403.

Xoplaki E., Luterbacher J., Paeth H., Dietrich D., Steiner N., Grosjean M. and Wanner H., 2005. European spring and autumn temperature variability and change of extremes over the last half millennium. Geophysical Research Letters 32, L15713, doi: 10.1029/2005GL023424.

Yang B., Braeuning A., Johnson K.R. and Yafeng S. 2002. General characteristics of temperature variation in China during the last two millennia. Geophysical Research Letters 29, doi: 10.1029/2001GL014485.

Zhang J. and Crowley T.J. 1989. Historical climate records in China and reconstruction of past climates. Journal of Climate 2: 833–849.

Using the Past to Understand the Present Land Use and Land Cover

Matthias Bürgi[1*], Anna M. Hersperger[1], Marcus Hall[2], Emiliy W. B. (Russell) Southgate[3] and Nina Schneeberger[1]

[1] Swiss Federal Research Institute WSL, Zürcherstrasse 111, CH-8903 Birmensdorf, Switzerland
[2] Institut of Environmental Sciences, University of Zürich, Winthherthurerstr. 190, CH-8057 Zürich, and Dept. of History, College of Humanities, The University of Utah, Salt Lake City, UT 84112
[3] Notre Dame Academy, Middleburg, VA 20117, U.S.A.
[*] Corresponding author: e-mail: matthias.buergi@wsl.ch

Abstract
Landscapes must be understood as dynamic time-dependent entities rather than static associations of biotic and abiotic elements. In particular, former human activities must be better appreciated and better incorporated in descriptions of landscape processes. To describe past landscapes, oral, written, (carto-)graphic and ecological sources can be used. Combinations of these sources usually provide reliable historical information, if based on a critical analysis of the quality and background of the data, including cross-checking information from the different data sources. The general public, planners, politicians, land managers, ecological modelers, and restoration ecologists are just some of the potential users of landscape history.

Keywords: landscape history, land use change, source types, historical ecology

F. Kienast, O. Wildi & S. Ghosh (eds.), A Changing World. Challenges for Landscape Research, 133–144.

Relevance and Methods of Landscape History

The potential contribution of landscape history – the study of the evolution of landscapes and ecosystems over centuries – to provide a better understanding of the present land use and land cover is increasingly recognized (Russell 1997; Swetnam *et al.* 1999). Most landscapes are cultural landscapes, shaped over time, in an interactive process linking human needs with natural resources in a specific topographic and spatial setting. Periods of distinct use and management of the land can often be distinguished, with some human activities leaving only a short imprint on the scenery and ecosystems, whereas others remaining visible over thousands of years depending on their level of impact. Thus, current pattern and processes of landscapes as well as ecosystem functions should be interpreted and understood with an integrative historic-ecological approach (Swetnam *et al.* 1999; Bürgi and Russell 2001).

It is not easy to incorporate the temporal dimension – which is inevitably process-oriented – into landscape studies, as the concept of landscape is usually rather static (Cosgrove 1984). Several authors discuss the problems of historic, process oriented approaches in landscape studies, such as Hobbs (1997), who state that in many landscape ecological studies processes are often afforded less attention than landscape pattern or may even be ignored altogether. Crumley (1998) points out that landscape ecologists either tend to ignore people or assume their effect to be negative. Meine (1999) states that if natural scientists and historians look at the same landscape they see different things and draw different lessons from what they see. Some of the differences in their landscape perception arise from the differences in the academic cultures of science and humanities. Whereas natural scientists are interested in finding generalized processes forming patterns of predictable events, historians typically focus their work on the particularities of a locality. In landscape ecology, human activities are regarded at best as one factor among many that have an impact on the system under study. Similarly in history, the spatial setting of historical events is just one among many aspects considered. For an integrative understanding of landscape changes, the two perspectives of science and humanities have to be combined.

Several fields of research, such as historical geography, environmental history, human ecology and historical ecology have a long tradition of considering humans as a biotic factor (McDonnell and Pickett 1993). In these fields, methods and approaches have been developed for combining information from different sources (Sheail 1980; Russell 1997; Egan and Howell 2001). In the following sections we provide an overview of the sources of information about past landscapes, as well as a survey of the uses and applications of this information.

Voices from the Past

Written and oral information

Many public and privately owned archives are replete with documents containing information about past land use and land management. A farmers' diary may be as valuable as contemporary newspaper reports or official agricultural statistics in contributing towards the reconstruction of past land-use practices and human impacts on the land (Russell 1997; Edmonds 2001). By combining different source types, often a more complete picture of landscape evolution can be gained.

Whereas it may be sufficient for a historian or a folklorist to describe and document a specific land use, a historical ecologist is interested in intensity, frequency and spatial extent, i.e. the disturbance regime of a specific land use. Only by collecting such detailed information, it is possible to incorporate the human impact fully in a study of ecosystem change (e.g., Wohlgemuth *et al.* 2002). However, even then, it is often hard or even impossible to meet the rigorous needs of quantification that is characteristic of ecological studies (Bürgi and Russell 2001). In many cases, researchers have to estimate and make informed guesses in order to fill in the lacuna in data left by discontinuous documentation or ways have to be found to incorporate qualitative information.

Written sources contain information that was regarded as relevant in the past. The content of these sources cannot be changed or extended – we have to take whatever there is. In contrast, this limitation does not exist when contemporary witnesses are asked about how, when, and where they have used and shaped the land. Therefore, oral history bears the potential to provide valuable information about the human impact on the land (Fogerty 2001). An obvious limitation of oral history is that the temporal span is limited by life expectancy. Furthermore, any interpretation of oral histories must address the question of faulty memories, biases etc (Perks and Thomson 1998).

Combining oral and written information often provides further potential to reconstruct past landscapes. These can then be combined with historical maps and pictures, which are described in the next section.

Historical maps and pictures

Historical maps are useful for retrospective analysis of landscape patterns and their change over time (e.g., Kienast 1993; Petit and Lambin 2002). Comparisons of old and modern maps highlight the major changes in land use (Fig. 1). However, the mapping criteria of the past need not have been the same as today and it is often impossible to find the historic mapping instructions. Thus, direct comparisons of old and modern maps in a Geographical Information System (GIS) often requires a procedure called "rubber sheeting" to correct for spatial mismatches and generate an estimate of the comparability of the maps.

Pictures, especially photographs, often provide greater detail and realism about a landscape than maps. All the same, they are subject to distortion, since photographs are taken for a purpose, and this purpose may bias the information provided in the photograph. The use of historical aerial photographs to study landscape change (Fig. 2) is a well-established method in vegetation science (e.g., Swetnam *et al.* 1999) and has also been applied in erosion monitoring (e.g., Thee *et al.* 1990). In more recent times, landscape change studies increasingly made use of satellite imagery and remote sensing (e.g., Serneels and Lambin 2001; Turner *et al.* 2001; Nagendra *et al.* 2004). This is described in detail in Zimmermann *et al.* (2007).

Qualitative interpretations of repeated terrestrial photographs (Fig. 3) have been used in several studies of landscape change (e.g., Tanner 1999; Nüsser 2001). Photographs, whether of aerial or terrestrial origin, are generally easier to interpret for studies of landscape change than works of art. Some authors have successfully evaluated paintings and etchings from the 17th and the 18th century (e.g., Zumbühl 1980), but the appropriate interpretation of these sources in the context of landscape change remains highly challenging.

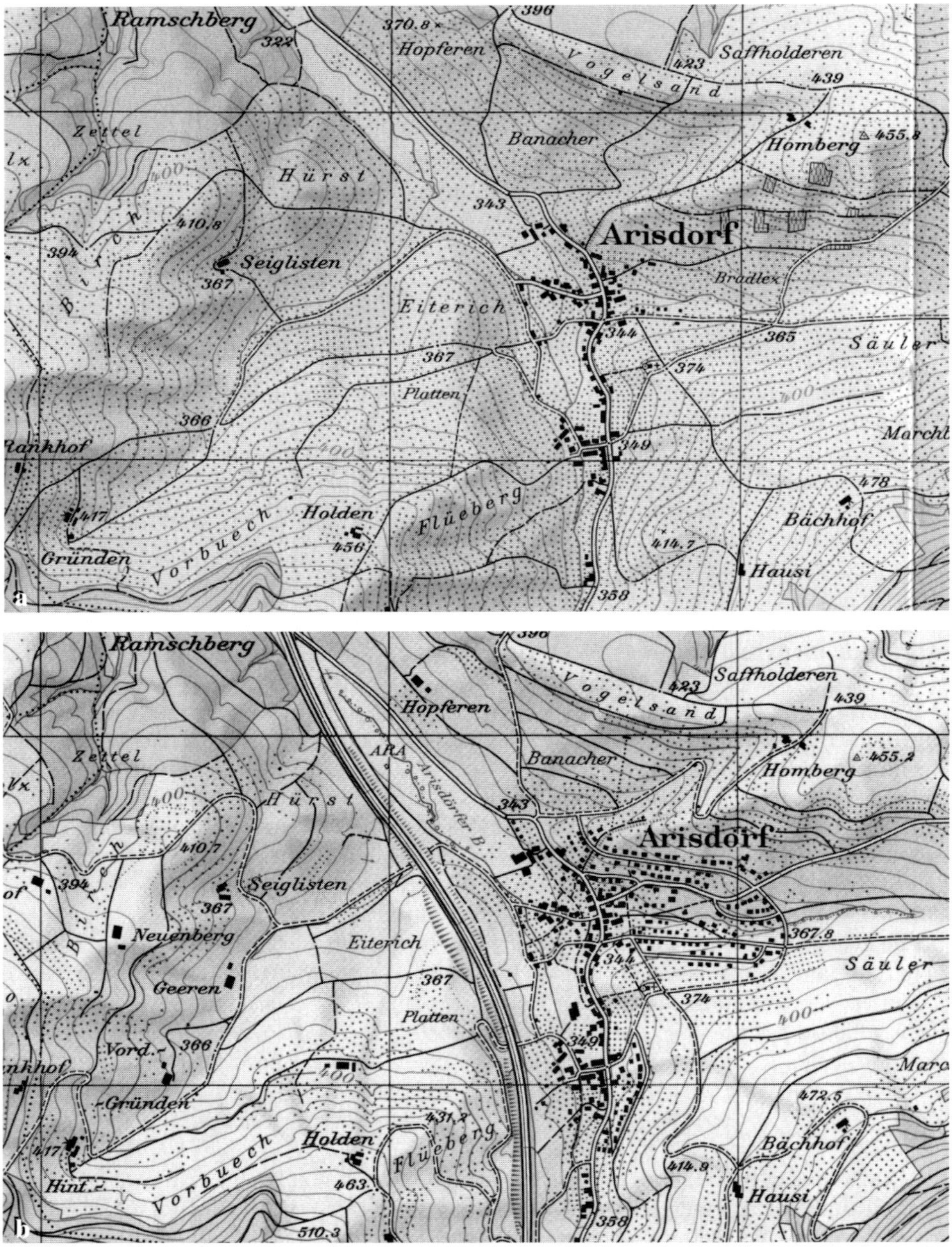

Fig. 1. Landscape change in the Swiss community of Arisdorf as shown in maps from 1955 (a, top) and 1988 (b, bottom). The landscape, in 1955 dominated by orchards (depicted as points) today is dominated by the newly constructed highway. (Source: Swiss National Maps 1:25 000, 1068 Sissach, reproduced with permission of the Federal Office of Topography, swisstopo (BA046410)).

Fig. 2. Landscape change in the Swiss community of Arisdorf as shown in aerial photographs from 1953 (a, top) and 1994 (b, bottom). Aerial photographs allow the analysis of field pattern. In the present case, the average size of the fields increased significantly. (Source: (a) Federal Office of Topography, swisstopo, SA 28, Aufn. 1839, 23.3.1953 (b) Federal Office of Topography, swisstopo, Linie 64, Aufn. 6662, 27.7.1994. Reprinted in Tanner 1999).

Fig. 3. Landscape change in the Swiss community of Arisdorf as shown in terrestrial photographs from 1941 (a, top) and 1999 (b, bottom). Unlike maps and aerial photographs, terrestrial photographs allow evaluating landscapes from a human perspective. Repeat terrestrial photography therefore is a powerful tool to analyse landscape change including qualitative and aesthetic aspects. In the example of Arisdorf the points on the map (Fig. 1a) become visible as three dimensional fruit trees, making the village be set in a forest of orchards. (Source: (a) Photo by Kling/Hans Eppens, Denkmalverzeichnis des Kantons Baselland 1941/1942, Photoarchiv Denkmalpflege des Kantons Basel-Landschaft, Liestal, Switzerland, reproduced in Tanner 1999, (b) Photo by Karl Martin Tanner, Seltisberg, Switzerland).

Reading the landscape

For landscape history, the most important physical evidence is the landscape as it presents itself today. As a result of century-long human activities, the current landscape contains a wealth of information about these activities, their intensity and spatial context in which they were conducted. Stone walls, for example, even if standing in a dense forest give exact information about how a former pasture land was spatially organized (e.g., Raup and Carlson 1941). Terraces, now in pastureland, show where more intensive agriculture was once common (Zimmermann 1972). In forests, the stand structure often reflects a long history of forest use and management.

The prime sources of information about forest development are tree rings (Esper *et al.* 2007). They contain information not only about climate, but also about fire frequency and intensity (e.g., Swetnam 1993), natural or anthropogenic changes in water regime (e.g., Rigling *et al.* 2002), forest use and management (e.g., Rigling and Schweingruber 1997), and the timing of forest succession after land abandonment (e.g., Iseli and Schweingruber 1990). Like many written documents, tree rings allow precise dating of events. Therefore, combining dendrochronological information with historical documents is of high interest for landscape history and ecological history in general.

Not only the trees, but also the ground they stand on contain information about past land use and land cover. One might utilize traces of plough horizons, or anomalies in nutrient content due to agricultural impacts that depleted the soils over a long period of time (Compton and Boone 2000; Koerner *et al.* 1999). Other natural archives include the sediments of ponds and lakes, peat in mires and small hollows, and soils, where pollen and macrofossils have been deposited and preserved. The pollen and macrofossil record can allow the reconstruction of the local and regional vegetation composition, and can reflect a region's changing human activities (e.g., Davis 1973; Russell 1993; Fuller *et al.* 1998; Lotter 1999).

Using a source-critical approach

All these sources of information about past landscapes are subject to several levels of interpretation. Historical sources, after all, reflect the past but are not the past itself. Natural evidence (such as tree rings) and human evidence (such as early meteorological data) (Egan and Howell 2001) therefore depend on the accuracy of the measurements, as well as on the attitude, bias, and agenda of the measurers. A measurement may vary according to tool, season, or calibration; a description of the former landscape may be scientific or lyric, detailed or general, but perhaps very pertinent to the questions being asked. Researchers using historical sources must also pay special attention to correlations masked as causations: in the example of measuring medieval climate change by tracing archival dates of annual grape harvests, the causal chain will depend on such factors as, 1) the accuracy and consistency of recording the harvest date, 2) the recorder's possible motives for falsely listing earlier or later dates, 3) the species of grapes or the height of vines, and 4) the synergy of precipitation, temperature, soil, sunlight, and wind on grape maturity (Ladurie 1972).

To help compensate for the subjectivity of historical documents, researchers will often combine several sources of information. Moreover, because most archival records provide only partial answers to historic questions, researchers often speak in terms of likelihood, probability, and plausibility (Bürgi and Russell 2001). By paying attention to the source, noting both its strengths and weaknesses for answering each question, researchers can address the post-modernist critique that former landscapes can never be described precisely.

Why Study the History of Landscapes

Landscape changes are traditionally studied in a descriptive way (e.g., Ewald 1978). Despite today's more analytical approach in research, such descriptive studies are still relevant. They satisfy the public's general interest in landscape history, which is shown by the success of many popular publications (e.g., Schama 1995; Tanner 1999), or in the relevance of traditional cultural landscapes for tourism, such as in the popularity of hiking along historic traffic routes or irrigation conduits (e.g., Crook and Jones 1999). Thus, descriptive studies of landscape history enables to locate society and individuals in time and space.

In planning processes, information about past landscape states is relevant on three different levels. First, a thorough analysis of the driving forces of landscape change in the past (e.g., Bürgi and Turner 2002; Bürgi *et al.* 2004) results in a deeper understanding of the changing relationship between societies and landscapes and it enables us to analyze the legacy of past land uses in present-day landscapes. Therefore, landscape history has explicitly been proposed as a tool for landscape and conservation planning (Marcucci 2000; Foster 2000). Second, studying landscape history in a municipality can be a good starting point for participatory planning processes, as public discussions about where the landscape is coming from provide a baseline for reasoning about where it might be heading (e.g., HSR Rapperswil 2002). Third, the development of new policy goals clearly requires information about long-term changes in the targeted system. This is especially true for land-use planning, including the development of urban areas, agriculture, forestry, infrastructure, but also topics such as sustainable development and biodiversity.

Why Study the History of Ecosystems

Not only landscapes, but also ecosystems with a history of human impacts can only be properly understood if the changing history of these impacts is considered (e.g., McDonnell and Pickett 1993; Vitousek *et al.* 1997). Several international long-term research programs, such as the MAB-Program (http://www.unesco.org/mab/), LTER (http://lternet.edu/) or PAGES (http://www.pages.unibe.ch/), especially with its focus 5 "Past Ecosystems Processes and Human-Environment Interactions" and the therein located activity "Human impacts on terrestrial ecosystems" (HITE, http://www.liv.ac.uk/geography/hite), have implemented this history oriented approach on the ecosystem level. The integration of history into ecologically oriented studies is especially crucial for systems characterized by slow changes and long time lags between impacts and effects, such as forests (e. g., Magnuson 1990; Foster and Aber 2004), or in other ecosystems dominated by human impacts, such as anthropogenic grasslands (Cousins *et al.* 2002; Foster and Motzkin 2003).

Ecological studies that include historical aspects also find a growing application in the field of restoration ecology, the science of assisting in the recovery of damaged ecosystems (Society for Ecological Restoration International 2004). With damaged and abandoned land being the world's fastest growing land type (Wali 1992), restoration ecologists are utilising historical landscape studies to provide descriptions of former conditions to be restored (Hall 2001). Landscape historians therefore find restoration ecologists to be an eager audience seeking available information about past natural systems (Egan and Howell 2001).

Restoration, furthermore, is not limited to efforts at bringing back relatively untouched, 'pristine' systems (such as wild forests or wetlands), but is also concerned with bringing back desirable historic anthropogenic ecosystems (such as managed forests, pastureland or gardens). Indeed, research about former natural ecological processes (e.g., fire frequency, flooding events, predator-prey interaction) or about former human land use and management (e.g., grazing pressure, forestry practices, wildlife harvest) may be more useful to restoration practitioners than static descriptions of past landscape states. The rise of the concept of historical variability (Landres *et al.* 1999) reflects a growing need for precise information about historical changes in disturbance regimes (e.g., Hellberg 2004).

Using Landscape History

We surveyed the various pitfalls of working with historical sources for landscape history, and how they can be accounted for in a source-critical approach. Generally, information about the past does not automatically provide answers about how to manage such ecosystems today or in the future (e.g., Hellberg 2004). The decisions about which historical questions we pursue must be based on our current needs and value systems. While certain restoration projects may require very specific descriptions of the past (e.g., Egan and Howell 2001), in most cases, more general information about former land use and management will already help to better understand current ecosystem functions as well as landscape patterns and processes.

Modelling studies in particular may be prone to promoting unduly accurate forecasts, projections or predictions (Veldkamp and Lambin 2001). The more complex the system and the more factors involved, the more difficult or even impossible it is to make accurate predictions. Often, scenario analyses will be a good way to mitigate the tension between predictability and uncertainty (Lambin *et al.* 2000). That said, the authors also believe that retrospective analyses can reduce unexpected behaviour (Middelham 2001) to a certain extent and under specific conditions. In any case, historical research should inform but not dictate future land-use decisions (Umbricht 2003).

Outlook

The diversity of source types providing data about the forms and functions of past landscapes clearly requires an interdisciplinary dialogue in order to develop new methods and approaches for combining a wide range of information with different reliabilities. It will remain impossible to rigorously test hypotheses regarding the potential link between changes in environmental features and changes in human activities, but experimentation and modelling are valuable tools to gain additional insights into their plausibility and the dynamics of the interactions studied. In any case, landscape ecologists must incorporate circumstantial evidence and inferential reasoning in applying such integrative methods. Not including historical information in landscape ecological studies frequently leads to misinterpretation of the observed environmental change (McDonnell and Pickett 1993).

The long list of uses of information about former landscapes reflects the growing awareness of the interconnectedness of societal and environmental development. It seems likely that landscape historians, historical ecologists and environmental historians will play increasingly crucial roles in basic landscape and ecological research as well as in related practical applications such as restoration ecology, planning processes and outreach programs.

References

Bürgi M. and Russell E.W.B. 2001. Integrative methods to study landscape changes. Land Use Policy 18: 9–16.

Bürgi M. and Turner M. 2002. Factors and processes shaping land cover and land cover changes along the Wisconsin River, U.S.A. Ecosystems 5: 184–201.

Bürgi M., Hersperger A. and Schneeberger B. 2004. Driving forces of landscape change – current and new directions. Landscape Ecology 19: 857–868.

Compton J.E. and Boone R.D. 2000. Long-term impacts of agriculture on soil carbon and nitrogen in New England forests. Ecology 81: 2314–2330.

Cosgrove D.E. 1984. Social formation and symbolic landscape. University of Wisconsin Press, Madison.

Cousins S.A.O., Eriksson A. and Franzen D. 2002. Reconstructing past land use and vegetation patterns using palaeogeographical and archaeological data – A focus on grasslands in Nynas by the Baltic Sea in south-eastern Sweden. Landscape and urban planning 61: 1–18.

Crook D.S. and Jones A.M. 1999. Traditional irrigation and its importance to the tourist landscape of Valais, Switzerland. Landscape Research 24: 49–65.

Crumley C.L. 1998. Foreword. In Balée, W. (ed.). Advances in historical ecology, pp. ix-xiv. Columbia University Press, New York.

Davis M.B. 1973. Pollen evidence of changing land use around the shores of Lake Washington. Northwest Science 47: 133–148.

Edmonds M. 2001. The pleasures and pitfalls of written records. In: Egan D. and Howell E.A. (eds.) 2001. The historical ecology handbook. Island Press, Washington, 73–99.

Egan D. and Howell E.A. (eds.) 2001. The historical ecology handbook. Island Press, Washington.

Ewald K.C. 1978. Der Landschaftswandel. Zur Veränderung schweizerischer Kulturlandschaften im 20. Jahrhundert. Berichte Eidgenössische Anstalt für das forstliche Versuchswesen, 191.

Esper J., Frank D.C. and Luterbacher J. 2007. On selected issues and challenges in dendroclimatology. In: Kienast F., Wildi O. and Ghosh S. (eds.). A Changing World. Challenges for Landscape Research, Vol. 8: 113–132. Springer Landscape Series, Dordrecht.

Fogerty J.E. 2001. Oral history: A guide to its creation and use. In: Egan D. and Howell E.A. (eds.) 2001. The historical ecology handbook. Island Press, Washington, 101–120.

Foster D.R. 2000. From bobolinks to bears: interjecting geographical history into ecological studies, environmental interpretation, and conservation planning. Journal of Biogeography 27: 27–30.

Foster D.R. and Aber J.D. 2004. Forests in Time. The Environmental Consequences of 1000 Years of Change in New England. Yale University Press, New Haven.

Foster D.R. and Motzkin G. 2003. Interpreting and conserving the openland habitats of coastal New England: insights from landscape history. Forest Ecolology and Management 185: 127–150.

Fuller J.L., Foster D.R., McLachlan J.C. and Drake N. 1998. Impact of Human Activity on Regional Forest Composition and Dynamics in Central New England. Ecosystems 1: 76–95.

Hall M. 2001. Repairing Mountains: Restoration, Ecology, and Wilderness in Twentieth-Century Utah. Environmental History 6: 574–601.

Hellberg E. 2004. Historical variability of deciduous trees and deciduous forests in Northern Sweden. Effects of forest fires, land-use and climate. Acta Universitatis Agriculturae Sueciae. Silvestria 308.

Hobbs R. 1997. Future landscapes and the future of landscape ecology. Landscape and urban planning 37: 1–9.

HSR Rapperswil (ed.) 2002. Werkzeugkasten LEK.

Iseli M. and Schweingruber F.H. 1990. Baumalter als Ausdruck der Bestandesdynamik in Brachlandflächen. Schweizerische Zeitschrift für Forstwesen 7: 581–593.

Kienast F. 1993. Analysis of historic landscape patterns with a Geographical Information System a methodological outline. Landscape Ecology 8: 103–118.

Koerner W., Dambrine E., Dupouey J.L. and Benoit M. 1999. 15N of forest soil and understorey vegetation reflect the former agricultural land use. Oecologia 121: 421–425.

Ladurie E. Le Roy. 1972. Times of feast, times of famine: a history of climate since the year 1000. Allen and Unwin, London.

Lambin E.F., Rounsevell M.D.A. and Geist H.J. 2000. Are agricultural land-use models able to predict changes in land-use intensity? Agriculture, Ecosystems and Environment 82: 321–331.

Landres P.B., Morgan P. and Swanson F.J. 1999. Overwiew of the use of natural variability concepts in managing ecological systems. Ecological Applications 9: 1179–1188.

Lotter A.F. 1999. Late-glacial and Holocene vegetation history and dynamics as shown by pollen and plant macrofossil analyses in annually laminated sediments from Soppensee, central Switzerland. Vegetation History Archaeobotany 8: 165–184.

Magnuson J.J. 1990. Long-term ecological research and the invisible present. BioScience 40: 495–501.

Marcucci D.J. 2000. Landscape history as a planning tool. Landscape and Urban Planning 49: 67–81.

McDonnell M.J. and Pickett S.T.A. 1993. Humans as components of ecosystems – the ecology of subtle human effects and populated areas. Springer.

Meine C. 1999. It's about time: conservation biology and history. Conservation Biology 13: 1–3.

Middelham F. 2001. Traffic simulation – Predictability: some thoughts on modeling. Future Generation Computer Systems 17: 627–636.

Nagendra H., Munroe D.K. and Southworth, J. 2004. From pattern to process: landscape fragmentation and the analysis of land use/land cover change. Agriculture, Ecosystems and Environment 101: 111–115.

Nüsser M. 2001. Understanding cultural landscape transformation: a re-photographic survey in Chitral, eastern Hindukush, Pakistan. Landscape and urban planning 57: 241–255.

Perks R. and Thomson A. (eds.) 1998. The oral history reader. Routledge.

Petit C.C. and Lambin E.F. 2002. Long-term land-cover changes in the Belgian Ardennes (1775–1929): model-based reconstruction vs. historical maps. Global Change Biology 8: 616–630.

Raup H.M. and Carlson R.E. 1941. The history of land use in the Harvard Forest. Harvard Forest Bulletin, 20.

Rigling A. and Schweingruber F.H. 1997. Entwicklung waldföhrenreicher Wälder im Gebiet Brienz-Wiesen (GR). Eine historisch-dendroökologische Studie. Schweizerische Zeitschrift für Forstwesen 148: 173–196.

Rigling A., Brühlhart H., Bräker O.U., Forster T. and Schweingruber F.H. 2002. Effects of irrigation on diameter growth and vertical resin duct production in Pinus sylvestris L. on dry sites in the central Alps, Switzerland. Forest Ecology and Management 175: 285–296.

Russell E.W.B. 1993. Early stages of secondary succession recorded in soil pollen on the North Carolina Piedmont. The American Midland Naturalist 129: 384–396.

Russell E.W.B. 1997. People and the land through time. Linking ecology and history. Yale University Press, New Haven.

Schama S. 1995. Landscape and memory. Harper Collins. London.

Serneels S. and Lambin E.F. 2001 Proximate causes of land-use change in Narok District, Kenya: a spatial statistical model. Agriculture, Ecosystems and Environment 85: 65–81.

Sheail J. 1980. Historical ecology: the documentary evidence. Huntington, Natural Environment Research Council, Institute of Terrestrial Ecology. 21 pp.

Society for Ecological Restoration International 2004. http://www.ser.org/content/ecological_restoration_primer.asp#3.

Swetnam T.W. 1993. Fire history and climate change in Giant Sequoia groves. Science 262: 886–889.

Swetnam T.W., Allen C.D. and Betancourt J.L. 1999. Applied historical ecology: using the past to manage for the future. Ecological Applications 9: 1189–1206.

Tanner K.M. 1999. Augen-Blicke: Bilder zum Landschaftswandel im Baselbiet. Verlag des Kantons Basel-Landschaft, Liestal. 264 pp.

Thee P., Zeller J. and Hägeli M. 1990. Wildbachverbau: Photogrammetrische Geländeauswertungen. Berichte Eidgenössische Anstalt für das forstliche Versuchswesen 324: 45 pp.

Turner B.L., Villar S.C., Foster D., Geoghegan J., Keys E., Klepeis P., Lawrence D., Mendoza P.M., Manson S., Ogneva-Himmelberger Y., Plotkin A.B., Salicrup D.P., Chowdhury R.R., Savitsky B., Schneider L., Schmook B., Vance C. 2001. Deforestation in the southern Yucatan peninsular region: an integrative approach. Forest Ecology and Management 154: 353–370.

Umbricht M.J. 2003. Welche Landschaft wollen wir? Denkmodelle für die Landschaft der Zukunft. Dissertation ETH Zurich Nr. 15324.

Veldkamp A. and Lambin E.F. 2001. Prediciting land-use change. Agriculture, Ecosystems and Environment 85: 1–6.

Vitousek P.M., Mooney H.A., Lubchenco J. and Melillo J.M. 1997. Human domination of earth's ecosystems. Science 277: 494–499.

Wali M.K. (ed.) 1992. Ecosystem rehabilitation: Preamble to sustainable development. SPB Academic Publishing, The Hague.

Wohlgemuth T., Bürgi M., Scheidegger C. and Schütz M. 2002. Dominance reduction of species through disturbance – a proposed management principle for central European forests. Forest Ecology and Management 166: 1–15.

Zimmermann W. 1972. Die Flurwüstungen im Kanton Schaffhausen. Diss. Univ. Zürich. 142 pp.

Zimmermann N.E., Washington-Allen R.A., Ramsey R.D., Schaepman M.E., Mathys L., Kötz B., Kneubühler M. and Edwards T.M. 2007. Modern remote sensing for environmental monitoring of landscape states and trajectories. In: Kienast F., Wildi O. and Ghosh S. (eds.). A Changing World. Challenges for Landscape Research, Vol. 8: 65–91. Springer Landscape Series, Dordrecht.

Zumbühl H.J. 1980. Die Schwankungen der Grindelwaldgletscher in den historischen Bild- und Schriftquellen des 12. bis 19. Jahrhunderts. Birkhäuser Verlag, Basel.

Integrating Population Genetics with Landscape Ecology to Infer Spatio-temporal Processes

Rolf Holderegger[1*], Felix Gugerli[1], Christoph Scheidegger[1] and Pierre Taberlet[2]

[1] Swiss Federal Research Institute WSL, Zürcherstrasse 111, CH-8903 Birmensdorf, Switzerland
[2] Laboratoire d'Ecologie Alpine, Equipe Génomique des Populations et Biodiversité, UMR CNRS 5553, BP 53, Université Joseph Fourier, 38041 Grenoble Cedex 9, France
[*] Corresponding author: e-mail: rolf.holderegger@wsl.ch

Abstract
The last decade has seen the rise of the research fields of DNA analysis and population or ecological genetics. They have the potential to allow the revision of landscape ecological concepts such as habitat connectivity or fragmentation. In this chapter, we first ask how population genetics can support and extend landscape ecological research from analysing patterns to understanding processes, and we introduce the concepts of neutral and adaptive genetic diversity. We further outline relevant population genetic applications, provide corresponding examples and discuss the benefits and limitations of molecular techniques by referring to the topics of migration, dispersal and gene flow at various spatial and temporal scales. The discussion highlights migration patterns of species in the postglacial landscape, the assessment of historical dispersal and gene flow among populations, a description of how the current movement of plant propagules can be measured and a brief reassessment of the metapopulation concept and of what would be required to provide unequivocal empirical data on metapopulations. Despite the evident usefulness of molecular methods in landscape ecological research and the spatial context inherent to both landscape ecology and population genetics, the two fields have hitherto virtually stayed apart. Thus, we claim that the emerging field of landscape genetics, a marriage of population genetics with landscape ecology, should be given research priority, both theoretically and empirically. We expect a significant contribution of this field to basic and applied topics such as fragmentation and the management of natural ecosystems.

Keywords: adaptive genetic variation, connectivity, gene flow, landscape genetics, metapopulation, neutral genetic variation, population genetics

F. Kienast, O. Wildi & S. Ghosh (eds.), A Changing World. Challenges for Landscape Research, 145–156.

Population Genetics as "A Bridge Over Troubled Water" in Landscape Ecology?

Landscape ecology is a rapidly evolving and flourishing field of ecological research (Li and Wu 2004). The qualitative and quantitative description of landscapes is a key issue in applying ecological knowledge to current environmental and socio-cultural problems (Turner *et al.* 2001). From a biological point of view, it has long been recognised that the landscape influences the ecological and evolutionary processes that took, take or will place in it. Terrestrial habitat patches embedded in a landscape matrix can theoretically be treated as "islands in the sea" according to the classical island biogeography model of MacArthur and Wilson (1967). However, the real world often does not function along the rules of theoretical models with spatial idealisations and multiple assumptions (Baquette 2004). The matrix of the landscape may dramatically alter the way the system behaves. For instance, dispersal corridors may locally interrupt the fundamental relationship of decreasing movement of individuals (or genes) with increasing distance among habitat patches. Two remote islands connected by "a bridge" will show higher exchange than two close islands separated by "troubled water". Connectivity as a "bridge over troubled water" could well act independent of pure spatial distance, depending on the quality of a landscape (Levin 1992). Hence, the evaluation of landscape structure is essential (Turner *et al.* 2001), especially in problem-orientated fields such as conservation biology or restoration ecology.

The above mentioned landscape structure is often described with spatial, temporal and ecological indices (Bolliger *et al.* 2007). Landscape indices that have been extensively used are, e.g., "connectivity", "contagion" or "fractal dimension" (Turner *et al.* 2001). In order to adequately determine landscape structure, more than one index is usually needed, since "connectivity" is only partly defined by "proximity" as illustrated above. However, do values of landscape indices really reflect an underlying ecological (or evolutionary) process? Li and Wu (2004) recently stated that the relationship between pattern and process has rarely been shown in landscape ecological research. Often, a close relationship is simply assumed but not proven. Spatial pattern analysis in landscape ecology would be of limited use if it could not help in explaining or predicting processes. This dilemma led Li and Wu (2004) to doubt the relevance of landscape indices (Bolliger *et al.* 2007).

How can the ecological meaning of landscape indices be verified? As we will show below, population (Hartl and Clark 1997) or ecological genetics (Lowe *et al.* 2004) offer powerful tools to investigate ecological processes. By studying an appropriate gene (or locus) and by using an appropriate sampling design, population genetics can generate data on ecological processes acting at various spatial and temporal scales (from single habitat patches to whole continents and from years to decades or even millennia; Lowe *et al.* 2004). Genetic data may thus give credit to landscape indices.

Using Neutral Genetic Diversity to Infer Landscape Processes

Genetic diversity is a major facet of biodiversity (Rio Convention; www.biodiv.org/convention/articles.asp). It is essential for the short- and long-term evolution of species and for their potential to react to environmental change (Lande and Shannon 1996). Thus, genetic investigations are often incorporated in conservation projects and practical management plans of endangered species. But is all the genetic diversity that we measure of such high importance?

There are two basic types of genetic diversity, namely neutral and adaptive genetic diversity (Holderegger *et al.* in press). Genetic diversity values obtained from most molecular genetic markers currently analysed in laboratories represent neutral genetic variation (Pearman 2001), because the corresponding genes are not subject to natural selection. Which gene variants or alleles an individual carries, does thus not affect its fitness. The corresponding genetic variation is, in effect, selectively neutral (Reed and Frankham 2001). Unfortunately, the identification of genes subject to natural selection, such as genes that influence juvenile survival, growth rates or disease resistance, needs labour-, time- and cost-intensive experiments under controlled environmental conditions (quantitative genetics; Latta 2003). The corresponding genetic variation is adaptive or selective. In the future, the new field of ecological genomics might open a way to directly investigate adaptive genes in the laboratory using molecular techniques (Jackson *et al.* 2002; Luikart *et al.* 2003).

If most measurements of genetic diversity based on the analysis of molecular markers are selectively neutral, what are then their meaning and practical benefit? Is there a correlation between the diversity of neutral genes and the diversity of genes upon which natural selection acts that would allow us to draw conclusions on the evolutionary potential (adaptability) of populations? This question is not of purely academic interest but central to conservation biology and the management of natural living resources, because neutral molecular genetic variation is often taken as a surrogate for the variation of adaptive genes (Pearman 2001; Reed and Frankham 2001).

Unfortunately, there seems to be no such direct correlation as shown by several recent reviews on animals and plants (Reed and Frankham 2001; Merilä and Crnokrak 2001; McKay and Latta 2002; Latta 2003; Holderegger *et al.* 2006). Thus, neutral genetic variation does not provide direct evidence for adaptation or the evolutionary potential of populations or species. Why should such a relationship between neutral and adaptive genetic variation have been expected anyway? Genetic variation and genetic differentiation (i.e. the dissimilarity among populations; Lowe *et al.* 2004) is not only caused by local adaptation, which would affect the selectively active genes, but also by population processes such as random sampling effects (genetic drift and bottlenecks; Hartl and Clark 1997), migration, dispersal and gene flow among populations, changes in population size, population fragmentation or mating behaviour (Frankham *et al.* 2002). All these processes act upon neutral genes in essentially the same way as they act on adaptive genes, but since selection does not interfere with neutral genetic variation, the latter truly reflects the action of these processes free of distortion. Neutral genetic markers are therefore excellent tools to study ecological processes such as dispersal and gene flow (see below), but the naive use of neutral genetic diversity as an indicator of biodiversity at the population level is questionable (Pearman 2001; Holderegger *et al.* in press; for a discussion of value systems see Buchecker *et al.* 2007 and Duelli *et al.* 2007).

Different Places, Different Times, Different Processes

Once upon a time: postglacial migration

Biogeographic patterns perfectly serve to show how molecular genetic methods allow the investigation of historical processes acting over thousands of years and on large spatial scales. Here, our example deals with the postglacial migration of species during the last 15 000 years. Molecular biogeography, i.e. phylogeography, has seen a tremendous rise during the last decade (Avise 2002). Phylogeography allows determining the evolutionary relationships between extant populations. Hence, postglacial (re-)immigration pathways can be inferred from genetic variation (Taberlet *et al.* 1998).

Cells contain organelles which possess their own DNA. Unlike genes from the cell nucleus, genes of these organelles are mostly uniparentally inherited, i.e. solely by the mother or the father. This is the case for mitochondria in both animals and plants and for chloroplasts in plants (Avise 2002; Holderegger *et al.* in press). In most plant species, chloroplasts are

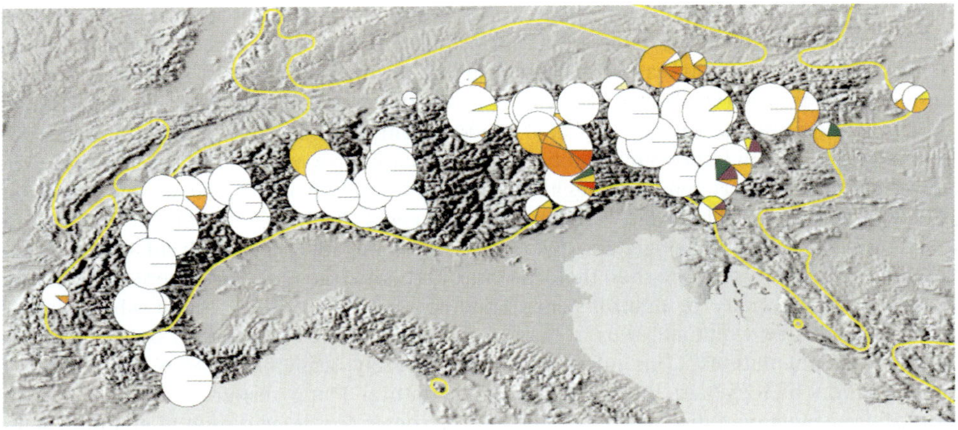

Fig. 1. Evidence of a founder effect in the western part of the Alpine range of Norway Spruce *(Picea abies)*. The relative frequencies of different mitochondrial DNA types (*nad*1 intron 2 sequence genotypes) are given by different colours, and sizes of circles are proportional to sample size per site. (Figure taken from Gugerli *et al.* (2001) with permission by Blackwell Science.)

inherited from the mother (maternally inherited), and the spatial distribution of chloroplast genotypes thus reflects seed dispersal. One can also distinguish dispersal by seed from gene flow by pollen if both nuclear genes and chloroplast and/or mitochondrial genes are investigated (for methods see Ennos 1994 and Lowe *et al.* 2004). This illustrates the elegance of molecular genetics with its large set of different genetic markers, each of which is potentially appropriate for a different question.

By making use of uniparentally inherited organelle markers (gene sequence variants from mitochondrial DNA), Gugerli *et al.* (2001) studied the postglacial re-colonisation of Central Europe by Norway Spruce *(Picea abies)*. The results suggest that Norway Spruce re-immigrated from eastern European glacial refugia. Northern and western Alpine populations were predominantly founded by immigrants originating from the Carpathian Mountains, while the Southern Alps appear to have been mainly colonised by individuals arriving from the Balkan Peninsula. Furthermore, Norway Spruce experienced a severe founder effect (i.e. colonisation by only a few colonising individuals; Frankham *et al.* 2002) during the postglacial re-colonisation of the Alps, which has led to a lower genetic diversity in the western than in the eastern Alps (Fig. 1). These phylogeographic findings are in good agreement with fossil pollen records (Lang 1994). The example shows that (1) molecular genetics can provide an empirical test of a given hypothesis by using an independent data source (here DNA markers versus fossil pollen) and (2) additional details about a dynamic process (here the origin and migration pathways of colonising individuals) can be inferred.

Intensive phylogeographic research has now been carried out, and reviews are provided, e.g., by Taberlet *et al.* (1998) and Hewitt (2004), as well as Petit *et al.* (2003) concentrating on European woody species and Schönswetter *et al.* (2005) recapitulating the glacial history of alpine plants. However, phylogeographic research has mainly dealt with single species. Ongoing research now aims at correlating multi-species phylogeographic patterns (i.e. comparative phylogeography; Taberlet *et al.* 1998) with several ecological parameters at large geographic scales. The latter is a typical subject of landscape ecological research. A corresponding example is the INTRABIODIV project on habitat, plant species and gene diversity in the European Alps and the Carpathians (http://intrabiodiv.vitamib.com).

Once just before: historical gene flow

Genetic differentiation among populations, e.g., expressed as F_{ST} (Frankham *et al.* 2002), is one of the most often calculated population genetic parameters. It ranges between 0 (no differentiation at all) and 1 (complete differentiation). Theoretically, high gene flow among populations should lead to genetic homogenisation of populations, while genetic isolation should cause pronounced genetic differentiation of populations. Based on this simple reasoning, the genetic differentiation of populations (under Wright's island model; Conner and Hartl 2004) can be used to indirectly infer the amount of past migration or gene flow among populations, *Nm,* using the equation $F_{ST} = 1/(4Nm + 1)$ (Frankham *et al.* 2002).

An example is provided by a molecular investigation (genetic fingerprinting with random amplified polymorphic DNAs, RAPDs) of populations of English Yew *(Taxus baccata)* in Switzerland by Hilfiker *et al.* (2004a). This marker type, mainly from the nuclear genome of the plants, is distributed to the next generation both by seed and pollen. In this study, *Nm*-values between population pairs of this conifer were calculated based on their pairwise genetic differentiation *(F_{ST})* and subsequently correlated with the geographic distances between them (Fig. 2). Swiss populations were connected by substantial past gene flow through pollen and/or seed, which decreased with increasing geographic distance (a pattern called "isolation by distance"; Lowe *et al.* 2004). However, Hilfiker *et al.* (2004b) found that the average genetic differentiation of small populations was higher than that of large populations, indicating that English Yew could well suffer from the negative population genetic consequences of small population size such as lower genetic variation or genetic drift (i.e. stochastic changes of the genetic composition of a population; Frankham *et al.* 2002). This example illustrates that the genetic structure of populations is a function of inter-population differences, but that the local random sampling effect of genetic drift within populations determines their genetic diversity.

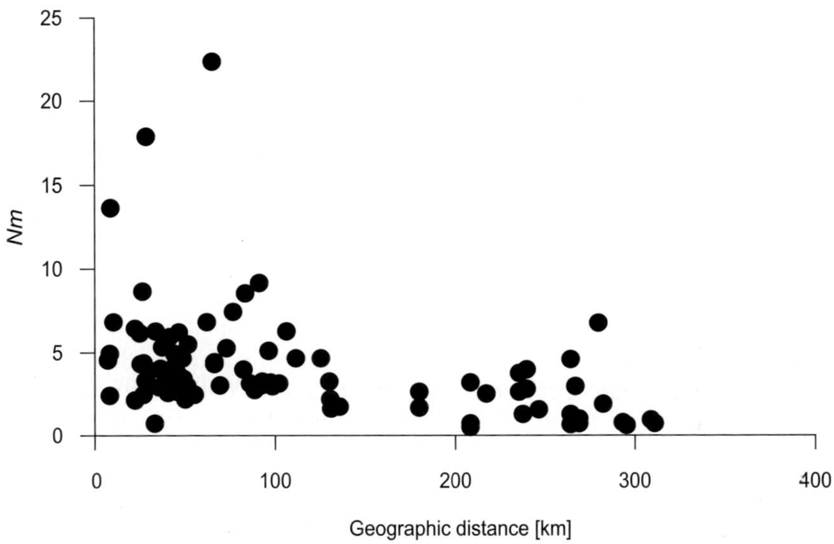

Fig. 2. Historical gene flow of English Yew *(Taxus baccata)* in Switzerland as estimated by *Nm*-values. The *Nm*-values are based on random amplified polymorphic DNAs (RAPDs). The data show decreasing gene flow with increasing geographic distance (Mantel test: $r_m = -0{,}166$, $P < 0{,}05$).

There are two important points to be stressed. (1) The calculation of *Nm*-values of gene flow does not take into account the spatial arrangement of populations. Hence, it is spatially not explicit and disregards any landscape feature. (2) The gene flow estimates presented in the example on English Yew (Fig. 2) are mainly a measure of past processes in the history of populations (Whitlock and McCauley 1999). They are based on population differentiation, which is a result of several processes acting in space *and* time. *Nm*-values are integrals over a certain, but usually unknown time period. This is an important fact, because *Nm*-values refer to historical patterns and do not necessarily reflect current processes. For instance, a molecular-genetic analysis of adults in the remaining populations of a long-lived tree species that recently experienced a substantial decrease in abundance due to habitat fragmentation might still indicate outcrossing and frequent gene exchange among populations. In reality, the populations may currently be characterised by inbreeding and almost complete genetic isolation in the recently fragmented landscape. However, it is the current processes that are most relevant for population persistence and of interest for conservation biologists (Frankham *et al.* 2002).

The present picture: current dispersal and gene flow

Population and ecological geneticists use two principal methodological approaches to refer to current migration and gene flow patterns: (1) progeny analysis (Smouse and Sork 2004) and (2) assignment tests (Manel *et al.* 2005).

In a progeny analysis of the scattered, insect-pollinated Wild Service Tree *(Sorbus torminalis)* in Switzerland by Hoebee *et al.* (WSL, Birmensdorf, unpubl. data), the positions of all adult individuals within a population were mapped and their genotypes determined using highly resolving genetic markers (microsatellites; Frankham *et al.* 2002). These trees represented all potential mates within the sampled area. Open pollinated seeds from several mother trees were then sampled, and the progenies (offspring) also submitted to genetic analysis. From the combined data set, it was possible to determine the fathers that sired the seeds of a given mother tree and to directly calculate current gene flow by pollen (Sork *et al.* 1999). As an example of the results of such a progeny analysis, Figure 3 shows that (1) many fathers, scattered over the ca. 20 ha area, sired the seeds of a chosen mother tree, (2) pollen was transported by insects over several hundred metres and (3) about 30% of the pollen entered the population from outside, i.e. none of the trees within the study population was identified as father. Hence, the amount 30% refers to gene immigration by pollen into the population. Similar results were obtained by Oddou-Muratorio *et al.* (2003, 2004) for a French population of the Wild Service Tree. A review of current gene flow in forest trees is provided by Smouse and Sork (2004).

In another example, Godoy and Jordano (2001) conducted a genetic study of the St. Lucy Cherry *(Prunus mahaleb)*, in which they made use of the fact that not only the genes of the offspring are transported in the seed of plants, but that the seed coat consists of maternal tissue and, thus, carries the genotype of the mother plant. Therefore, the authors trapped bird-dispersed seeds and allocated them to the genotype of their respective mother using a molecular-genetic method (microsatellites). In this way, these authors determined contemporary seed dispersal. In particular, they could also evaluate the potential for long-distance dispersal (as defined as dispersal among populations), which is otherwise difficult to estimate (Ouburg *et al.* 1999). Godoy and Jordano (2001) identified almost 18% of the sampled seeds as long-distance immigrants – an unexpectedly high value. The information gained from such a genetic analysis clearly exceeds that of traditional seed trapping experiments. In the latter, the exact location of the seed sources can often not be identified.

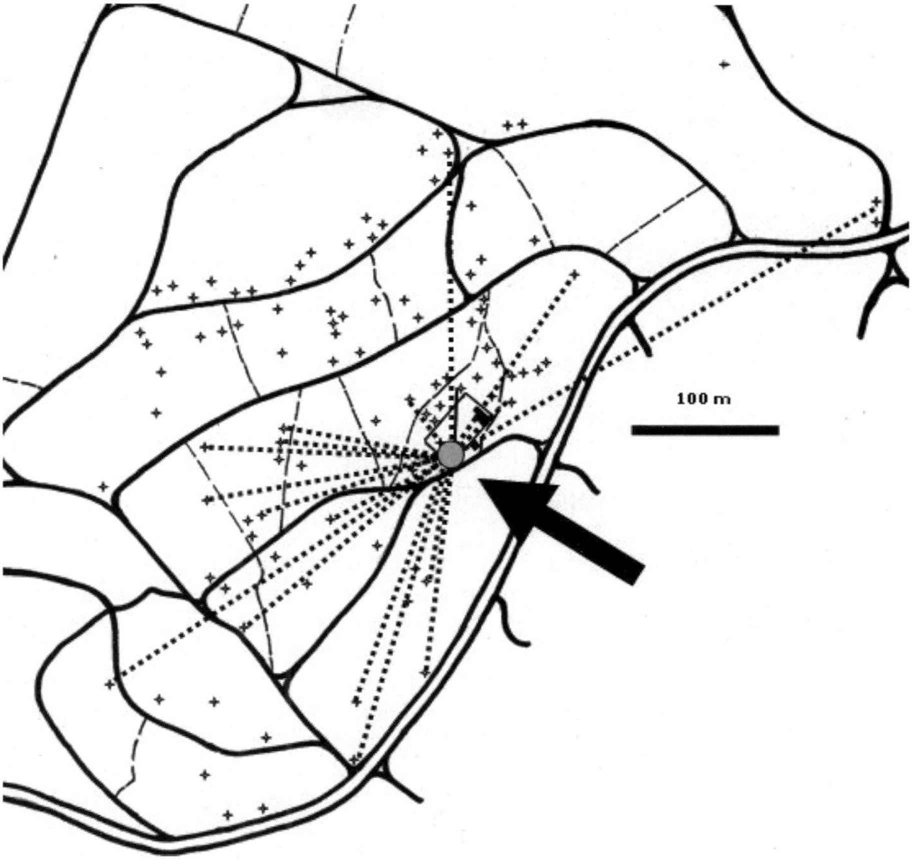

Fig. 3. Contemporary gene flow by pollen in a Swiss population of the Wild Service Tree *(Sorbus torminalis)* as revealed by genetic progeny analysis using microsatellite markers. The grey circle refers to a focal mother tree whose progeny was investigated, crosses refer to adult *S. torminalis* trees, broken black lines connect the mother with the fathers that sired the progeny, and the solid black arrow refers to 30% pollen immigration from outside the investigated population.

The second, widely used approach to infer contemporary migration and gene flow is the application of assignment tests (Rannala and Mountain 1997; Manel *et al.* 2005). Based on its genotype, an individual is probabilistically assigned to the population with which it has the greatest genetic similarity. If it is not assigned to the population from which it had been sampled, it is presumed to be a recent migrant (Frankham *et al.* 2002). Assignment tests require relatively strong genetic differentiation of populations to assign an immigrant individual to a distinct population with high probability, which can be a drawback in empirical studies.

Other methods to infer current gene flow exist. One is the TWOGENER approach, which does not require complete sampling of adult populations (Smouse and Sork 2004; Sork *et al.* 2005). It should also be stressed that for many organisms with minute diaspores such as fungi or lichens, molecular techniques are currently the only methods available that provide reliable estimates of propagule dispersal at the species level (Walser *et al.* 2001).

From the examples given above, it is evident that genetic methods can be used to get real-world estimates of migration, dispersal and gene flow at the landscape level. They may therefore be used to evaluate the ecological relevance of landscape indices of fragmentation or, otherwise, connectivity (Li and Wu 2004). Given the possibility to infer both historical and current patterns of gene flow, population and ecological genetics also offer the possibility to link changes in population connectivity with landscape changes in time (Bürgi *et al.* 2007). For instance, the isolating effect of motorways on wild animals could be estimated by (1) investigating the historical gene flow among populations that are nowadays separated by motorways and (2) by comparing historical gene flow with current gene flow among populations on both sides of the motorways. Corresponding research is conducted on Bobcat *(Lynx rufus)* and Coyote *(Canis latrans)* in California (Riley *et al.* 2004) and on Roe Deer *(Capreolus capreolus)* in Central Europe (Coulon *et al.* 2004; Hindenlang *et al.*; WSL, Birmensdorf, unpubl. data). However, in contrast to what is sometimes suggested by practitioners, population genetics cannot provide absolute measurements of population isolation (Whitlock and McCauley 1999).

Whatever will be: metapopulations

Strictly speaking, a metapopulation can be defined as a group of local populations, partly occupying a set of suitable local habitat patches, showing a turnover of local extinction and re-colonisation in each generation (Baquette 2004). The inclusion of "in each generation" in this definition is important, because it sets the time frame of metapopulation dynamics. By applying this strict definition, real metapopulation dynamics have only been proven for a hand-full of animal species of mainly short generation times such as several butterflies and some amphibians and passerines (Hanski 1998; Baquette 2004). The scientific debate on whether plant species show classical metapopulation dynamics is still continuing (Freckleton and Watkinson 2002; Ehrlén and Eriksson 2003). This is not surprising given the overlapping generation times and the long life span of many plant species (e.g., beyond 100 years for tree species). In plants, metapopulation dynamics have often been indirectly inferred from the genetic structure of populations, in cases where regional processes were shown to overrule local ones (Oddou-Muratorio *et al.* 2001; Hilfiker *et al.* 2004a). However, this genetic inference better fits to a relaxed definition of metapopulations as a group of local populations connected by gene flow (Freckleton and Watkinson 2002). Given the lack of sufficient empirical evidence, the strict metapopulation concept for plants is far from being generally accepted.

A short summary on the present state of the art in metapopulation research including metapopulation genetics could be: there is much theory, simulation and modelling, but little empirical data. What would be needed to prove that a group of populations shows metapopulation dynamics in a strict sense? (1) The suitable habitat of a species must coincide with a network of empty and occupied patches, and the local populations are located in discrete areas. This requires complete field surveys and sound analyses of habitat quality. It is not an easy task to show that a habitat patch is principally suitable but not occupied. In plants, one could achieve this task with seeding and/or transplantation experiments. (2) There must be extinction and re-colonisation events in each generation (Baquette 2004). This calls for long-term demographic data sets that are usually not available and almost impossible to be generated for long-lived species. For instance, metapopulation dynamics of a long-lived tree species could be expected to act within a time frame of several hundred years. Here, the typical solution is to take habitat history as a surrogate of population history using landscape historical methods (Landergott *et al.* 2001; Lindborg and Eriksson 2004;

Bürgi *et al.* 2007). (3) Local population dynamics need to be asynchronous (Baquette 2004) in order to verify that the system under study is not driven by a generally acting intrinsic or extrinsic ecological factor or process other than regional population dynamics. Again, detailed long-term demographic data would be required. (4) Dispersal among local populations has to be shown in each generation. By definition, local populations within metapopulations have to be loosely connected or partially isolated (Frankham *et al.* 2002). To show dispersal (and gene flow by pollen in plants), population genetics comes into play again.

An example is provided by a genetic investigation of an assumed metapopulation of the Water Clover Fern, *Marsilea strigosa,* in a French pond system (Vitalis *et al.* 2002). The classical metapopulation concept assumes that (re-)colonisation of an unoccupied patch takes place by founders form a random sample of all occupied habitat patches in the system. By using microsatellite markers, Vitalis *et al.* (2002) found near-patch recolonisation to be dominant. This leads to a distinct structuring of the genotypes within the assumed pond metapopulation. It thus takes a long time for a gene-variant (allele) to spread over the metapopulation. This rather small-scale dispersal pattern, in contradiction to the classical metapopulation theory, would have been impossible to detect without genetic methods. An overview of population genetics in metapopulation research is provided by Hanski and Gaggiotti (2004).

As Baquette (2004) points out, the metapopulation concept has had a huge conceptual effect on the thinking of both empirical and theoretical ecologists, because it has reorientated their perspectives of the importance of spatial patterns and migration or gene flow among populations. But, perhaps, the importance of the concept lies more in its heuristic value than in practical applications.

Que Sera: An Outlook

A glance at textbooks on landscape ecology shows that issues of population and ecological genetics (including conservation genetics) are either not treated at all (e.g., Turner *et al.* 2001) or treated only in a general way (e.g., Forman 1995). On the other hand, the same could be said about textbooks on population genetics with respect to landscape ecological issues (e. g., Frankham *et al.* 2002). However, the two fields are not as disparate as one might assume at first glance. They often have the same aim, but use different methods or concepts. Most population genetic concepts are at least spatially implicit, and spatial statistics such as autocorrelation analyses are standard methods in population genetics. Investigations of the evolution of species in space and time increasingly apply statistical or modelling approaches from landscape ecology, e.g., kriging (Hoffmann *et al.* 2003). It therefore seems a timely task to incorporate population genetics into landscape ecological studies and vice versa (Holderegger *et al.* in press). Both fields would benefit from this integration. So far, population genetic studies have often investigated single populations below or sets of populations at larger spatial scales beyond a landscape. Spatially explicit genetic investigations at the scale of real landscapes have meanwhile become necessary.

The integration of population genetics and landscape ecology has recently been named landscape genetics by Manel *et al.* (2003; for an extension to landscape genomics see Luikart *et al.* 2003). The field promises to bring population and ecological genetics closer to real-world problems. This asks for a spatially and often also temporarily explicit approach. Landscape ecology can provide the theoretical framework to interpret corresponding spatio-temporal genetic data. Literally, the comparison of isolines exhibiting a spatially explicit measure of habitat fragmentation with "genetic contour lines" based on the genetic differentiation of populations would allow a verification of the effects of dispersal barriers and/or dispersal corridors. On the other hand, molecular investigations provide independent tests for basic assumptions of landscape ecology (see above). Genetic studies can also be used to infer how historical processes have affected species diversity (Lindborg and Eriksson 2004). An example that integrates habitat quality and population demography in the field, landscape assessment by GIS and testing for historical (from museum specimen) and present population isolation using molecular genetics is given for Capercaillie *(Tetragalus urogallus)* in Suter *et al.* (2007). As shown by this example on an endangered bird species, landscape genetic research needs the collective knowledge of scientists from different fields such as landscape ecology, ecological modelling, biodiversity and population genetics: landscape genetics is cross-disciplinary by definition.

Acknowledgements

The authors thank all members of the Section of Ecological Genetics at the WSL, who provided some of the data presented here. Simon Egli, Sucharita Ghosh, Felix Kienast, Victoria Sork, Otto Wildi and an anonymous referee gave comments on earlier versions of the manuscript. Financial support by the European Commission and the Swiss Agency for Education and Science (projects BAFE/BIOTECH and INTRABIODIV), the Swiss Science Foundation (NCCR Plant Survival) and the Swiss Agency for the Environment, Forest and Landscape BUWAL (project preservation and use of forest genetic resources) is greatly acknowledged.

References

Avise J.C. 2002. Phylogeography: the history and formation of species. Harvard University Press, Cambridge.

Baquette M. 2004. The classical metapopulation theory and the real, natural world: a critical appraisal. Basic and Applied Ecology 5: 213–224.

Bolliger J., Wagner H.H. and Turner M.G. 2007. Identifying and quantifying landscape patterns in space and time. In: Kienast F., Wildi O. and Ghosh S. (eds.). A Changing World. Challenges for Landscape Research, Vol. 8: 177–194. Springer Landscape Series, Dordrecht.

Buchecker M., Kianicka S. and Junker B. 2007. Value systems: drivers of human-landscape interactions. In: Kienast F., Wildi O. and Ghosh S. (eds.). A Changing World. Challenges for Landscape Research, Vol. 8: 7–26. Springer Landscape Series, Dordrecht.

Bürgi M., Hersberger A., Hall M., (Russell) Southgate E.W.B. and Schneeberger N. 2007. Using the past to understand the present land use and land cover. In: Kienast F., Wildi O. and Ghosh S. (eds.). A Changing World. Challenges for Landscape Research, Vol. 8: 133–144. Springer Landscape Series, Dordrecht.

Conner J.K. and Hartl D.L. 2004. A primer to ecological genetics. Sinauer, Sunderland.

Coulon A., Cosson J.F., Angibault J.M., Cargnelutti B., Galan M., Morellet N., Petit E., Aulagnier S. and Hewison A.J.M. 2004. Landscape connectivity influences gene flow in a Roe Deer population inhabiting a fragmented landscape: an individual-based approach. Molecular Ecology 13: 2841–2850.

Duelli P., Baur P., Buchecker M., Gugerli F., Holderegger R. and Wohlgemuth T. 2007. The role of value systems in biodiversity research. In: Kienast F., Wildi O. and Ghosh S. (eds.). A Changing World. Challenges for Landscape Research, Vol. 8: pp. 27–34. Springer Landscape Series, Dordrecht.

Ennos R.A. 1994. Estimating the relative rates of pollen and seed migration among plant populations. Heredity 72: 250–259.

Ehrlén J. and Eriksson O. 2003. Large-scale spatial dynamics of plants: a response to Freckleton and Watkinson. Journal of Ecology 91: 316–320.

Forman R.T.T. 1995. Land mosaics. The ecology of landscapes and regions. Cambridge University Press, Cambridge.

Frankham R., Ballou J.D. and Briscoe D.A. 2002. Introduction to conservation genetics. Cambridge University Press, Cambridge.

Freckleton R.P. and Watkinson A.R. 2002. Large-scale spatial dynamics of plants: metapopulations, regional ensembles and patchy populations. Journal of Ecology 90: 419–434.

Godoy J.A. and Jordano P. 2001. Seed dispersal by animals: exact identification of source trees with endocarp DNA microsatellites. Molecular Ecology 10: 2275–2283.

Gugerli F., Sperisen C., Büchler U., Magni F., Geburek T., Jeandroz S. and Senn J. 2001. Haplotype variation in a mitochondrial tandem repeat of Norway Spruce *(Picea abies)* populations suggests a serious founder effect during postglacial re-colonisation of the western Alps. Molecular Ecology 10: 1255–1263.

Hanski I. 1998. Metapopulation dynamics. Nature 396: 41–49.

Hanski I. and Gaggiotti O. (eds.). 2004. Ecology, genetics and evolution of metapopulations. Elsevier, Amsterdam.

Hartl D.L. and Clark A.G. 1997. Principles of population genetics. Sinauer, Sunderland.

Hewitt G.M. 2004. Genetic consequences of climatic oscillations in the Quaternary. Philosophical Transactions of the Royal Society of London Series B Biological Sciences 359: 183–195.

Hilfiker K., Holderegger R., Rotach P. and Gugerli F. 2004a. Dynamics of genetic variation in *Taxus baccata:* local versus regional perspectives. Canadian Journal of Botany 82: 219–227.

Hilfiker K., Gugerli F., Schütz J.-P., Rotach P. and Holderegger R. 2004b. Low RAPD variation and female-biased sex ratio indicate genetic drift in small populations of the dioecious conifer *Taxus baccata* in Switzerland. Conservation Genetics 5: 357–365.

Hoffmann M.H., Glass A.S., Tomiuk J., Schmuths H., Fritsch R.M. and Bachmann K. 2003. Analysis of molecular data of *Arabidopsis thaliana* (l.) Heynh. (Brassicaceae) with geographical information systems (GIS). Molecular Ecology 12: 1007–1019.

Holderegger R., Kamm U. and Gugerli F. 2006. Adaptive vs. neutral genetic diversity: implications for landscape genetics. Landscape Ecology 21: 797–807.

Jackson R.B., Linder C.R., Lynch M., Purugganan M., Somerville S. and Thayer S.S. 2002. Linking molecular insight and ecological research. Trends in Ecology and Evolution 17: 409–114.

Lande R. and Shannon S. 1996. The role of genetic variation in adaptation and population persistence in a changing environment. Evolution 50: 434–437.

Landergott U., Holderegger R., Kozlowski G. and Schneller J.J. 2001. Historical bottlenecks decrease genetic diversity in natural populations of *Dryopteris cristata*. Heredity 87: 344–355.

Lang G. 1994. Quartäre Vegetationsgeschichte Europas. Gustav Fischer, Jena.

Latta R.G. 2003. Gene flow, adaptive population divergence and comparative population structure across loci. New Phytologist 161: 51–58.

Levin S.A. 1992. The problem of pattern and scale in ecology. Ecology 48: 1043–1967.

Li H. and Wu J. 2004. Use and misuse of landscape indices. Landscape Ecology 19: 389–399.

Lindborg R. and Eriksson O. 2004. Historical landscape connectivity affects present plant species diversity. Ecology 85: 1840–1845.

Lowe A., Harris S. and Ashton P. 2004. Ecological genetics. Design, analysis and application. Blackwell, Oxford.

Luikart G., England P., Tallmon D., Jordan S. and Taberlet P. 2003. The power and promise of population genomics: from genotyping to genome typing. Nature Reviews Genetics 4: 981–994.

MacArthur R.H. and Wilson E.O. 1967. The theory of island biogeography. Princeton University Press, Princeton.

Manel S., Schwartz M.K., Luikart G. and Taberlet P. 2003. Landscape genetics: combining landscape ecology and population genetics. Trends in Ecology and Evolution 18: 189–197.

Manel S., Gaggiotti O.E. and Waples R.S. 2005. Assignment methods: matching biological questions with appropriate techniques. Trends in Ecology and Evolution 20: 136–142.

McKay J.K. and Latta R.G. 2002. Adaptive population divergence: markers, QTL and traits. Trends in Ecology and Evolution 17: 285–291.

Merilä J. and Crnokrak P. 2001. Comparison of genetic differentiation at marker loci and quantitative traits. Journal of Evolutionary Biology 14: 892–903.

Oddou-Muratorio S., Petit R.J., Le Guerroue B., Guesnet D. and Demesure B. 2001. Pollen-versus seed-mediated gene flow in a scattered tree species. Evolution 55: 1123–1135.

Oddou-Muratorio S., Houot M.-L., Demesure-Musch B. and Austerlitz F. 2003. Pollen flow in the Wild Service Tree, *Sorbus torminalis* (L.) Crantz. I. Evaluating the paternity analysis procedure in continuous populations. Molecular Ecology 12: 3427–3439.

Odduo-Muratorio S., Demesure-Musch B., Pélissier R. and Gouyon P.H. 2004. Impacts of gene flow and logging history on the local genetic structure of a scattered tree species, *Sorbus torminalis* (L.) Crantz. Molecular Ecology 13: 3689–3702.

Ouborg N.J., Piquot Y. and Van Groenendael J.M. 1999. Population genetics, molecular markers and the study of dispersal in plants. Journal of Ecology 87: 551–568.

Pearman P.B. 2001 Conservation value of independently evolving units: sacred cow or testable hypothesis? Conservation Biology 15: 780–783.

Petit R.J., Aquinagalde I., De Beaulieu J.-L., Bittkau C., Brower S., Cheddadi R., Ennos R.A., Fineschi S., Grivet D., Lascoux M., Mohanty A., Müller-Starck G., Demesure-Musch B., Palmé A., Martin J.P., Martin J.P., Rendell S. and Vendramin G.G. 2003. Glacial refugia: hotspots but not melting pots of genetic diversity. Science 300: 1563–1565.

Rannala B. and Mountain J.L. 1997. Detecting immigration by using multilocus genotypes. Proceedings of the National Academy of Sciences USA 94: 9147–9201.

Reed D.H. and Frankham R. 2001. How closely correlated are molecular and quantitative measures of genetic variation? A meta-analysis. Evolution 55: 1095–1103.

Riley S.P.D., Pollinger J., Sauvajot R.M., York E.C., Bromley C., Fuller T.K. and Wayne R.K. 2004. Freeway reduces gene flow in wide-ranging mammalian carnivores. 18th Annual Meeting of the Society for Conservation Biology. Columbia University, New York, pp. 232.

Schönswetter P., Stehlik I., Holderegger R. and Tribsch A. 2005. Molecular evidence for glacial refugia of mountain plants in the European Alps. Molecular Ecology 19: 3597–3555.

Smouse P.E. and Sork V.L. 2004. Measuring pollen flow in forest trees: an exposition of alternative approaches. Forest Ecology and Management 197: 21–38.

Sork V.L., Nason J., Campbell D.R. and Fernandez J.F. 1999. Landscape approaches to historical and contemporary gene flow in plants. Trends in Ecology and Evolution 14: 219–224.

Sork V.L., Smouse P.E., Apsit V.J., Dyer R.J. and Westfall R.D. 2005. A two-generation analysis of pollen pool genetic structure in flowering dogwood, *Cornus florida* (Cornaceae), in the Missouri Ozarks. American Journal of Botany 92: 262–271.

Suter W., Bollmann K., Holderegger R. and Senn J. 2007. Landscape permeability: from individual dispersal to population persistence. In: Kienast F., Wildi O. and Ghosh S. (eds.). A Changing World. Challenges for Landscape Research, Vol. 8: 157–174. Springer Landscape Series, Dordrecht.

Taberlet P., Fumagalli L., West-Savey A.-G. and Cosson J.-F. 1998. Comparative phylogeography and postglacial colonisation routes in Europe. Molecular Ecology 7: 453–464.

Turner M.G., Gardner R.H. and O'Neill R.V. 2001. Landscape ecology in theory and practice. Springer, New York.

Vitalis R., Riba M., Colas B., Grillas P. and Olivieri I. 2002. Multilocus genetic structure at contrasted spatial scales of the endangered water fern *Marsilea strigosa* Willd. (Marsileaceae, Pteridophyta). American Journal of Botany 89: 1142–1155.

Walser J.-C., Zoller S., Büchler U. and Scheidegger C. 2001. Species-specific detection of *Lobaria pulmonaria* (lichenised ascomycete) diaspores in litter samples trapped in snow cover. Molecular Ecology 10: 2129–2138.

Whitlock M.C. and McCauley D.E. 1999. Indirect measures of gene flow and migration: $F_{ST} \neq 1/(4Nm + 1)$. Heredity 82: 117–125.

Landscape Permeability: From Individual Dispersal to Population Persistence

Werner Suter*, Kurt Bollmann and Rolf Holderegger

Swiss Federal Research Institute WSL, Zürcherstrasse 111, CH-8903 Birmensdorf, Switzerland
* Corresponding author: e-mail: werner.suter@wsl.ch

Abstract
Landscape permeability, usually known as connectivity in landscape ecology, defines to what degree organisms are capable of moving through a landscape. It is an important functional aspect linking landscape structure to the dynamics of populations, when these are at least in part determined by emigration and immigration of individuals. Landscape permeability is thus most relevant in landscapes with fragmented habitat and where populations are organised as metapopulations. This chapter briefly reviews how the two paradigms "habitat fragmentation" and "metapopulation dynamics" are related, which main effects of habitat fragmentation on populations have been found, and how structural landscape elements such as edges, matrices, barriers and corridors may determine landscape permeability for dispersing individuals. We use an example from our own work on capercaillie (*Tetrao urogallus*; Aves; Tetraonidae) to illustrate how the commonly used static approach of relating spatial population patterns to landscape structure is limited by the lack of empirical data on how dispersal actually takes place, and how the problem may be mitigated by studying the result of the unknown dispersal by genetic methods. We then follow the development leading from static distribution models to dynamic, spatially explicit population models and conclude that validation lags behind the theoretical development, again as a matter of lack of data on dispersal, particularly for vagile, large-bodied animal species. The chapter is concluded with some management-related observations regarding the restoration of landscape connectivity by means of movement corridors.

Keywords: connectivity, fragmentation, metapopulations, modelling, SEPM, dispersers, capercaillie, *Tetrao urgallus*

F. Kienast, O. Wildi & S. Ghosh (eds.), A Changing World. Challenges for Landscape Research, 157–174.

Introduction

Ecological processes operating within a landscape can be seen in their most basic form as the flow of matter, energy and information between habitats (Turner *et al.* 2001). Flow strength is controlled by the permeability of the landscape, i.e. by the spatial configuration of landscape elements supporting or restraining flow. In landscape ecology, permeability is usually known as connectivity, defined by the degree to which the landscape facilitates or impedes flow among resource patches (Bélisle 2005; Taylor *et al.* 1993; Tischendorf and Fahrig 2000). The notion of resource patchiness may thus be seen as an equivalent of habitat fragmentation. In this chapter, we address recent progress in linking individual dispersal to population persistence in fragmented habitats, focusing on animals (for plants see Holderegger *et al.* 2007). The theme has been extensively dealt with in the literature, including a number of books (e.g. Gutzwiller 2002; Hanski 1999; McCullough 1996) and paper series (e.g. Villard 2002). We concentrate on the question of how landscape structure influences the dispersal of individuals and thus affects (meta-)population dynamics, how this has been explored in theoretical and modelling approaches, and what empirical data are available.

We begin with a brief look at how the two paradigms "habitat fragmentation" and "metapopulation dynamics" are related, and at the main effects habitat fragmentation can have on populations. We then briefly discuss the role of structural landscape elements for landscape permeability to dispersing individuals. An example from our own work on capercaillie (*Tetrao urogallus;* Aves; Tetraonidae) illustrates how the commonly used static approach of relating spatial population patterns to landscape structure is limited by the lack of data on dispersing individuals. In the following chapter, we look at how this problem has been tackled so far, and which challenges remain. We end with some management-related conclusions.

Habitat Fragmentation and Metapopulation Dynamics: Two Basic Paradigms

We prefer the term «habitat fragmentation» to «landscape fragmentation», since it is the habitat rather than the landscape that become fragmented by human activity. Habitat fragmentation and its effects on populations have been a key issue in ecology since the 1970s, but the concept has proven multifaceted (Fahrig 2003). Implications from modelling and empirical results are similarly manifold, depending on which organisms, habitats, and biogeographical areas are studied (Haila 2002; Villard 2002), and which temporal and spatial scales are addressed (Debinski and Holt 2000; Urban 2005). In general, increasing habitat fragmentation is coupled with decreasing habitat patch size and increasing inter-patch distances across the «non-habitat» matrix (Turner *et al.* 2001). Fragmentation also results in an increase of boundary lines between habitat units, which can act as obstacles to dispersing individuals. Hence, habitat connectivity as seen from a biological point of view is reduced by fragmentation but in effect strongly depends on the dispersal ability (D'Eon *et al.* 2002) of the organism in question.

At about the same time as landscape ecology emerged as a discipline, the theory of metapopulation dynamics was developed by population ecologists. It aims at understanding the overall persistence of an array of local populations that are spatially separated but loosely linked by dispersal (Hanski and Gilpin 1997). Thus, metapopulation theory is easily applicable to animal populations in fragmented habitats (With 2004; for plants see Freckleton and Watkinson 2002). The rate at which individuals (or genes) are exchanged between local populations is central to the concept, and here metapopulation theory is bound to the idea of landscape connectivity.

The recent rapid development of tools for spatial analysis has directed much effort towards pattern analysis and development of landscape metrics or indices. As a result, landscape ecology has suffered from having neglected relationships between landscape patterns and ecological processes (Goodwin 2003; Li and Wu 2004; Wu and Hobbs 2002), although recent theoretical developments have helped to fill this gap (With 2002). Ironically, patch/matrix-oriented thinking in metapopulation theory has also given strong emphasis to patterns despite the fundamentally process-based nature of metapopulation models. Views thus differ on the general applicability of metapopulation models to natural systems (Baguette and Mennechez 2004; Shreeve *et al.* 2004). Most applications refer to small-bodied species such as arthropods and small mammals for which patch incidence is easily measured, but the concept should also be amenable to large mammals if dynamics of subpopulations are measured in terms of their demography (Elmhagen and Angerbjörn 2001). Modelling has steadily progressed in the last decade, but validation lags behind. There is an urgent need for empirical data on how organisms move through landscapes and thereby perceive and react to obstacles and resources (Bélisle 2005; Lidicker 2002; McGarigal and Cushman 2002), because metapopulation dynamics will ultimately be shaped by individual-based processes (see also paragraph "From structure to process: ...", this chapter).

Effects of Fragmentation

Declining population size due to splitting up of contiguous habitats may simply reflect the decrease in habitat area, without any superimposed fragmentation effects (Fahrig 2003). Yet the reduction of population size associated with habitat fragmentation is often disproportionately larger than the proportion of habitat lost, especially when populations in small patches become extinct (Andrén 1994; Beier *et al.* 2002). Such immediate consequences of habitat fragmentation have primarily nurtured the scientific interest in this topic. Processes resulting in reduced fitness within isolated and small populations (Frankham *et al.* 2002; Young and Clarke 2000) are often related to increased amounts of boundaries (edge effects), which may alter regimes of predation, parasitism or disturbance (Hansson *et al.* 1995). Thus, habitat quality also decreases with decreasing patch size, and this may affect the persistence of populations. The probability that individuals from some other (source) patches immigrate will at the same time decrease, mainly because distances between patches have become larger and more difficult to overcome. Hence, size and number of habitat fragments and their spatial arrangement in a landscape play important roles in mortality/extinction and migration/colonisation rates. Matrix composition and quality are currently also being discovered as important factors in metapopulation research. Spatial aspects are now routinely incorporated in conservation modelling including viability analyses of vertebrate species with the aid of specialized software (Reed *et al.* 2002).

Genetic effects in small and isolated populations may additionally reduce reproductive success and lower population persistence (Lande 1999; Saccheri *et al.* 1998). Conservation genetic theory predicts that small population size tends to increase the probability of genetic drift (i.e. the random sampling of genetic variants), decreases genetic diversity (i.e. genetic erosion) and heterozygosity, increases breeding among related individuals (inbreeding), and reduces the fitness of inbred individuals (inbreeding depression; Young and Clarke 2000; Frankham *et al.* 2002). The viability of small populations may become reduced within only a few generations. On the other hand, diminished gene exchange among populations may lead to population differentiation (see paragraph "From structure to process: ...", this chapter). Potential evolutionary differentiation as a consequence of habitat fragmentation reflects the fact that habitat fragmentation does not necessarily have a negative connotation *per se.*

Habitat islands can temporarily become safe from predators, parasitoids and diseases (Frankham *et al.* 2002) or temporarily support species that commute between different habitats, be it on a daily (e.g. feeding – shelter) or seasonal basis, or once in a life time (e.g. different juvenile and adult habitats). Generally, experimental fragmentation studies mainly show the immediate negative effects of habitat loss and habitat change during the first few years, i.e. the actual process of fragmentation (Zschokke *et al.* 2000), while long-term stabilising or even positive effects that might only come into play after years of community restructuring are often missed by shorter empirical studies.

Edges, Matrix, Barriers and Corridors: Requisites of Landscape Permeability

Because of increased edge length, the proportion of fragment area subject to influences originating in the surrounding land (i.e. the matrix) increases. Edge effects include both abiotic factors, such as wind or fire, and biotic factors, such as predators, parasites, disease vectors, or man (Laurance *et al.* 2002). Edge effects can often not be generalised (Lahti 2001). While edges may be penetrable from the outside, they may still act as barriers for many organisms living in the fragmented habitat but ready to disperse (Cale 2003). However, the strength of this effect may be more related to the contrast between habitat and matrix (Collinge and Palmer 2002; Holmquist 1998; Ricketts 2001) and the extent and structure of the matrix that has to be crossed (Haynes and Cronin 2003) than to the edge itself.

Habitat patches and surrounding matrices are usually considered to be binary systems of suitable and unsuitable habitat (Turner *et al.* 2001). Crossing ability and propensity are species- and often also sex-specific, but generally increase with body mass (Grubb and Doherty 1999). Many species are reluctant to cross even small expanses of unsuitable land, although they would be able to do so physically (Bélisle *et al.* 2001; Laurance *et al.* 2002; Gobeil and Villard 2002; Creegan and Osborne 2005). Independent of animal size, (behavioural) barriers may thus restrict movements across the landscape (Harris and Reed 2002). However, a rising number of studies on the dispersal behaviour of a wide range of animals provide evidence that the matrix does not simply consist of unsuitable «desert» between habitat islands, but that it can have some habitat quality of its own (Enoksson *et al.* 1995; Haynes and Cronin 2004; Bender and Fahrig 2005). Such quality will enhance the permeability of the matrix. Permeability may also depend on the spatial structure of the matrix, with some species using certain elements in preference to others when dispersing (Cale 2003). In heterogeneous landscapes, dispersal patterns are significantly affected by edge mediated behaviour (Ovaskainen 2004). When moving across boundaries, the strength of the contrast at boundaries determines the direction of movements of animals: low contrast boundaries exhibit net immigration, and high contrast boundaries experience net emigration (Collinge and Palmer 2002). While moving through different habitat types, the animals experience high variation in mortality risk and thus highly variable matrix permeability (Hein *et al.* 2003).

Linear landscape structures such as roads or rivers often act as barriers for many terrestrial animals; whereas for aquatic organisms, rivers function as dispersal corridors. Such linear landscape structures can therefore either reduce or increase species-specific landscape permeability (Forman and Alexander 1998). Animals may not cross a linear structure because of either physical or behavioural inability, or they simply avoid areas bordering linear man-made structures (Nellemann *et al.* 2001). Thus, road networks simultaneously fragment habitats and produce edge effects (Mech *et al.* 1988; Forman *et al.* 2003). Strips of remnant (or restored) habitat linking larger expanses of the same habitat (e.g. forest patches in an agricultural matrix) can work as corridors enhancing dispersal or migrational movements, particularly for larger ground-dwelling animals. There is still some controversy about how effectively corridors enhance connectivity of habitat patches (Beier and Noss 1998), as applications of the corridor concept in biological conservation have hardly been evaluated with respect to their capacity of defragmenting habitats at regional scales (Vos *et al.* 2002). Additionally, corridors are likely to favour groups of species with particular life histories more than other groups (Hudgens and Haddad 2003; see also paragraph "Managing landscape permeability", this chapter). For smaller organisms such as insects, strips of (semi-)natural vegetation often have significance as habitat *per se* rather than as movement corridor, and may even represent the last refuges for rare species (e.g. «road reserves», Saunders and Hobbs 1991). On the other hand, corridors, *sensu stricto*, may not be a prerequisite for successful dispersal of mobile species as long as a series of smaller habitat patches can function as «stepping stones» between larger habitat expanses and thus become a functional corridor (With 2002).

Landscape Permeability: Lessons from Capercaillie

The capercaillie *(Tetrao urogallus)*, one of the largest forest-dwelling galliform birds, serves to illustrate the importance of a detailed understanding of spatial landscape aspects in the conservation biology of endangered species. Habitats of capercaillie in Central Europe have become strongly fragmented, and most populations show recent decreasing trends (Klaus and Bergmann 1994; Storch 2000). Population decline and fragmentation have been particularly severe in the Swiss Alps after 1970 (Mollet *et al.* 2003). The remnants apparently form metapopulations (Segelbacher and Storch 2002). Here the contemporary spatial pattern of occupied forest areas resembles islands within a matrix of unsuitable woodland or open montane-subalpine farmland (Fig. 1). Conservation measures to halt the overall decline have rarely succeeded, probably because they were local in scope and neglected population processes operating at the landscape scale (i.e. hundreds of km^2; Storch 2002). A complex interplay of local habitat quality at the forest stand scale, regional habitat fragmentation, and large-scale climatic factors such as cold and wet summers is suspected to have caused the decline (Lindström *et al.* 1996). Since capercaillie possesses characteristics of an umbrella species, conserving the capercaillie will also benefit a wider array of bird species (Suter *et al.* 2002) and generally improve the habitat of subalpine forest communities.

Ecological factors driving population processes in capercaillie have been shown to operate on different spatial scales, from local stands to entire regions (10 ha to 100 km^2; Andrén 1994; Kurki *et al.* 2000; Storch 1997, 2002, 2003). For example, at least three spatial scales have to be considered to understand habitat requirements (Storch 1997): vegetation structure (small scale), stand mosaic (intermediate scale) and spatial arrangement of forests in the landscape (large scale). A habitat area described at the intermediate scale usually comprises stands both used and unused by capercaillie. Used stands are characterised by intermediate canopy cover, rich field-layer and low-branched solitary trees (Bollmann *et al.* 2005a). Graf *et al.* (2005) applied a habitat suitability model of the capercaillie for the central and eastern Prealps and Alps and produced a map of discrete distribution patches resembling a metapopulation pattern (Fig. 1). The total area of the patches (1,187 km^2) was much larger than the area actually occupied by capercaillie, but coincided with the former distribution of capercaillie in the seventies, when total Swiss population size was at least twice as large as it is today (Mollet *et al.* 2003).

According to a stochastic model of Grimm and Storch (2000), the minimum viable size for an isolated capercaillie population in the Alps is 470 individuals. This figure corresponds

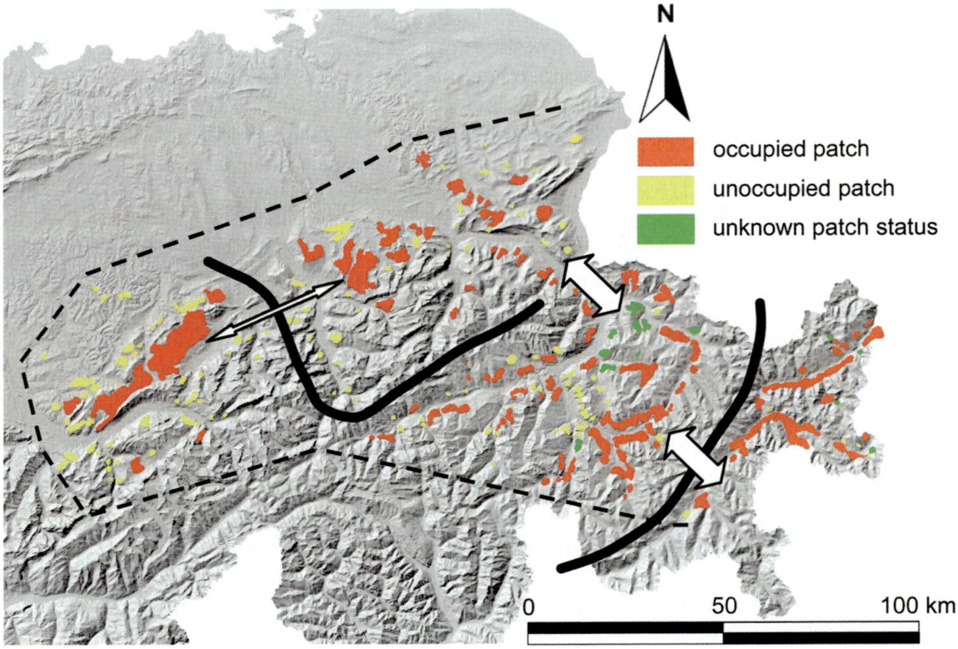

Fig. 1. Distribution and occupation of potential habitat patches by capercaillie *(Tetrao urogallus)* in the Central and Eastern Alps of Switzerland. Gene flow (arrows) among regions is derived from genetic analyses (Jacob *et al.*, WSL Birmensdorf, unpubl. data). Arrow sizes represent the relative amount of gene flow. The patches were predicted and derived from the habitat models of Graf *et al.* (2006). Areas believed to act as barriers to dispersal (Mollet *et al.* 2003) are depicted with thick lines. The dashed line denotes the study area.

to a minimum area requirement of 250 km^2, although in a metapopulation arrangement, patches of 50–100 km^2 might suffice as long as they allow the exchange of dispersers (Grimm and Storch 2000). Only four of the Alpine populations in Switzerland inhabit forest areas ≥100 km^2. We assume that these forests are the hubs of a habitat network of core populations which are spatially linked with smaller populations via dispersing individuals. Such metapopulations may thus manage with considerably less than 250 km^2 of forest area to be viable, but the dispersal rate between occupied patches would therefore be a key factor for the persistence of local capercaillie populations (Segelbacher and Storch 2002; Storch and Segelbacher 2000). Dispersal rate should depend on parameters related to the patches (patch area, patch quality and local population size) as well as parameters defining the gap (inter-patch distance, matrix quality, topography, anthropogenic barriers etc.). In the Swiss Alps, patches were occupied only if they were larger than 54 ha and less than 10 km away from the next occupied patch (Bollmann *et al.* 2005b). These results are in line with dispersal distances of 5–10 km determined by field observations and telemetric studies (review in Storch and Segelbacher 2000). Currently no empirical data are available for parameters that potentially influence dispersal rates in capercaillie.

Population genetic methods have the potential to significantly improve estimates of dispersal, where traditional field methods reach their limits. An evaluation of genetic differentiation of capercaillie populations in the European Alps revealed only low to moderate structuring, although populations at the northern edge of the Alps were clearly distinct from central Alpine populations (Segelbacher and Storch 2002). Genetic distances between different edge populations correlated significantly with geographic distances (isolation by distance), while central Alpine populations showed a higher degree of genetic differentiation from each other, irrespective of distance (Segelbacher and Storch 2002). Our own data (G. Jacob *et al.* unpubl., WSL Birmensdorf) provide evidence of some dispersal across mountain ridges dividing the central Swiss Alps and between the central Alps and the northern edge of the Alps. Surprisingly, there is much less exchange of individuals between central and eastern edge populations (Fig. 1). This is contrary to the common belief that the mountain chain stretching between the central Alpine populations and edge populations in eastern Switzerland forms a major barrier to dispersal, whereas the gap between the edge populations is only of recent origin. In summary, there seems to be still a considerable but varying level of dispersal, and hence connectivity, in the Swiss Alps. Finding discrepancies between estimates of dispersal made by telemetric methods and genetic investigations is a common experience. It may reveal the fact that telemetry measures the contemporary gene flow and genetic methods provide surrogates of the historic dispersal ability of the species. The discrepancies also illustrate well the need for a better understanding of how landscape structures promote or impede dispersal events. Judging landscape permeability is crucial for the effective conservation of capercaillie, but important questions remain unanswered, for example: which topographical features really function as dispersal barriers (see above); what role matrix characteristics play in dispersal; or how patch size and quality influence the dispersal ability and propensity of individual birds.

From Structure to Process: Current and Future Challenges

Understanding, modelling and predicting population dynamics and persistence in a spatially explicit context is a challenging task and a hot topic of current and possibly, future research. In Figure 2, we have tried to illustrate how we perceive linkages between approaches that are mainly landscape-, individual-, or population-based.

A starting point on the long way to understand the dynamics of populations in space is the understanding of landscapes and populations in terms of their structures. Landscape ecological work has originally been concerned with landscape metrics to describe, analyse and measure the structural units which form a landscape. Along with a growing interest in the functional understanding of landscape structure (Tischendorf and Fahrig 2000), more complex metrics such as measures of fragmentation or connectivity/cohesion (e.g. Chardon *et al.* 2003; Jaeger 2000; Opdam *et al.* 2003; Fig. 2 A1) were developed.

Landscape metrics, together with spatial information on species occurrence, are the raw material for modelling species distributions in relation to habitat quality and configuration, matrix resistance, corridors or barriers, and for modelling landscape connectivity and suitability for given species (Fig. 2 B1). Some studies have modelled ecological connectivity at the regional scale without referring to particular organisms (e.g. Marulli and Mallarach 2005), but in general the concept of landscape functioning needs an organismal reference in order to be meaningful. Most studies use empirical data on the occurrence of species (or entire communities; e.g. Perault and Lomolino 2000), whereas in the assessment of landscape permeability for a focal species, assumptions on how the species perceive landscape elements such as barriers might be used (Hunter *et al.* 2003). There are a wide range of modelling approaches for a wealth of species. These models are generally static and often probabilistic in nature (Burgman *et al.* 2005; Guisan and Zimmermann 2000). They have a

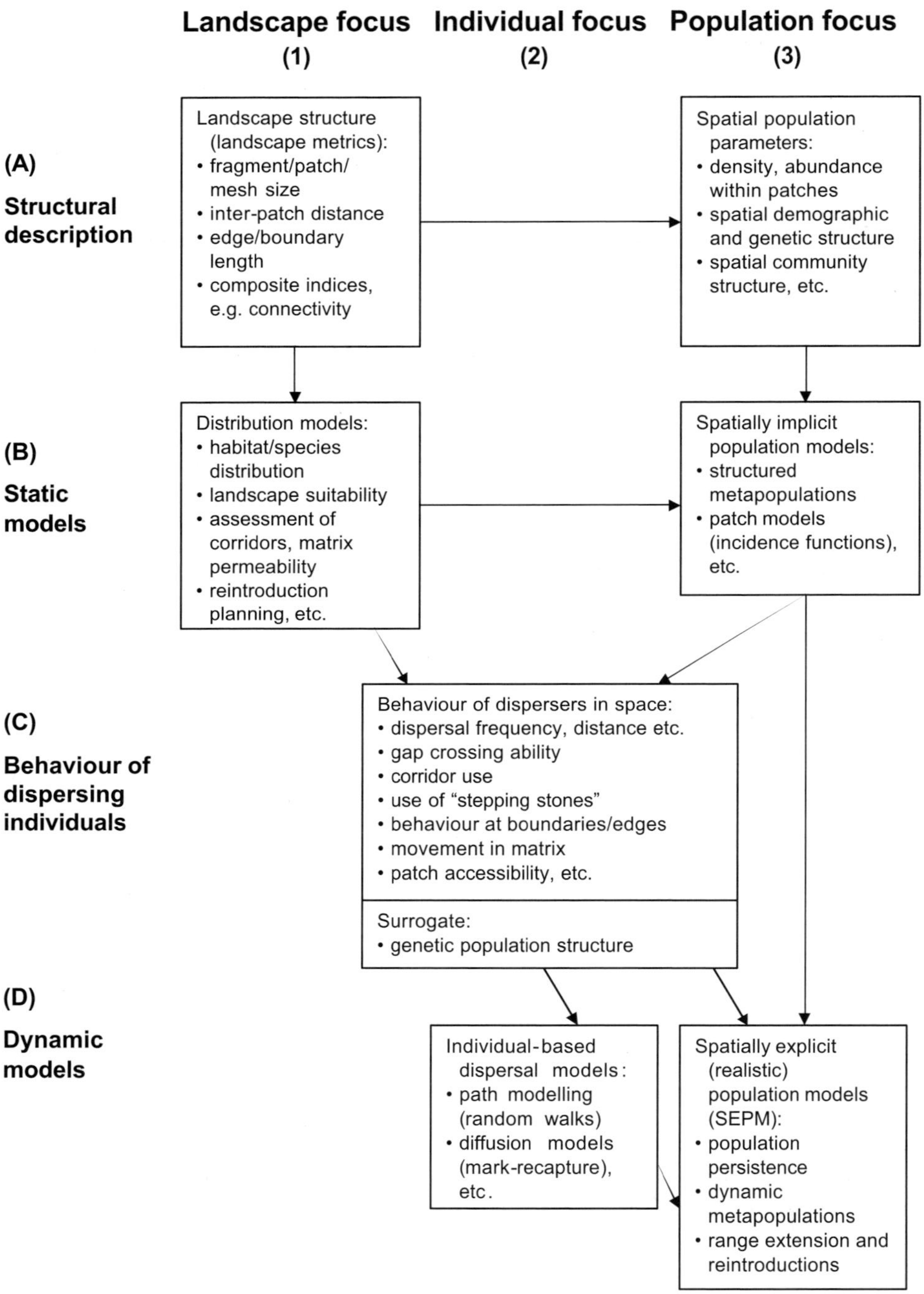

Fig. 2. Graphical representation of ways and linkages leading from structural information on landscapes and animal populations to spatially explicit (= realistic; Hanski and Gaggiotti 2004) population models (see text for explanations).

strong emphasis on landscape structural data, while data regarding the focal species are often rather crude measurements, such as presence/absence (incidence; e.g. Verbeylen *et al.* 2003). This is why we list them as "landscape-based" models. Many of these studies, however, are done in a conservation-related context. Our model of capercaillie distribution summarised above is a typical example, and it exemplifies the limits of this type of approach for understanding the viability of fragmented populations when demographic data are not available.

The way to progress beyond static models and eventually to understand how population dynamics are moulded by their landscape context starts by linking population parameters to landscape metrics (Fig. 2 A3). The variety of population parameters lending themselves to analysis ranges from measures of species abundance to temporally explicit demographic data but may also include approaches of studying the genetic structure of populations in space (see below). There is now a plethora of papers exploring spatial population structure in relation to landscape metrics, although a preponderance of studies using smaller vertebrates (small mammals, birds, amphibians etc.) is obvious. However, working with invertebrates on smaller spatial scales allows experimental approaches to be used (Grez *et al.* 2004; Haddad and Baum 1999) and provides insight into scale-dependence of species responses to landscape structure (Krawchuk and Taylor 2003).

Analogous to the landscape-based approach, modelling the distribution of populations in space will first produce a static model, which for fragmented landscapes is usually within the metapopulation framework (Hanski 1999; Hanski and Gaggiotti 2004; Hanski and Gilpin 1997). Our label "static" for structural metapopulation models in Figure 2 B3 simply means that the inherently dynamic aspect of patch extinction/colonisation is often based on assumptions about the dynamics of subpopulations rather than empirical data. In this case, the distribution pattern of a metapopulation as it is seen ("snapshot") is the outcome not only of population processes within the patch but also of unknown behavioural decisions of individual dispersers moving across a landscape (Andreassen *et al.* 2002; Sutherland 1996). The distinction between approaches classified as B1 or B3 in Figure 2 is thus small in many published examples.

From this point onwards, however, progress in understanding how populations behave under the constraints set by real landscapes cannot be achieved without taking the behaviour of dispersing individuals into account. Dispersal, particularly in an evolutionary context or in terms of geographical patterns (Bullock *et al.* 2002; Clobert *et al.* 2001), has been well studied. However, dispersal as an individual-based process, in which organisms move through a landscape, navigate, and make decisions in response to habitat and landscape structures encountered (Fig. 2 C2), has received much less attention (Lidicker 2002). Empirical studies have focused on small species with limited mobility such as arthropods (often butterflies and beetles), small mammals (almost exclusively small rodents), reptiles, amphibians, and some birds (see Bowne and Bowers 2004 for a review of interpatch movements). When empirical data are lacking, one can look at the genetic structure in a metapopulation and infer movement rates between patches and the role that barriers, corridors, and the matrix might play therein. A number of studies already used this landscape-genetic approach (Coulon *et al.* 2004; Manel *et al.* 2003). We complemented our static capercaillie distribution model with a genetic analysis and found that gene flux between patches was not entirely in agreement with conclusions drawn from interpreting the patch distribution map and intermediary corridors and barriers as they presented themselves to the human eye (previous chapter).

Another way to explore behaviour of dispersers in the landscape is by means of individual-based models (Fig. 2 D2). Movement analysis often employs random walk techniques or related diffusion modelling (Ovaskainen 2004), but several other approaches have also been taken. Recent simulation studies using virtual organisms have investigated patch reachability as a function of various landscape metrics as well as behavioural tradeoffs for dispersers (Hein *et al.* 2004; Tischendorf *et al.* 2003; Zollner and Lima 2005). Several authors have concluded that dispersal may be adequately captured by simple depictions or algorithms (King and With 2002; Zollner and Lima 2005) whereas others have pointed out that modelling outcome regarding patch accessibility or colonisation probability may heavily depend on how movement is modelled, at what grain size landscape is represented, and how realistically behaviour is implemented in the model (Jepsen *et al.* 2005). Results from data-driven individual-based models support the notion that the main challenge lies in dealing with behavioural complexity rather than the spatial structure of the landscape (Morales and Ellner 2002; Ovaskainen 2004; Whittington *et al.* 2004).

The final step in exploring of how population dynamics work in a specific landscape configuration leads to spatially explicit population models (SEPM; Fig. 2 D3). These models differ from other landscape models in that they allow modelling movements of individuals within a heterogeneous landscape and an estimation of how this movement influences (meta) population dynamics (Dunning *et al.* 1995). SEPMs are increasingly used by both researchers and managers (Dunning *et al.* 1995; Etienne *et al.* 2004). They have tremendous potential for application in species conservation and management and are able to address questions regarding population viability in fragmented landscapes, effects of landscape change or direct human impact on species distribution, whether parks can sustain populations of focal species, or how to plan reintroduction schemes, among others. Recent examples include models for space-demanding medium- to large-sized wildlife species such as raptors (Lawler and Schumaker 2004) or mammalian carnivores (Carroll *et al.* 2003, 2006; Macdonald and Rushton 2003; Wiegand *et al.* 2004a). Many of these models addressing large spatial scales use relatively coarse parameters for estimating dispersal movements. Concern has repeatedly been expressed that uncertainties due to assumptions or overt simplifications in parameterising movements or patch-specific demography may severely affect the usefulness of SEPMs for management purpose (Beissinger *et al.* 2006; Doak and Mills 1994; Wiegand *et al.* 2004b; Zollner and Lima 2005). Current progress with theoretical SEPMs has shown ways of how to mitigate a lack of empirical data in some cases (Parvinen *et al.* 2003; Wiegand *et al.* 2004b), but ultimately, there is no substitution for real data when it comes to validate these "data-hungry" models.

Most of the papers cited above make some mention of which types of field data are especially scarce, and we subscribe to most of them. Nonetheless, we would like to conclude this section with an eclectic list of aspects that are particularly data-deficient in our view.

1. *Motivation and results of emigration:* How does emigration rate relate to patch characteristics and quality (which includes population size or density in a patch), i.e. what patch parameters make some subpopulations to sources and others to sinks? To what degree does immigration enhance subpopulation persistence? Are there thresholds of self-sustaining populations (Alderman *et al.* 2005)? How important is long-distance dispersal in animals (and plants; Nathan *et al.* 2002)?
2. *Behaviour and navigation of dispersers:* How do dispersing individuals navigate in the landscape (Andreassen *et al.* 2002; Schooley and Wiens 2003), and how do they interact with landscape features or quality gradients between patches and matrix (Haynes and Cronin 2004)? How do connectivity measurements relate to dispersal rate and distances in different organisms? There is much space for better understanding individual paths across the landscape in terms of behavioural tradeoffs (Zollner and Lima 2005) or travel costs (Bélisle 2005).
3. *Measurement of multiple aspects in the same study:* Almost all papers focus on one or two major aspects such as dispersal, patch occupancy or patch demography in relation to

landscape features. For very few species have multiple aspects such as reproduction, survivorship and movement of individuals between habitat patches been measured simultaneously (Smith and Hellmann 2002). This field is wide open to future research.

4. *Data for large-bodied, vagile species are needed:* We have repeatedly pointed to the fact that much work on metapopulations has used small organisms at small spatial scales (m^2 to hectares rather than hundreds or thousands of km^2). Size matters (Lawton 1999), and results obtained from small organisms cannot easily be upscaled to yield meaningful results in population viability models of far-dispersing, large-bodied organisms. Data for such species generally remain in short supply, although these species are often of great conservation concern. For example, most existing models on large carnivores so far had to substitute inexistent data on dispersal behaviour with simple assumptions on landscape permeability.

Managing Landscape Permeability

In man-altered landscapes, connectivity can be restored. This fact has quickly been taken up by managers and conservationists, and fragmentation theory has found its way into conservation practice. Since – as we have argued earlier in this chapter – empirical evidence is rather scarce at the scale relevant to landscape management (Niemelä 2001; Simberloff *et al.* 1992), it probably was the intuitive appeal of landscape connectivity that made legislation and conservation practice overtake empirical research (Soulé and Terborgh 1999). In fact, if people ranging from grass-root conservationists to international NGO leaders have one common idea about what is needed most in conservation, it is the conviction that habitats have to be reconnected. The idea is more and more converted into action at local, regional, or even continental scales (Soulé and Terborgh 1999). Local solutions often are technical measures such as passage-ways or crossing structures to mitigate barrier effects of roads (mostly for mammals) or barrages (for fish). On a slightly larger scale, habitat patches may be linked by habitat corridors, which is the underlying idea for building networks of hedgerows in Central Europe (Schuller *et al.* 2000). At an international and continental scale, networks of protected areas are created to ensure long-term survival of populations of animal species demanding extensive spaces (Bennett 1994; Large Herbivore Initiative LHI). The European Union's «Natura 2000» program now attempts to achieve "connectedness" across different scales (Kleyer *et al.* 1996). While the contribution of technical constructions to enhance landscape permeability is firmly established in terms of passage rates of wildlife (Clevenger and Waltho 2000; McDonald and St. Clair 2004), their roles in actually maintaining long-distance migration (Berger 2004) or dispersal are less clear. Regional or even nation-wide schemes to restore wildlife migration routes, at least in the form as they are currently *en vogue* in Central Europe (Woess *et al.* 2002), are often based on expert opinion rather than solid data on wildlife movements. The same is true for habitat corridors, i.e. linear habitat structures such as hedges or forest strips that are also created for enhancing connectivity. Criticism on the basis of a possibly unfavourable cost-benefit ratio has pointed out the paucity of empirical support (Simberloff *et al.* 1992; Noss and Beier 2000), but there are now examples which tested and confirmed the functionality of large corridors for long-ranging species (Dixon *et al.* 2006).

"Ecological compensation" or set-aside schemes that come into effect for farmland in many European countries can potentially improve the landscape for dispersing organisms by creating more habitat patches, reducing inter-patch distances, and ameliorating matrix hostility. With respect to biodiversity of plants and taxa of small animals, these schemes, in general, are improvements over the former situation (van Buskirk and Willi 2004), but there is usually little concern for spatial aspects of restored habitats accounting for the needs of moving individuals. However, proper placement of restored patches can improve restoration success for focal species (Huxel and Hastings 1999), but best practice may not follow from easy rules of thumb (Schultz and Crone 2004). Thorough assessments of schemes that have been implemented for several years are still few and recent results show that success is often limited or untestable, either because schemes had no proper planning or because ecological data prior to instalment are not available (Vos *et al.* 2002). There is an urgent need for controlled, quasi-experimental set-ups in order to come up with recommendations and measures with clearly stated objectives. Success of directed management has to be assessed in order to evaluate the effectiveness of measures taken and, if necessary, to adapt future management.

Acknowledgements

Writing this chapter has been supported by the two WSL research programs "Forest – Wildlife – Landscape" and "Periurban Landscapes", while the capercaillie research projects have been made possible by additional grants from the Swiss Foundation for Scientific Research SNF (grant no. 3100–65199.01), from the Swiss Federal Office for the Environment FOEN, and from MAVA foundation. The authors are grateful for conceptual discussions with Peter Duelli and Josef Senn, and for substantial critical input by David Frank, Brett Goodwin and Eric Matthysen that helped to considerably improve an earlier version of the manuscript.

References

Alderman J., McCollin D., Hinsley S.A., Bellamy P.E., Picton P. and Crockett R. 2005. Modelling the effects of dispersal and landscape configuration on population distribution and viability in fragmented habitat. Landscape Ecology 20: 857–870.

Andreassen H.P., Stenseth N.C. and Ims R.A. 2002. Dispersal behaviour and population dynamics of vertebrates. In: Bullock J.M., Kenward R.E. and Hails R.S. (eds.), Dispersal Ecology. Blackwell Science, Malden, pp. 237–256.

Andrén H. 1994. Effects of habitat fragmentation on birds and mammals in landscapes with different proportions of suitable habitat: a review. Oikos 71: 355–366.

Baguette M. and Mennechez G. 2004. Resource and habitat patches, landscape ecology and metapopulation biology: a consensual viewpoint. Oikos 106: 399–403.

Beier P. and Noss R.F. 1998. Do habitat corridors provide connectivity? Conservation Biology 12: 1241–1252.

Beier P., Van Drielen M. and Kankam B.O. 2002. Avifaunal collapse in West African forest fragments. Conservation Biology 16: 1097–1111.

Beissinger S.R., Walters J.R., Catanzaro D.G., Smith K.G., Dunning J.B. Jr., Haig S.M., Noon B.R. and Stith B.M. 2006. Modeling approaches in avian conservation and the role of field biologists. Ornithological Monographs 59: 1–56.

Bélisle M. 2005. Measuring landscape connectivity: the challenge of behavioural landscape ecology. Ecology 86: 1988–1995.

Bélisle M., Desrochers A. and Fortin M.-J. 2001. Influence of forest cover on the movements of forest birds: a homing experiment. Ecology 82: 1893–1904.

Bender D.J. and Fahrig L. 2005. Matrix structure obscures the relationship between interpatch movement and patch size and isolation. Ecology 86: 1023–1033.

Bennett G. (ed.) 1994. Conserving Europe's Natural Heritage. Graham and Trotman, London.

Berger J. 2004. The last mile: How to sustain long-distance migration in mammals. Conservation Biology 18: 320–331.

Bollmann K., Weibel P. and Graf R.F. 2005a. An analysis of central Alpine capercaillie spring habitat at the forest stand scale. Forest Ecology and Management 215: 307–318.

Bollmann K., Graf R.F. and Suter W. 2005b. Assessing the roles of patch area, isolation, and connectivity for predicting capercaillie occurrence in the Alps. In: Pohlmeyer K. (ed.) Extended abstracts of the XXVIIth Congress of the International Union of Game Biologists. DSV-Verlag, Hamburg, pp. 61–62.

Bowne D.R. and Bowers M.A. 2004. Interpatch movements in spatially structured populations: a literature review. Landscape Ecology 19: 1–20.

Bullock J.M., Kenward R.E. and Hails R.S. 2002. Dispersal Ecology. Blackwell Science, Malden.

Burgman M.A., Lindenmayer D.B. and Elith J. 2005. Managing landscapes for conservation under uncertainty. Ecology 86: 2007–2017.

Cale P.G. 2003. The influence of social behaviour, dispersal and landscape fragmentation on population structure in a sedentary bird. Biological Conservation 109: 237–248.

Carroll C., Phillips M.K., Schumaker N.H. and Smith D.W. 2003. Impacts of landscape change on wolf restoration success: planning a reintroduction program based on static and dynamic spatial models. Conservation Biology 17: 536–548.

Carroll C., Phillips M.K., Lopez-Gonzalez C.A. and Schumaker N.H. 2006. Defining recovery goals and strategies for endangered species: the wolf as a case study. BioScience 56: 25–37.

Chardon J.P., Adriaensen F. and Matthysen E. 2003. Incorporating landscape elements into a connectivity measure: a case study for the Speckled wood butterfly (*Pararge aegeria* L.). Landscape Ecology 18: 561–573.

Clevenger A.P. and Waltho N. 2000. Factors influencing the effectiveness of wildlife underpasses in Banff National Park, Alberta, Canada. Conservation Biology 14: 47–56.

Clobert J., Danchin E., Dhondt A.A. and Nichols J.D. (eds.) 2001. Dispersal. Oxford University Press, Oxford.

Collinge S.K. and Palmer T.M. 2002. The influences of patch shape and boundary contrast on insect response to fragmentation in California grasslands. Landscape Ecology 17: 647–656.

Coulon A.F., Cosson J.F., Angibault J.M., Cargnelutti B., Galan M., Morellet N., Petit E., Aulagnier S. and Hewison A.J.M. 2004. Landscape connectivity influences gene flow in a roe deer population inhabiting a fragmented landscape: an individual-based approach. Molecular Ecology 13: 2841–2850.

Creegan H.P. and Osborne P.E. 2005. Gap-crossing decisions of woodland songbirds in Scotland: an experimental approach. Journal of Applied Ecology 42: 678–687.

Debinski D.M. and Holt R.D. 2000. A survey and overview of habitat fragmentation experiments. Conservation Biology 14: 342–355.

D'Eon R.G., Glenn S.M., Parfitt I. and Fortin J.-M. 2002. Landscape connectivity as a function of scale and organism vagility in a real forested landscape. Conservation Ecology 6, 2: 10.

Dixon J.D., Oli M.K., Wooten M.C., Eason T.H., McCown J.W. and Paetkau D. 2006. Effectiveness of a regional corridor in connecting two Florida black bear populations. Conservation Biology 20: 155–162.

Doak D.F. and Mills L.S. 1994. A useful role for theory in conservation. Ecology 75: 615–626.

Dunning J.B., Stewart D.J., Danielson B.J., Noon B.R., Root T.L., Lamberson R.H. and Stevens E.E. 1995. Spatially explicit population models – current forms and future uses. Ecological Applications 5: 3–11.

Elmhagen B. and Angerbjörn A. 2001. The applicability of metapopulation theory to large mammals. Oikos 94: 89–100.

Enoksson B., Angelstam P. and Larsson K. 1995. Deciduous forest and resident birds: the problem of fragmentation within a coniferous forest landscape. Landscape Ecology 10: 267–275.

Etienne R.S., ter Braak C.J.F. and Vos C.C. 2004. Application of stochastic patch occupancy models to real metapopulations. In: Hanski I. and Gaggiotti O.E. (eds.), Ecology, Genetics, and Evolution of Metapopulations. Elsevier Academic Press, Burlington, MA, pp. 105–132.

Fahrig L. 2003. Effects of habitat fragmentation on biodiversity. Annual Review of Ecology and Systematics 34: 487–515.

Forman R.T.T. and Alexander L.E. 1998. Roads and their major ecological effects. Annual Review of Ecology and Systematics 29: 207–231.

Forman R.T.T., Sperling D., Bissonette, J.A., Clevenfer A.T., Cutshall C.D., Dale V.H., Fahrig L., France R., Goldman C.R., Heanu K., Jones J.A., Swanson F.J., Turrentine T. and Winter T.C. 2003. Road Ecology – Science and Solutions. Island Press, Washington D.C.

Frankham R., Ballou J.D. and Briscoe D.A. 2002. Introduction to Conservation Genetics. Cambridge University Press, Cambridge.

Freckleton R.P. and Watkinson A.R. 2002. Large-scale spatial dynamics of plants: metapopulations, regional ensembles and patchy populations. Journal of Ecology 90: 419–434.

Grimm V. and Storch I. 2000. Minimum viable population size of capercaillie *Tetrao urogallus:* results from a stochastic model. Wildlife Biology 6: 219–225.

Gobeil J.-F. and Villard M.-A. 2002. Permeability of three boreal forest landscape types to bird movements as determined from experimental translocations. Oikos 98: 447–458.

Goodwin B.J. 2003. Is landscape connectivity a dependent or independent variable? Landscape Ecology 18: 687–699.

Graf R.F., Bollmann K., Suter W. and Bugmann H. 2005. The importance of spatial scale in habitat models: capercaillie in the Swiss Alps. Landscape Ecology 20: 703–717.

Graf R.F., Bollmann K., Sachot S., Suter W. and Bugmann H. 2006. On the generality of habitat distribution models: a case study of capercaillie in three Swiss regions. Ecography: 29: 319–328.

Grez A.A., Zaviezo T. and Reyes S. 2004. Short-term effects of habitat fragmentation on the abundance and species richness of beetles in experimental alfalfa microlandscapes. Revista Chilena de Historia Natural 77: 547–558.

Grimm V. and Storch I. 2000. Minimum viable population size of capercaillie *Tetrao urogallus:* results from a stochastic model. Wildlife Biology 6: 219–225.

Grubb T.C. and Doherty P.F. 1999. On home-range gap-crossing. Auk 116: 618–628.

Guisan A. and Zimmermann N.E. 2000. Predictive habitat distribution models in ecology. Ecological Modelling 135: 147–186.

Gutzwiller K. (ed.) 2002. Applying landscape ecology in biological conservation. Springer, New York, 518 pp.

Haddad N.M. and Baum K.A. 1999. An experimental test of corridor effects on butterfly densities. Ecological Applications 9: 623–633.

Haila Y. 2002. A conceptual genealogy of fragmentation research: From island biogeography to landscape ecology. Ecological Applications 12: 321–334.

Hanski I. 1999. Metapopulation Ecology. Oxford University Press, Oxford, 313 pp.

Hanski I. and Gaggiotti O.E. (eds.) 2004. Ecology, Genetics, and Evolution of Metapopulations. Elsevier Academic Press, Burlington, MA, 696 pp.

Hanski I. and Gilpin M. eds. 1997. Metapopulation biology. Academic Press, London, 512 pp.

Hansson L., Fahrig L. and Merriam G. (eds.) 1995. Mosaic landscapes and ecological processes. Chapman and Hall, London.

Harris R.J. and Reed J.M. 2002. Behavioral barriers to non-migratory movements of birds. Annales Zoologici Fennici 39: 275–290.

Haynes K.J. and Cronin J.T. 2003. Matrix composition affects the spatial ecology of a prairie planthopper. Ecology 84: 2856–2866.

Haynes K.J. and Cronin J.T. 2004. Confounding of patch quality and matrix effects in herbivore movement studies. Landscape Ecology 19: 119–124.

Hein S., Gombert J., Hovestadt T. and Poethke H.-J. 2003. Movement patterns of the bush cricket *Platycleis albopunctata* in different types of habitat: matrix is not always matrix. Ecological Entomology 28: 432–438.

Hein S., Pfenning B., Hovestadt T. and Poethke H.-J. 2004. Patch density, movement pattern, and realised dispersal distances in a patch-matrix landscape – a simulation study. Ecological Modelling 174: 411–420.

Holderegger R., Gugerli F., Scheidegger C. and Taberlet P. 2007. Integrating population genetics with landscape ecology to infer spatio-temporal processes. In: Kienast F., Wildi O. and Ghosh S. (eds.). A Changing World. Challenges for Landscape Research, Vol. 8: 145–156. Springer Landscape Series, Dordrecht.

Holmquist J.G. 1998. Permeability of patch boundaries to benthic invertebrates: influences of boundary contrasts, light level, and faunal density and mobility. Oikos 81: 558–566.

Hudgens B.R. and Haddad N.M. 2003. Predicting which species will benefit from corridors in fragmented landscapes from population growth models. The American Naturalist 161: 808–820.

Hunter R.D., Fisher R.N. and Crooks K.R. 2003. Landscape-level connectivity in coastal southern California, U.S.A, as assessed through carnivore habitat suitability. Natural Areas Journal 23: 302–314.

Huxel G.R. and Hastings A. 1999. Habitat loss, fragmentation, and restoration. Restoration Ecology 7: 309–315.

Jaeger J.A.G. 2000. Landscape division, splitting index, and effective mesh size: new measures of landscape fragmentation. Landscape Ecology 15: 115–130.

Jepsen J.U., Baveco J.M., Topping C.J., Verboom J. and Vos C.C. 2005. Evaluating the effect of corridors and landscape heterogeneity on dispersal probability: a comparison of three spatially explicit modelling approaches. Ecological Modelling 181: 445–459.

King A.W. and With K.A. 2002. Dispersal success on spatially structured landscapes: when do spatial pattern and dispersal behaviour really matter? Ecological Modelling 147: 23–39.

Klaus S. and Bergmann H.-H. 1994. Distribution, status and limiting factors of capercaillie in central Europe, particularly in Germany, including an evaluation of reintroductions. Gibier Faune Sauvage 11: 57–80.

Kleyer M., Kaule G. and Settele J. 1996. Landscape fragmentation and landscape planning with focus on Germany. In: Settele J., Margules C. R., Poschlod P. and Henle K. (eds.), Species Survival in Fragmented Landscapes. Kluwer Academic Publishers, Dordrecht, pp. 138–151.

Krawchuk M.A. and Taylor P.D. 2003. Changing importance of habitat structure across multiple spatial scales for three species of insects. Oikos 103: 153–161.

Kurki S., Nikula A., Helle P. and Linden H. 2000. Landscape fragmentation and forest composition effects on grouse breeding success in boreal forests. Ecology 81: 1985–1997.

Lahti D.C. 2001. The "edge effect on nest predation" hypothesis after twenty years. Biological Conservation 99: 365–374.

Lande R. 1999. Extinction risks from anthropogenic, ecological, and genetic factors. In: Landweber L.F. and Dobson P. (eds.), Genetics and the Extinction of Species. Princeton University Press, Princeton, pp. 1–22.

Laurance W.F., Lovejoy T.E., Vasconcelos H.L., Bruna E.M., Didham R.K., Stouffer P.C., Gascon C., Bierregaard R.O., Laurance S.G. and Sampaio E. 2002. Ecosystem decay of Amazonian forest fragments: A 22-year investigation. Conservation Biology 16: 605–618.

Lawler J.L. and Schumaker N.H. 2004. Evaluating habitat as a surrogate for population viability using a spatially explicit population model. Environmental Monitoring and Assessment 94: 85–100.

Lawton J. 1999. Size matters. Oikos 85: 19–21.

Li H. and Wu J. 2004. Use and misuse of landscape indices. Landscape Ecology 19: 389–399.

Lidicker W.Z. 2002. From dispersal to landscapes: progress in the understanding of population dynamics. Acta Theriologica 47, Suppl. 1: 23–37.

Lindström J., Ranta E. and Linden H. 1996. Large-scale synchrony in the population dynamics of capercaillie, black grouse and hazel grouse populations in Finland. Oikos 76: 221–227.

Macdonald D.W. and Rushton S. 2003. Modelling space use and dispersal of mammals in real landscapes: a tool for conservation. Journal of Biogeography 30: 607–620.

Manel S., Schwartz M.K., Luikart G. and Taberlet P. 2003. Landscape genetics: combining landscape ecology and population genetics. Trends in Ecology and Evolution 18: 189–197.

Marulli J. and Mallarach J.M. 2005. A GIS methodology for assessing ecological connectivity: application to the Barcelona Metropolitan Area. Landscape and Planning 71: 243–262.

McCullough D.R. 1996. Metapopulations and Wildlife Conservation. Island Press, Washington, 429 pp.

McDonald W. and St. Clair C.C. 2004. Elements that promote highway crossing structure use by small mammals in Banff National Park. Journal of Applied Ecology 41: 82–93.

McGarigal K. and Cushman S.A. 2002. Comparative evaluation of experimental approaches to the study of habitat fragmentation effects. Ecological Applications 12: 335–345.

Mech L.D., Fritts S.H., Radde G.L. and Paul W.J. 1988. Wolf distribution and road density in Minnesota. Wildlife Society Bulletin 16: 85–87.

Mollet P., Badilatti B., Bollmann K., Graf R.F., Hess R., Jenny H., Mulhauser B., Perrenoud A., Rudmann F., Sachot S. and Studer J. 2003. Verbreitung und Bestand des Auerhuhns *Tetrao urogallus* in der Schweiz 2001 und ihre Veränderungen im 19. und 20. Jahrhundert. Der Ornithologische Beobachter 100: 67–86.

Morales J.M. and Ellner S.P. 2002. Scaling up animal movements in heterogeneous landscapes: the importance of behaviour. Ecology 83: 2240–2247.

Nathan R., Katul G.G., Horn H.S., Thomas S.M., Oren R., Avissar R., Pacala S.W. and Levin S.A. 2002. Mechanisms of long-distance dispersal of seeds by wind. Nature 418: 409–413.

Nellemann C., Vistnes I., Jordhöy, P. and Strand O. 2001. Winter distribution of wild reindeer in relation to power lines, roads and resorts. Biological Conservation 101: 351–360.

Niemelä, J. 2001. The utility of movement corridors in forested landscapes. Scandinavian Journal of Forest Research Suppl. 3: 70–78.

Noss R.F. and Beier P. 2000. Arguing over little things: response to Haddad *et al.* Conservation Biology 14: 1546–1548.

Opdam P., Verboom J. and Pouwels R. 2003. Landscape cohesion: an index for the conservation potential of landscapes for biodiversity. Landscape Ecology 18: 113–126.

Ovaskainen O. 2004. Habitat-specific movement parameters estimated using mark-recapture data and a diffusion model. Ecology 85: 242–257.

Parvinen K., Dieckmann U., Gyllenberg M. and Metz J.A.J. 2003. Evolution of dispersal in metapopulations with local density dependence and demographic stochasticity. Journal of Evolutionary Biology 16: 143–153.

Perault D.R. and Lomolino M.V. 2000. Corridors and mammal community structure across a fragmented, old-growth forest landscape. Ecological Monographs 70: 401–422.

Reed J.M., Mills L.S., Dunning J.B., *et al.* 2002. Emerging issues in population viability analysis. Conservation Biology 16: 7–19.

Ricketts T.H. 2001. The matrix matters: Effective isolation in fragmented landscapes. The American Naturalist 158: 87–99.

Saccheri I., Kuussaari M., Kankare M., Vikman P., Fortelius W. and Hanski I. 1998. Inbreeding and extinction in a butterfly metapopulation. Nature 392: 491–494.

Saunders D.A. and Hobbs R.J. (eds.) 1991. Nature Conservation 2: The Role of Corridors. Surrey Beatty, Chipping Norton.

Schooley R.L. and Wiens J.A. 2003. Finding habitat patches and directional connectivity. Oikos 102: 559–570.

Schuller D., Brunken-Winkler H., Busch P., Forster M., Janiesch P., von Lemm R., Niedringhaus R. and Strasser H. 2000. Sustainable land use in an agriculturally misused landscape in northwest Germany through ecotechnical restoration by a 'Patch-Network-Concept'. Ecological Engineering 16: 99–117.

Schultz C.B. and Crone E.E. 2005. Patch size and connectivity thresholds for butterfly habitat restoration. Conservation Biology 19: 887–896.

Segelbacher G. and Storch I. 2002. Capercaillie in the Alps: genetic evidence of metapopulation structure and population decline. Molecular Ecology 11: 1669–1677.

Shreeve T.G., Dennis L.H. and Van Dyck H. 2004. Resources, habitats and metapopulations – wither reality? Oikos 106: 404–408.

Simberloff D., Farr J.A., Cox, J. and Mehlman D.W. 1992. Movement corridors – conservation bargains or poor investments? Conservation Biology 6: 493–504.

Smith J.N.M. and Hellmann J.J. 2002. Population persistence in fragmented landscapes. Trends in Ecology and Evolution 17: 397–399.

Soulé M.E. and Terborgh J. 1999. Continental conservation: Scientific foundations of regional reserve networks. Island Press, Washington, D.C., 238 pp.

Storch I. 1997. The importance of scale in habitat conservation for an endangered species: the capercaillie in Central Europe. In: Bissonette J.A. (ed.) Wildlife and Landscape Ecology: Effects of Pattern and Scale. Springer-Verlag, New York, pp. 310–330.

Storch I. (ed.) 2000. Grouse: Status Survey and Conservation Action Plan 2000–2004. IUCN, Gland.

Storch I. 2002. On spatial resolution in habitat models: can small-scale forest structure explain Capercaillie numbers? Conservation Ecology 6(1): 6 [online] URL: http://www.consecol.org/vol6/iss1/art6.

Storch I. 2003. Linking a multiscale habitat concept to species conservation. In: Bissonette J.A. and Storch I. (eds.). Landscape Ecology and Resource Management: Linking Theory with Practice. Island Press, Washington DC, pp. 302–320.

Storch I. and Segelbacher G. 2000. Genetic correlates of spatial population structure in central European capercaillie *Tetrao urogallus* and black grouse *T. tetrix*: a project in progress. Wildlife Biology 6: 305–310.

Suter W., Graf R.F. and Hess R. 2002. Capercaillie (Tetrao urogallus) and avian biodiversity: Testing the umbrella-species concept. Conservation Biology 16: 778–788.

Sutherland W.J. 1996. From Individual Behaviour to Population Ecology. Oxford University Press, Oxford, 213 pp.

Taylor P.D., Fahrig L., Henein K. and Merriam G. 1993. Connectivity is a vital element of landscape structure. Oikos 68: 571–573.

Tischendorf L. and Fahrig L. 2000. On the usage and measurement of landscape connectivity. Oikos 90: 7–19.

Tischendorf L., Bender D.J. and Fahrig L. 2003. Evaluation of patch isolation metrics in mosaic landscapes for specialist vs. generalist dispersers. Landscape Ecology 18: 41–50.

Turner M.G., Gardner R.H. and O'Neill R.V. 2001. Landscape Ecology in Theory and Practice. Springer, New York.

Urban D. 2005. Modeling ecological processes across scales. Ecology 86: 1996–2006.

van Buskirk J. and Willi Y. 2004. Enhancement of farmland biodiversity within set-aside land. Conservation Biology 18: 987–994.

Verbeylen G., De Bruyn L., Adriaensen F. and Matthysen E. 2003. Does matrix resistance influence Red squirrel (*Sciurus vulgaris* L. 1758) distribution in an urban landscape? Landscape Ecology 18: 791–805.

Villard M.A. 2002. Habitat fragmentation: Major conservation issue or intellectual attractor? Ecological Applications 12: 319–320.

Vos C.C., Baveco H. and Grashof-Bokdam C.J. 2002. Corridors and species dispersal. In: Gutzwiller K.J. (ed.). Applying Landscape Ecology in Biological Conservation. Springer, New York, pp. 84–104.

Wiegand T., Knauer F., Kaczensky P. and Naves J. 2004a. Expansion of brown bears *(Ursus arctos)* into the eastern Alps: a spatially explicit population model. Biodiversity and Conservation 13: 79–114.

Wiegand T., Revilla E. and Knauer F. 2004b. Dealing with uncertainty in spatially explicit population models. Biodiversity and Conservation 13: 53–78.

Whittington J., Cassady St. Clair C. and Mercer G. 2004. Path tortuosity and the permeability of roads and trails to wolf movement. Ecology and Society 9(1): 4. [online] URL: http://www.ecologyandsociety.org/vol9/iss1/art4.

With K.A. 2002. Using percolation theory to assess landscape connectivity and effects of habitat fragmentation. In: Gutzwiller K.J. (ed.). Applying Landscape Ecology in Biological Conservation. Springer, New York, pp. 105–130.

With K.A. 2004. Metapopulation dynamics: perspectives from landscape ecology. In: Hanski I. and Gaggiotti O.E. (eds.). Ecology, genetics, and evolution of metapopulations, Elsevier, Amsterdam pp. 23–44.

Woess M., Grillmayer R. and Voelk F.H. 2002. Green bridges and wildlife corridors in Austria. Zeitschrift für Jagdwissenschaften 48, Suppl.: 25–32.

Wu J. and Hobbs R. 2002. Key issues and research priorities in landscape ecology: An idiosyncratic synthesis. Landscape Ecology 17: 355–365.

Young A.G. and Clarke G.M. (eds.) 2000. Genetics, Demography and Viability of Fragmented Populations. Cambridge University Press, Cambridge, 456 pp.

Zollner P.A. and Lima S.L. 2005. Behavioral tradeoffs when dispersing across a patchy landscape. Oikos 108: 219–230.

Zschokke S., Dolt C., Rusterholz H.P., Oggier P., Braschler B., Thommen G.H., Lüdin E., Erhardt A. and Baur B. 2000. Short-term responses of plants and invertebrates to experimental small-scale grassland fragmentation. Oecologia 125: 559–572.

Spatial Pattern Recognition,
Time Series Analysis and Dynamic Modeling

Identifying and Quantifying Landscape Patterns in Space and Time

Janine Bolliger[1*], Helene H. Wagner[1] and Monica G. Turner[2]

[1] Swiss Federal Research Institute WSL, Zürcherstrasse 111, CH-8903 Birmensdorf, Switzerland
[2] Department of Zoology, University of Wisconsin-Madison, Madison, Wisconsin 53706, U.S.A
[*] Corresponding author: e-mail: janine.bolliger@wsl.ch

Abstract
In landscape ecology, approaches to identify and quantify landscape patterns are well developed for discrete landscape representations. Discretisation is often seen as a form of generalisation and simplification. Landscape patterns however are shaped by complex dynamic processes acting at various spatial and temporal scales. Thus, standard landscape metrics that quantify static, discrete overall landscape pattern or individual patch properties may not suffice when viewing landscapes as gradients or when quantifying spatially dynamic response surfaces resulting from model simulations of landscapes. The spatio-temporal dynamics of patterns can be quantified using various approaches originating in different fields and ranging from geography, geology, engineering, physics, plant community ecology and complex systems theory. This book chapter provides an overview on quantitative measures that may be used as indicators to assess landscape patterns in space and time for discrete and continuous landscape representations and discusses promising avenues for addressing the most pressing needs for spatial analysis of gradient-dominated and dynamic landscapes.

Keywords: landscape analysis, landscape metrics, landscape pattern, spatial pattern, temporal pattern

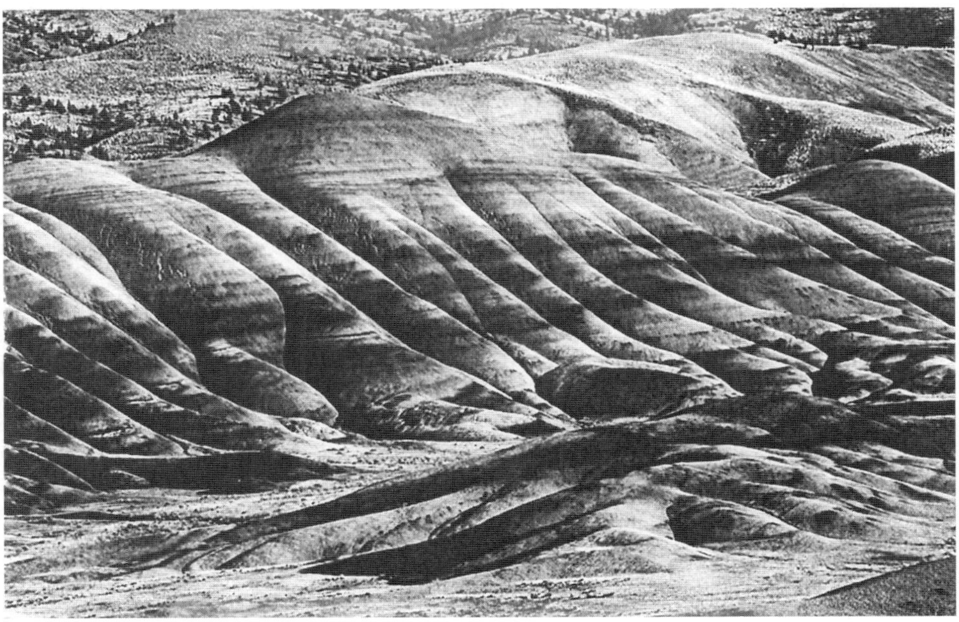

F. Kienast, O. Wildi & S. Ghosh (eds.), A Changing World. Challenges for Landscape Research, 177–194.

Landscape Patterns

A landscape may be defined as an area that is spatially heterogeneous in at least one factor of interest (Turner *et al.* 2001). This minimum definition stresses heterogeneity as a key concept. In fact, most ecological systems are heterogeneous: environmental factors vary in space and time, and most species are unevenly distributed. Landscape ecology focuses on quantifying heterogeneity and investigating its causes and ecological consequences across ranges of scales (Turner 2005). When dealing with heterogeneity however it is important to discriminate between different types of heterogeneity, to recognise its sources and to consider scale (Levin 1978, 1992; Wiens 2000).

Types of landscape patterns

An ecological system or system property of interest may be heterogeneous in space and in time. In landscape ecology, spatial heterogeneity is usually referred to as landscape pattern or landscape structure, whereas landscape dynamics generally refers to changes in landscape patterns through time. To date, landscape structure has received more attention from landscape ecologists than landscape dynamics (Gustafson 1998).

Both landscape structure and landscape dynamics can be studied using discrete or continuous models of heterogeneity. Many approaches assume that the landscape consists of discrete, non-overlapping objects or patches that belong to mutually exclusive classes or system states. Patches are either embedded in an assumedly homogeneous matrix or form a mosaic (Forman and Godron 1986; Forman 1995). Such discrete landscape representations are widespread and are helpful to simplify and quantify complex landscapes. Discrete landscape representations have been extremely successful in a wide range of landscape ecological topics such as habitat fragmentation (Hargis *et al.* 1998), landscape descriptions (Haines-Young and Chopping 1996), monitoring of landscape change (Lausch and Herzog 2002) or conservation (Thompson and McGarigal 2002). It has been argued however that many natural phenomena may be primarily continuous in character and exist as gradients rather than as features with discrete boundaries (Regan *et al.* 2000; Bolliger and Mladenoff 2005; McGarigal and Cushman 2005). Contrary to discrete boundaries, gradients describe gradual transitions of feature properties. Gradient-based concepts of representing landscape patterns may be more appropriate for many landscape properties than the spatially discrete patch-mosaic concept (Bolliger and Mladenoff 2005; McGarigal and Cushman 2005) as classification of spatially continuous features into discrete units may result in information loss. However, whether a phenomenon appears as relatively discrete or continuous will often depend on the scale of the study, especially the spatial resolution or grain, the measurement resolution and the hierarchical scale (Gosz 1993; McGarigal and Cushman 2005).

Conceptual discussions about whether natural features are of discrete or continuous nature are not new, and date back to Gleason's and Clement's discourses during the first half of the 20th century (Keller and Golley 2000). Methodologically, the continuity of natural features can be assessed using fuzzy logic. The method was originally introduced by Zadeh (1965) and developed further (Bezdek 1981; McBratney and Odeh 1997; Minasny and McBratney 2002). In this approach each unit (e.g., grid cells) may be assigned one or more types and its degree of belonging to each type is expressed as a membership function. Partial memberships are quantified for each type, indicating that some types are compositionally distinct whereas others may share common characteristics. Thus, the membership information may serve as baseline information to assess structural uncertainties, i.e., the degree of certainty with which landscape patterns can be discretised (Brown 1998; Bolliger and Mladenoff 2005).

Pattern-generating processes

Landscape patterns result from various interacting processes (Levin 1978; Forman and Godron 1986; Turner *et al.* 2001; Turner 2005). When modelling landscape dynamics, it is essential to distinguish between biotic and abiotic processes. While both types of processes affect the behaviour of a system, abiotic processes are assumed to depend only on external drivers, whereas biotic processes may themselves be affected by the system (Lischke *et al.* 2007). In empirical studies however the distinction between biotic and abiotic processes is often implicit.

Abiotic processes typically affect landscape patterns via interactions with externally imposed, oftentimes environmental factors, such as climate, soil, topography, or disturbance (fire, wind). For example, climatic parameters play a significant role for spatial distributions of trees. Studies assessing scenarios of elevated temperatures along an altitudinal gradient in the Swiss Alps show that major reorganisation in forests including species shifts are expected along an altitudinal gradient (Kienast *et al.* 1998; Bolliger *et al.* 2000; Bolliger 2002).

Biotic processes include e.g., dispersal or competition between or within species, but also disturbance (insect outbreak). Dispersal and population spread have been of continuing interest (Levin 1979; Webb 1987; Clark 1998; Clark *et al.* 1999), and dispersal in particular has proven to be important in structuring e.g., forest community composition (Jacquemin *et al.* 2001), or governing migration rates (Clark 1998). Biotic processes shape landscape patterns through internal interactions between the individual system components, e.g., organisms that may cause organisation by accumulation of small changes (Bak *et al.* 1987; Bak 1996), generating patchiness even in the absence of environmental heterogeneity.

Processes of both biotic and abiotic nature include disturbance and human impact. Disturbance may be considered abiotic in cases of e.g., fire or wind or biotic if disturbance is caused by e.g., pathogens. Whether human impact is viewed as biotic or abiotic depends on perspective and judgement. If effects of humans on the landscape are perceived as part of an ecosystem, they may be referred to as biotic. If human impacts are perceived as external drivers shaping the ecosystem, they may be referred to as abiotic. Human impact is most important in shaping landscapes and it includes various socio-economic policy- and management driven processes (Bürgi *et al.* 2007).

In most natural systems both biotic and abiotic processes are relevant for shaping landscape patterns. However, system interactions in environmentally extreme habitats e.g., desert, arctic, high/low pH values in soils) are likely to be primarily driven abiotically. Environments with little temporal heterogeneity (e.g., rain forests) are more likely to be dominated by biotic processes (Solé and Manrubia 1995; Solé *et al.* 2002).

As landscapes are the result of interacting biotic and abiotic processes, they pose great challenges for the quantitative assessment of primary processes shaping the patterns. Since the same process may produce many different patterns, two landscape patterns will rarely be identical, making statistical comparison between different landscapes or between different time steps difficult (Fortin *et al.* 2003), and methodological problems arise from the combined effects of several biotic and abiotic processes (Wagner and Fortin 2005).

Process interactions and scale

Processes may interact linearly (unidirectionally), or in nonlinear ways, including mutual, self-reinforcing (positive feedback), and self-inhibiting interactions (negative feedback). Nonlinear interactions may be a primary source of patterns in many systems (Bascompte and Solé 1995; Farkas *et al.* 2002; Green and Sadedin 2005).

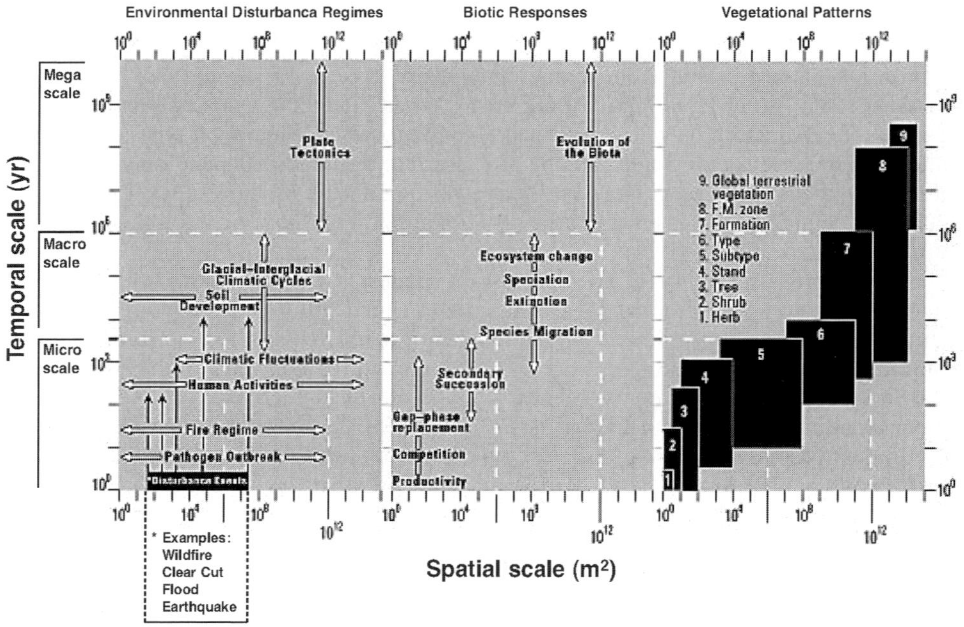

Fig. 1. Relationships between spatio-temporal scales and levels of organization. (Figure redrawn from Turner *et al.* (2001), modified from Delcourt *et al.* (1983)).

Pattern and processes operate on broad ranges of spatial and temporal scales and their characteristics are associated with scale (Levin 1978, 1992; Wu and Loucks 1995). Thus, effects of processes on pattern need to be considered with their characteristic temporal and spatial scale (Levin 1978, 1992; Wu and Loucks 1995). The space-time scaling of many biotic or abiotic processes is such that as a rule, fast processes operate on smaller spatial scales than slow processes (Fig. 1). In order to achieve high predictability, the spatial and temporal scales of a study need to match those of the process (Wiens 1989). For instance, the scale of interest for a research project investigating processes in a grassland that affect individuals of a particular insect species is defined by the organism's home range, and the time scale is identified by its life-history attributes (e.g., reproduction rate, mortality rate; Addicott *et al.* 1987; Wiens 1989).

If one process operates at a much slower rate than the other, an essentially non-linear interaction may appear unidirectional. The faster process is constrained by the quasi-static pattern created by the slower process. For instance, landscape ecological studies often assume the habitat mosaic to be constant in order to study its effect on the movement of organisms, population dynamics, or species distributions.

Indicators

Indicators are qualitative descriptors or quantitative measures that report key information to assess structure, function, or composition of a system (Dale 2001) with the most efficient use of available resources. They identify a system based on selected criteria that are monitored through time or space to inform about the system's state or condition (static), its changes, or trends (dynamic) (Dale 2001). Indicators facilitate decisions about whether intervention is desirable or necessary, which interventions might yield the best results, or how successful interventions have been. Hall and Grinnell (1919) were among the first to use the indicator concept by attributing animal species to specific life zones that are identified by geographical areas with comparable structure and composition.

Indicators are currently widely applied in many areas of research, environmental management, policy and decision making ranging from, for example, environmental pollution (Mal *et al.* 2002) to ecosystem integrity (Grove 2002). Here, we refer to indicators as quantitative measures derived from various methodologies in order to quantify and assess aspects of landscape and landscape-pattern properties.

Ecological indicators: Various types of indicators serve different applications. Ecological indicators, for example, are typically very specific for particular environments or taxonomic groups at specific locations (Fig. 2). They usually rely on expert knowledge and are derived from field observations. Ecological indicators encompass, for example Ellenberg's values (Ellenberg 1988) that benchmark details on specific plant species requirements (e.g., light thresholds), individual species, or communities. They are used to estimate species richness (Duelli and Obrist 1998, 2003), monitor land-use change (Cousins and Lindborg 2002), or assess the influence of disturbance and management (Dale *et al.* 2002a). Other examples of ecological indicators, in particular with respect to biodiversity, are discussed in Duelli *et al.* (2007). Advantages of ecological indicators thus include very specific information of a particular species or population at a particular location (Fig. 2), and are indicative of a process or response that may be too costly or difficult to measure directly. However, the information derived from ecological indicators does not necessarily allow up-scaling or generalisation to larger spatial or temporal scales.

Landscape indicators: Indicators that characterise properties at the landscape scale however supplement ecological indicators by providing information about, e.g., the amount and spatial arrangement of different land-cover types (Jones *et al.* 2000; Gergel *et al.* 2002; Wade *et al.* 2003), or environmental quality (Forman and Alexander 1998). Indicators for landscapes inform on the state of relevant landscape properties or their changes. Since landscapes are higher-level aggregations of individual patch properties, the information quantified by landscape indicators is more general. Indicators at the landscape scale can be derived

Fig. 2. Ecological indicators and landscape indicators.

from GIS databases, aerial photographs, or remotely sensed images. Advantages thus include that they may be much cheaper and more easily obtained for large areas. Also, they usually rely on standardised approaches and thus allow generalisation at larger spatial and temporal scales. Details on the systems however are usually not accounted for.

A broad variety of landscape characteristics can be quantified with indicators. Primary goals of landscape indicators are to quantify the amount and spatial arrangement of land cover, and to ensure comparability between different landscapes. However, the purpose of quantifying landscape patterns is to capture key features that matter, thus depending on the research question, objectives, allowable errors, and data availability.

Early landscape characterisations relied on two indicators: typology refers to the number of elements, and chorology measures the number of landscape elements or habitat types (Snacken and Antrop 1983). Today, landscapes are commonly described by composition and configuration (Gustafson 1998; Turner *et al.* 2001). Landscape composition can be assessed using indicators such as percent area. Configuration refers to the spatial arrangement of the individual elements. The overall spatial organisation of elements identifies how the landscape types are arranged in relation to each other, involving indicators derived from information theory (O'Neill *et al.* 1988), fractal geometry (Milne 1988; Milne *et al.* 1992; With 1994), or percolation theory (O'Neill *et al.* 1988; Gustafson and Parker 1992; Johnson and Milne 1992; Milne 1998; Wickham *et al.* 1999). Additionally, it has been stressed that the degree and type of interactions (connectivity) between the landscape elements play an important role in shaping ecological systems and landscapes (Taylor *et al.* 1993; With 1997; Bolliger *et al.* 2003; Bolliger 2005; Green and Sadedin 2005).

Landscape Indicators

Landscapes can be characterised discretely or continuously in space and time. Static landscape descriptions are conducted at particular time steps ("snapshots"), whereas dynamic characterisations are performed continuously through time ("movies").

Static, snapshot-like landscape descriptions usually rely on empirical observations (e.g., field measurements, GIS databases, aerial photographs) that represent landscapes at specific time-steps. The state and conditions of smaller-scale systems (e.g., dry meadow) can be reported or monitored through regular visits, e.g., weekly, annually). Empirical observations may provide information that dates back a few years or decades. For larger-scale systems such as landscapes, the system states can be monitored through time series obtained e.g., from temporal series of GIS data, aerial photographs, or remotely sensed images. As a rule however long-term effects of environmental change in the future or the past cannot be assessed with empirical data only. For dynamic, movie-type landscape assessments, models provide a suitable tool to improve the understanding of observed system functions, patterns, or diversity, and to assess consequences of changes of individual system components or of the environment on particular system properties. Models thus allow evaluations of alternative scenarios and help generate hypotheses for states and conditions of systems not only under current, but also under changing future or past conditions (e.g., temperature change, management change) (Lischke *et al.* 2007). Thus, indicators for model simulations inform on the system's state or condition under various scenarios of environmental changes.

In the following sections we present quantitative indicators relying on various methodological approaches that can be used to characterise landscapes or landscape-pattern properties. The indicators include measures to characterise particular periods (static) or time series (dynamic) based on discrete or continuous landscape representations (Fig. 3).

	Discrete (space)	Continuous (space)
Dynamic (time)	Transition matrix + * Markov chains + *	Spatio-temporal entropy + Diagnostics from non-linear dynamics + Power spectra + *
Static (time)	Landscape metrics + * Fractal dimension, lacunarity + * Algorithmic complexity + *	Surface metrology indices + * Fractal dimension, lacunarity + * Geostatistics + * Wavelet + *

Fig. 3. Quantitative landscape indicators used to characterise landscapes in space and time. Indicators representing landscapes based on empirical (field or GIS) data are labelled with *, whereas + refer to indicators that characterise systems using model simulations.

Static, discrete landscape indicators

Algorithmic complexity: Computers use specific algorithms for the compression of graphic files or images, e.g., of maps representing landscapes. The size of an optimally compressed file or image can be interpreted as the condensation of the entire set of interactions between the digital components of the image or file, e.g., a landscape. The file or image size can thus be interpreted as a measure of landscape complexity based on compression algorithms (Kaspar and Schuster 1987; Manson 2001; Sprott *et al.* 2002; Bolliger *et al.* 2003). Algorithmic complexity is easy to calculate and provides comparability between different images representing, e.g., the same landscape at different times or the results of different simulation runs. The metric however does not provide details on which landscape element contributes more or less to the size of the file, and it is a relative measure.

Landscape metrics: Landscape metrics statistically represent landscapes or individual patch-type properties and are standard tools to analyse questions regarding the composition and configuration of landscapes and individual patches (Turner *et al.* 2001; Cardille and Turner 2002; McGarigal *et al.* 2002).

Metrics for landscape composition identify and describe the landscape pattern, whereas landscape configuration refers to their spatial arrangement of the landscape elements. Landscape composition is assessed using metrics such as landscape diversity (Shannon-Weaver diversity), or the proportion of area occupied by habitat types (Turner *et al.* 2001). Metrics for landscape configuration involve e.g., probabilities of patch adjacency, patch shape, or connectivity between patches (Turner *et al.* 2001). Such metrics are of great value to investigations dealing with habitat fragmentation, where patch isolation may cause extinction of entire populations because dispersal or colonisation rates are reduced (With 1997, 2002), or where disaggregation of landscapes may foster persistence of populations in reducing the probability of some disturbances such as fire (Franklin and Forman 1987). Patch size and shape may influence a variety of ecological properties, e.g., flows between patches in animal foraging strategies (Zollner and Lima 1997). The shape characteristics can be directly related to the overall heterogeneity of the landscape, whereas the area of an individual patch is of great ecological relevance in that it determines the space to support viable populations.

Advantages of landscape metrics include that many metrics are easily calculated and widely known to landscape ecologists (e.g., FRAGSTATS, McGarigal *et al.* 2002). However, the ecological relevance of the broad range of available metrics may be difficult to asses and may lead to misleading conclusions if not analysed carefully regarding concept and limitations (Li and Wu 2004). For example, it has been shown that some metrics, though calculated differently, are highly correlated (Riitters *et al.* 1995; Gustafson 1998; Turner *et al.* 2001; Neel *et al.* 2004). Other studies have shown that the comparability of metrics across scales may be problematic (Wu *et al.* 2002), so that different conclusions are drawn on the ecology depending on the scale of the study (Turner *et al.* 2001; Greenberg *et al.* 2002; Thompson and McGarigal 2002; Li and Wu 2004). Furthermore, it has been stressed that current methods to quantify landscape properties are more advanced in comparison to our ability to interpret the landscape properties with respect to ecologically relevant processes (Turner *et al.* 2001). Thus, the search for relationships between patterns and processes requires careful evaluation (Turner *et al.* 2001; Li and Wu 2004), especially since we are currently lacking thorough understanding of the required degree of landscape change to provoke ecologically relevant implications (Turner *et al.* 2001; Wu and Hobbs 2002).

Static, continuous landscape indicators

Indicators derived from geostatistics: Geostatistics is a method to quantify continuous surfaces (e.g., landscapes) and to assess the degree and extent of spatial autocorrelation. Spatial autocorrelation is an indicator that measures the common phenomenon that nearby observations tend to be more similar than distant ones. The distance at which spatial autocorrelation levels off indicates the spatial scale of organisation in a system, e.g., the size of an ecological neighbourhood, so that observations beyond this distance may be considered as ecologically and statistically uncorrelated. Positive spatial autocorrelation is assumed to result from a spatial process, e.g., dispersal. The major approaches to quantify spatial autocorrelation differ in their practical objectives: geostatistical methods focus on the estimation of the spatial covariance structure of a variable (e.g., variogram modelling) in order to estimate population parameters from spatially dependent observations (block kriging) or to interpolate values at unobserved locations (e.g., kriging). Spatial statistics developed in geography on the other hand, aim at testing for the presence of a spatial process in order to model this process or to account for spatial autocorrelation when assessing the correlation between spatially structured variables (Cliff and Ord 1981; Fortin *et al.* 2001; Liebhold and Gurevitch 2002).

However, all these methods require an assumption of stationarity, i.e., the spatial autocorrelation structure must be the same throughout the study area. It is often sufficient to assume weak or second-order stationarity, where the mean is constant, the autocorrelation depends only on the geographic distance between sampling units, and the variance is finite and constant (Burrough 1995). Local spatial statistics can be calculated within a moving-window, thus assuming stationarity only within the extent of the window, in order to identify anomalous subareas or delimit boundaries (Boots 2002; Pearson 2002).

Variogram modelling is widely used for assessing the spatial structure of a continuous variable (Isaaks and Srivastava 1989; Haining 1997). An empirical variogram is a plot of the semivariance between pairs of observations, averaged for each distance class, against geographic distance. The plot can be interpreted visually to assess how the variance of the variable changes with distance and, possibly, direction in space. This spatial covariance structure can then be summarised by fitting a theoretical variogram model. More complicated cases require the fitting of model parameters as a function of direction or the combination of several basic variogram functions, and the spatial cross-correlation between pairs of variables can be modelled in a similar way to allow for multivariate analysis (Wackernagel 1998).

Any variance can be partitioned by the distance between observations (Wagner 2004) so that, e.g., the results of multivariate analysis such as ordination can be plotted in variogram form. This allows additional insights into the spatial structure due to different processes, e.g., the spatial structure of a community that is induced by a specific environmental factor or the scale of patchiness of the residual variance after accounting for environmental heterogeneity (Wagner 2004; Wagner and Fortin 2005).

Indicators derived from spectral analysis: Spectral analysis is probably among the best known methods to characterise temporal data. When analysing time-series, spectral analysis using Fourier transform subdivides the series into individual sine and cosine waves, assuming an overlay of periodic structures with different wave lengths (frequencies) and amplitudes. To examine the spectrum, the logarithm of the squared amplitude is plotted against the logarithm of the frequency. The analysis of spatial surfaces relies on the periodogram, where a measure of variance (instead of amplitude) is plotted against distance (instead of frequency) (Dale 2000).

Results from spectral analysis can be graphically displayed by periodograms. These can be used as indicators to assess whether any particular time or space scale is singled out. For example, if the log-log spectrum exhibits straight lines (power laws), no particular spatial or temporal scale is singled out, and the properties of a given frequency or distance stand for all frequencies or distances. The phenomenon is referred to as scale invariance and has been observed throughout a broad range of natural phenomena and research fields including earthquakes (Ceva 1998), avalanches in sand piles (Bak *et al.* 1987; Bak 1996), chemical reactions (Simoyi *et al.* 1982), population dynamics (Solé *et al.* 1993; Perry 1995), landscape pattern (Milne 1998; Bolliger *et al.* 2003; Bolliger 2005), or evolutionary ecology (Solé *et al.* 1999).

Applied to landscape ecology, spectral analysis may help to identify particular spatial or temporal scales that are typical for a landscape. However, although scaling relationships offer clues to how the fundamental processes of biology give rise to empirical evidence is still largely missing (Levin 1998). Also, the indicator is likely most suitable to analyse model simulations since long time series are required that may not be empirically available.

Fractals and lacunarity: Fractals (Mandelbrot 1982) are mathematical representations of the complexity. Many objects relevant to landscape ecological research have fractal qualities either in two or three dimensional space (e.g., coastlines).

Fractals have many properties (Hastings and Sugihara 1993; Sprott 2003). For example, they have geometries that are too irregular to be represented by ordinary geometry, or they are self-similar, meaning that individual elements of the object are similar to the whole. Infinite copies of the elements make up the landscape. Self-similarity implies that the object is independent of the scale of observation (scale invariance), i.e., no characteristic scale is singled out and large-scale patterns can be predicted from small-scale pattern properties and vice versa. The upper boundary of the scale is determined by the extent of the object itself (e.g., landscape). The lower boundary is identified by the grain of the data. In landscape research, fractal geometry can be used to quantify the spatial complexity apparent in landscapes (Mandelbrot 1982; Gardner *et al.* 1987; Milne 1988, 1991; Milne *et al.* 1992; With 1994; Sprott *et al.* 2002; Bolliger *et al.* 2003).

Fractals may be described using algorithms that quantify the proportion of the geometrical space that is occupied by the fractal and are expressed as fractal dimensions. Fractals with identical fractal dimension may have greatly different appearances (Plotnick *et al.* 1993). In discrete patterns these differences are determined by the size of the gaps. The gaps in geometric structures can be measured using lacunarity indices (Mandelbrot 1982; Plotnick *et al.* 1993) to quantify the texture associated with patterns of spatial dispersion. Lacunarity is a useful index of surface structure for continuous landscape data where it measures the distributions of local maxima (peaks) and minima (valleys) of continuous landscape data (McGarigal and Cushman 2005). Additionally, lacunarity captures multiple dimensions of segregation across multiple scales (Wu and Sui 2001). First developed by Mandelbrot (1982), a variety of algorithms to calculate lacunarity are available (Allain and Cloitre 1991).

Indicators derived from wavelet analysis: Wavelet analysis is similar to spectral analysis, but instead of representing a pattern by a linear combination of sinusoid functions at different scales, it uses more flexible wavelet functions (Percival 2001). Wavelet analysis provides a promising alternative for characterising and partitioning landscapes in the presence of multiple, overlapping processes, and the method can easily handle large data sets such as remote sensing data (Bradshaw and Spies 1992; Csillag and Kabos 2002; McGarigal and Cushman 2005). Wavelet analysis has the advantage that it preserves the hierarchical information about the structure of a surface and allows pattern decomposition at the same time (Bradshaw and Spies 1992). The results can be interpreted in two ways: the wavelet variance identifies the scales that contribute most strongly to the pattern (similar to a periodogram in spectral analysis), whereas the degree of matching of the wavelet function can be mapped directly onto the data, thus identifying the spatial location of a specific structure (similar to local spatial statistics). Wavelets can thus be used as explanatory variables to predict a biotic response (Keitt and Urban 2005).

Surface metrology indices: Continuous surfaces have more properties than can be assessed using for example geostatistical approaches that quantify correlation distances, or spectral analysis that identifies particular spatial or temporal scales of periodic structures. Additional properties of continuous surfaces include e.g., roughness, skewness, curvature, or local peaks. These characteristics can be assessed by surface metrology indices based on methods that have been developed in microscopy and molecular physics (Barbato *et al.* 1995; SPIP 2001). The indices have recently become of interest to the landscape ecological community (McGarigal and Cushman 2005) and offer promising grounds for exploring their use and limitation for landscape ecological research.

Dynamic, discrete landscape indicators

The first step towards an empirical quantification of landscape dynamics is often a comparison between at least two time steps (Fig. 3). For instance, land-use/land-cover classifications derived from remote sensing may be available for different years. The resulting transition frequencies may be referred to as an indicator of landscape change representing the degree and spatial location of change. Obviously, such a comparison is only valid if the methodology is comparable between classifications, i.e., the same georeferencing, resampling and classification algorithms or rules were applied. Comparison may be based on map properties (comparison of landscape metrics that quantify composition or configuration for each time step) or on pixels (assessment of transitions from one state to another) (Jenerette and Wu 2001).

With transition matrices and Markov chains, the change of each pixel state between two time steps can be summarised in a transition probability matrix (Dale *et al.* 2002b) (Fig. 3). This is a square matrix with as many rows and columns as there are states (e.g., land-use/land-cover classes). A cell in row A and column B contains the estimated transition probability for a pixel with initial state A to switch to state B in one time step, thus representing an indicator of the likelihood of change. The transition probability is estimated by the observed number of transitions from A to B divided by the number of pixels with initial state A. For a series of *k* time steps with constant interval, *k-1* transition matrices can be estimated. If the underlying process is assumed constant (stationary), the *k-1* matrices are expected to be identical and may be averaged. The pooled transition matrix summarises the amount of change between all land-use types per average time step. The equilibrium state of the system, if such a state exists, is defined by a vector that, when multiplied by the transition matrix, yields the same vector again (eigen vector of the transition matrix; Usher 1992).

Under the assumption that the transition of a pixel between time steps $t_1 < t_2$ depends solely on the state of the pixel in t_1, the transition probability matrix describes a first-order Markov chain (Nicheva 2001). In ecology, this assumption is sometimes replaced by introducing holding-time requirements, i.e., a pixel needs to persist in state A for at least x time steps before transition to state B is possible (Acevedo *et al.* 1995; Yemshanov and Perera 2002). In some situations, a Markov chain will converge to an equilibrium state, independent of initial conditions (Usher 1992). A traditional Markov chain represents a spatially implicit temporal process. The explicit introduction of a spatial dimension results in spatio-temporal Markov chains (STMC), which combine a Markov chain with a cellular automaton (Balzter *et al.* 1998).

Dynamic, continuous landscape indicators

Indicators for continuous, dynamic landscapes can be found in dynamic systems or information theory. Currently, they rely on model simulations since empirical data (e.g., field-derived, aerial photographs) rarely provide long time series. One example of a dynamic, continuous indicator is spatiotemporal complexity that quantifies patchy vegetation dynamics (Parrott 2005). The indicator is similar to the information-based Shannon entropy and distinguishes between ordered, random, and aggregated patchiness (Parrott 2005). Other examples of dynamic, continuous indicators include measures to assess equilibria and their stability to gain impressions on how a trajectory of a continuous-time differential equation behaves throughout the entire state space (e.g., landscape) (Lischke *et al.* 2007). If the trajectory spirals around a point equilibrium, a spiral point is observed. Saddle points are found where some trajectories are attracted to equilibria, and others are repelled from them. In non-linear dynamics, periodic attractors (or repellors) are referred to as limit cycles. For example, the size of a basin of attraction of some equilibrium and its limits, within which the state variable may be perturbed before the system switches to other basins of attraction, may provide measures of system resilience (Pykh 2002).

Lyapunov exponents are indicators to assess the predictability of a system by measuring the extent to which small changes are amplified. It thus quantifies how sensitive a system is to perturbations (changes). In most cases perturbations tend to be amplified until they grow large, no matter how tiny the initial perturbations were. This behaviour, where small perturbations are amplified, is called sensitivity dependence on initial conditions. The Lyapunov exponents reflect the average rate at which perturbations increase or decrease. There are as many Lyapunov exponents as there are dimensions of the state space, and each exponent indicates whether the perturbation will increase or decrease in a particular direction

(Eckmann and Ruelle 1985). Applications of Lyapunov exponents in landscape ecology may thus involve assessments on the predictability of models. Additionally, since dependence on initial conditions is one of the hallmarks of chaos (Sprott 2003), Lyapunov exponents can be used to assess whether a system is chaotic or not.

Assessments of the aesthetic appeal of art and nature (Hunziker *et al.* 2007) reveal that a balance of simplicity and complexity, order and unpredictability, is preferred by humans (Aks and Sprott 1996). Results showed that the correlations between Lyapunov exponents (representing the unpredictability of the dynamic process) and people's aesthetical preferences for patterns had on average Lyapunov exponents that corresponded to those of many natural objects (Aks and Sprott 1996).

Conclusions and Challenges

Identifying and quantifying the structure and dynamics of landscape patterns is central to many questions in basic and applied research, and indicators provide a useful framework to assess key properties of landscapes in space and time at most efficient use of the available resources. Selection of the appropriate suite of indicators with respect to the research objective and scale is a key to successful application. Ideally, indicators should inform about a system as comprehensively as possible with respect to the research question of interest. It is thus necessary to choose appropriate indicators that mirror adequately the system's structure, function, and composition including assessments on the indicator's use and limitation.

There are several important challenges:

1. *Characterising landscapes dynamically:* To date, landscapes have been characterised mostly spatially. However, it is widely recognised that many landscape elements exhibit non-equilibrium dynamic-transient behaviour (Lischke *et al.* 2007). Thus, future challenges in landscape ecology include increasing focus on dynamic assessments of landscapes and landscape change, especially since appropriate data will be increasingly available through remote sensing (Zimmermann *et al.* 2007).
2. *Characterising landscapes continuously:* Approaches to assess landscapes continuously include geostatistical approaches that indicate distances of correlation, spectral analysis to assess whether there are any particular spatial or temporal scales, and fractals allow analysis of the spatial complexity of a surface. However, continuous surfaces exhibit many more properties. Surface metrology indices and wavelets are promising tools to overcome this shortcoming, although their suitability for application in landscape ecology remains to be tested.
3. *Comparability of landscape indicators across scales and statistical tests:* Use of indicators for landscape ecological questions involves thorough testing to ensure comparability within and across different landscapes and scales. Quantifying and interpreting differences between landscapes/landscape patterns is a future challenge, since routine statistical tests are hardly applicable. Furthermore, not only pattern, but also driving processes need to be compared (Fortin *et al.* 2003).

Acknowledgements

We would like to thank the three editors of this book Felix Kienast, Rita Ghosh and Otto Wildi and two anonymous reviewers for their helpful and constructive suggestions and comments.

References

Acevedo M.F., Urban D.L. and Ablan M. 1995. Transition and gap models of forest dynamics. Ecological Applications 5: 1040–1055.

Addicott J.F., Aho J.M., Antolin M.F., Padilla D.K., Richardson J.S. and Soluk D.A. 1987. Ecological neighborhoods: scaling environmental patterns. Oikos 49: 340–346.

Aks D.J. and Sprott J.C. 1996. Quantifying aesthetic preference for chaotic patterns. Empirical Studies of the Arts 14: 1–16.

Allain C. and Cloitre M. 1991. Characterising the lacunarity of random and deterministic fractal sets. Physical Review A 44: 3552–3558.

Bak P. 1996. How nature works. Springer Verlag, New York.

Bak P., Tang C. and Wiesenfeld K. 1987. Self-organised criticality: an explanation of 1/f noise. Physical Review Letters 59: 381–384.

Balzter H., Braun P.W. and Kohler W. 1998. Cellular automata models for vegetation dynamics. Ecological Modelling 107: 113–125.

Barbato W.L., Carneiro K., Cuppini D., Garnaes J., Gori G., Hughes G., Jensen C.P., Jorgensen J.F., Jusko O., Livi S., McQuoid H., Nielsen L., Picotto G.B. and Wilening G. 1995. Scanning tunnelling microscopy methods for the characterisation for roughness and micro hardness measurements. Synthesis report for research contract with the European Union under its programme for applied metrology. In: European Commission Catalogue number, Brussels, Luxemburg.

Bascompte J. and Solé R.V. 1995. Rethinking complexity – modelling spatiotemporal dynamics in ecology. Trends in Ecology & Evolution 10: 361–366.

Bezdek J.C. 1981. Pattern recognition with fuzzy objective function algorithms. Plenum Press, New York.

Bolliger J. 2002. Schweizer Wälder und Klimaveränderungen: Vergleich von Simulationen quantitativer Vegetationsmodelle. Schweizerische Zeitschrift für Forstwesen 153: 167–175.

Bolliger J. 2005. Simulating complex landscapes with a generic model: sensitivity to qualitative and quantitative classifications. Ecological Complexity 2: 131–149.

Bolliger J., Kienast F. and Zimmermann N.E. 2000. Risks of global warming on montane and subalpine forests in Switzerland. Regional Environmental Change 1: 99–111.

Bolliger J. and Mladenoff D.J. 2005. Quantifying spatial classification uncertainties of the historical Wisconsin landscape (U.S.A). Ecography 28: 141–156.

Bolliger J., Sprott J.C. and Mladenoff D.J. 2003. Self-organisation and complexity in historical landscape patterns. Oikos 100: 541–553.

Boots B. 2002. Local measures of spatial association. Ecoscience 9: 168–176.

Bradshaw G.A. and Spies T.A. 1992. Characterising canopy gap structure in forests using wavelet analysis. Journal of Ecology 80: 205–215.

Brown D.G. 1998. Classification and boundary vagueness in mapping presettlement forest types. International Journal of Information Science 12: 105–129.

Bürgi M., Hersperger A., Hall M., (Russell) Southgate E.W.B. and Schneeberger N. 2007. Using the past to understand the present land use and land cover. In: Kienast F., Wildi O. and Ghosh S. (eds.). A Changing World. Challenges for Landscape Research, Vol. 8: 133–144. Springer Landscape Series, Dordrecht.

Burrough P.A. 1995. Spatial aspects of ecological data. In: van Tongeren O.F.R. (ed.), Data Analysis in Community and Landscape Ecology, pp. 213–251. Cambridge University Press, Cambridge.

Cardille J.A. and Turner M.G. 2002. Understanding landscape metrics I. In: Gergel S.E. and Turner M.G. (eds.). Learning landscape ecology, pp. 85–100. Springer, New York

Ceva H. 1998. On the asymptotic behvaieur for an earthquake model. Physics Letters A 245: 413–418.
Clark J.S. 1998. Why trees migrate so fast: Confronting theory with dispersal biology and the paleorecord. American Naturalist 152: 204–224.
Clark J.S., Kern R., Macklin E. and HilleRisLambers J. 1999. Seed dispersal near and far: Patterns across temperate and tropical forests. Ecology 80: 1475–1494.
Cliff A.D. and Ord J.K. 1981. Spatial Processes: Models and Applications. Pion, London.
Cousins S.A.O. and Lindborg R. 2002. Assessing changes in plant distribution patterns – indicator species versus plant functionl types. Ecological Indicators 1: 17–27.
Csillag F. and Kabos S. 2002. Wavelets, boundaries, and the spatial analysis of landscape pattern. Ecoscience 9: 177–190.
Dale M.R.T. 2000. Spatial pattern analysis in plant ecology. Cambridge University Press, Cambridge.
Dale V.A. 2001. Challenges in the development and use of ecological indicators. Ecological Indicators 1: 3–10.
Dale V.A., Beyeler S.C. and Jackson B. 2002a. Understory vegetation indicators of anthropogenic disturbance in longleaf pine forests at Fort Benning, Georgia, U.S.A. Ecological Indicators 1: 155–170.
Dale V.H., Fortes D.T. and Ashwood T.L. 2002b. A landscape transition matrix approach for land management. In: Liu J. and Taylor W. (eds.). Integrating Landscape Ecology into Natural Resource Management, pp. 265–293. Cambridge University Press, Cambridge.
Duelli P., Baur P., Buchecker M., Gugerli F., Holderegger R. and Wohlgemuth T. 2007. The role of value systems in biodiversity research. In: Kienast F., Wildi O. and Ghosh S. (eds.). A Changing World. Challenges for Landscape Research, Vol. 8: 27–34. Springer Landscape Series, Dordrecht.
Duelli P. and Obrist M.K. 2003. Biodiversity indicators: the choice of values and measures. Agriculture Ecosystems and Environment 98: 87–98.
Duelli P. and Obrist M.K. 1998. In search of the best correlates for local organismal biodiversity in cultivated areas. Biodiversity and Conservation 7: 297–309.
Eckmann J.-P. and Ruelle D. 1985. Ergodic theory of chaos and strange attractors. Rev. Mod. Phys 57: 617–656.
Ellenberg H. 1988. Vegetation ecology of Central Europe, 4th ed. Cambridge University Press.
Farkas I., Derenyi I., Jeong H., Neda Z., Oltvai Z.N., Ravasz E., Schubert A., Barabasi A.-L. and Vicsek T. 2002. Networks in life: scaling properties and eigenvalue spectra. Physica A 314: 25–34.
Forman R.T.T. 1995. Land mosaic: the ecology of landscapes and regions. Cambridge University Press, Cambridge, UK.
Forman R.T.T. and Alexander I.F. 1998. Roads and their major ecological effects. Ann. Rev. Ecol. Syst. 29: 207–231.
Forman R.T.T. and Godron M. 1986. Landscape Ecology. John Wiley & Sons, New York.
Fortin M.J., Boots B., Csillag F. and Remmel T.K. 2003. On the role of spatial stochastic models in understanding landscape indices in ecology. Oikos 102: 203–212.
Fortin M.-J., Dale M.R.T. and ver Hoef J. 2001. Spatial analysis in ecology. In: Piegorsch W.W. (ed.). The Encyclopedia of Environmetrics, pp. 2051–2058. John Wiley and Sons Ltd.
Franklin J.F. and Forman R.T.T. 1987. Creating landscape patterns by forest cutting: Ecological consequences and principles. Landscape Ecology 1: 5–18.
Gardner R.H., Milne B.T., Turner M.G. and O'Neil R.V. 1987. Neutral models for the analysis of broad-scale landscape pattern. Landscape Ecology 1: 19–28.
Gergel S.E., Turner M.G., Miller J.R., Melack J.M. and Stanley H.E. 2002. Landscape indicators of human impacts to riverine systems. Aquatic Science 64: 118–128.
Gosz J.R. 1993. Ecotonoe hierarchies. Ecological Applications 3: 369–376.
Green D.G. and Sadedin S. 2005. Interactions matter – complexity in landscapes and ecosystems. Ecological Complexity 2: 117–130.
Greenberg J.D., Gergel S.E. and Turner M.G. 2002. Understanding landscape metrics II. In: Gergel S.E. and Turner M.G. (eds.). Learning landscape ecology, pp. 101–111. Springer, New York.

Grove S.J. 2002. Tree basal area and dead wood as surrogate indicators of saproxylic insect faunal integrity: a case study from the Australian lowland tropics. Ecological Indicators 1: 171–188.

Gustafson E.J. 1998. Quantifying landscape spatial pattern: what is the state of the art? Ecosystems 1: 143–156.

Gustafson E.J. and Parker G.R. 1992. Relationship between landcover proportion and indices of landscape spatial pattern. Landscape Ecology 7: 101–110.

Haines-Young R. and Chopping M. 1996. Quantifying landscape structure: a review of landscape indices and their application to forested landscapes. Progress in Physical Geography 20: 418–445.

Haining R. 1997. Spatial Data Analysis in the Social and Environmental Sciences. Cambridge University Press, Cambridge.

Hall H.M. and Grinnell J. 1919. Life-zone indicators in California. Proc. Calif. Acad. Sci 37–67.

Hargis C.D., Bissonnette J.A. and David J.L. 1998. The behaviour of landscape metrics commonly used in the study of habitat fragmentation. Landscape Ecology 13: 167–186.

Hastings H.M. and Sugihara G. 1993. Fractals: a User's Guide for the Natural Sciences. Oxford University Press, Oxford.

Hunziker M., Buchecker M. and Hartig T. 2007. Space and place – two aspects of the human-landscape relationship. In: Kienast F., Wildi O. and Ghosh S. (eds.). A Changing World. Challenges for Landscape Research, Vol. 8: 47–62. Springer Landscape Series, Dordrecht.

Isaaks E.H. and Srivastava R.M. 1989. Applied Geostatistics. Oxford University Press, New York.

Jacquemin H., Butaye J. and Hermy M. 2001. Forest plant species richness in small, fragmented mixed deciduous forest patches: The role of area, time and dispersal limitations. Journal of Biogeography 28: 801–812.

Jenerette G.D. and Wu J.G. 2001. Analysis and simulation of land-use change in the central Arizona-Phoenix region, U.S.A. Landscape Ecology 16: 611–626.

Johnson A.R. and Milne B.T. 1992. Diffusion in fractal landscapes: simulations and experimental studies of tenebrionid beetle movement. Ecology 73: 1968–1983.

Jones K.B., Neale A.C., Nash M.S., Riitters K.H., Wickham J.D., O'Neill R.V. and Van Remortel R.D. 2000. Landscape correlates of breeding bird richness across the United States mid-atlantic region. Environmental Monitoring and Assessment 63: 159–174.

Kaspar F. and Schuster H.G. 1987. An easily calculable measure for the complexity of spatio-temporal patterns. Phys. Rev. A 36: 842.

Keitt T.H. and Urban D. 2005. Scale-specific inference using wavelets. Ecology 86: 2497–2504.

Keller D.R. and Golley F.B. 2000. The philosophy of ecology. The University of Georgia Press, Athens GA, U.S.A.

Kienast F., Wildi O. and Brzeziecki B. 1998. Potential impacts of climate change on species richness in mountain forests – an ecological risk assessment. Biological Conservation 83: 291–305.

Lausch A. and Herzog F. 2002. Applicability of landscape metrics for the monitoring of landscape change: Issues of scale, resolution and interpretability. Ecological Indicators 2: 3–15.

Levin S.A. 1998. Ecosystems and the bioshpere as complex adaptive systems. Ecosystems 1: 431–436.

Levin S.A. 1978. Pattern formation in ecological communities. In: Steele J.H. (ed.), Spatial pattern in plankton communities, pp. 433–465. Plenum Press, New York.

Levin S.A. 1979. Non-uniform stable solutions to reaction-diffusion equations: applications to ecological pattern formation. In: Haken H. (ed.). Pattern formation by dynamic systems and pattern recognition, pp. 210–222. Springer Verlag, Heidelberg, New York.

Levin S.A. 1992. The problem of pattern and scale in ecology. Ecology 73: 1943–1967.

Li H. and Wu J. 2004. Use and misuse of landscape indices. Landscape Ecology 19: 389–399.

Liebhold A.M. and Gurevitch J. 2002. Integrating the statistical analysis of spatial data in ecology. Ecography 25: 553–557.

Lischke H., Bolliger J. and Seppelt R. 2007. Dynamic spatio-temporal landscape models. In: Kienast F., Wildi O. and Ghosh S. (eds.). A Changing World. Challenges for Landscape Research, Vol. 8: 273–296. Springer Landscape Series, Dordrecht.

Mal T.K., Uveges J.L. and Turk K.W. 2002. Fluctuating asymmetry as an ecological indicator of heavy metal stress in *Lythrum salicaria*. Ecological Indicators 1: 189–195

Mandelbrot B. 1982. The fractal geometry of nature. Freeman.
Manson S.M. 2001. Simplifying complexity: a review of complexity theory. Geoforum 32: 405–414.
McBratney A.B. and Odeh I.O.A. 1997. Application of fuzzy sets in soil science: fuzzy logic, fuzzy measurements, and fuzzy decisions. Geoderma 77: 85–113.
McGarigal K. and Cushman S.A. 2005. The gradient concept of landscape structure. In: J. Wiens and Moss M. (eds.). Issues and perspectives in landscape ecology, pp. 112–119. Cambridge University Press, Cambridge.
McGarigal K., Cushman S.A., Neel M.C. and Ene E. 2002. FRAGSTATS: Spatial pattern analysis program for categorial maps. Computer software program produced by the authors at the University of Massachusetts, Amherst, MA, U.S.A. http://www.umass.edu/landeco/research/fragstats/fragstats.html.
Milne B.T. 1991. Lessons from applying fractal models to landscape patterns. In: Turner M.G. and Gardner R.H. (eds.). Quantiative methods in landscape ecology, pp. 199–235. Springer.
Milne B.T. 1988. Measung the fractal geometry of landscapes. Appl. Math. Computation 27: 67–79.
Milne B.T. 1998. Motivation and benefits of complex systems approaches in ecology. Ecosystems 1: 449–456.
Milne B.T., Turner M.G., Wiens J.A. and Johnson A.R. 1992. Interactions between the fractal geometry of landscapes and allometric herbivory. Theoretical Population Biology 41: 337–353.
Minasny B. and McBratney A.B. 2002. FuzME Version 3.0 http://www.usyd.edu.au/su/agric/acpa/fkme/program.html. The University of Sydney, Sydney, Australia, http://www.usyd.edu.au/su/agric/acpa.
Neel M.C., McGarigal K. and Cushman S.A. 2004. Behaviour of class-level landscape metrics across gradients of class aggregation and area. Landscape Ecology 19: 435–455.
Nicheva D. 2001. Encyclopedia of Environmetrics. In: El-Shaarawi A.H. and Piegorsch W.W. (eds.). Markov chains, pp. 1207–1208. John Wiley & Sons, Chichester.
O'Neill R.V., Milne B.T., Turner M.G. and Gardner R.H. 1988. Resource utilisation scales and landscape pattern. Landscape Ecology 2: 63–69.
Parrott L. 2005. Quantifying the complexity of simulated spatiotemporal population dynamics. Ecological Complexity 2: 175–184.
Pearson D.M. 2002. The application of local measures of spatial autocorrelation for describing pattern in north Australian landscapes. Journal of Environmental Management 64: 85–95.
Percival D.B. 2001. Wavelets. In: Piegorsch W.W. (ed.). The Encyclopedia of Environmetrics, pp. 2338–2351. John Wiley and Sons Ltd., New York.
Perry D.A. 1995. Self-organising systems across scales. Trends in Ecology and Evolution 10: 241–244.
Plotnick R.E., Gardner R.H. and O'Neill R.V. 1993. Lacunarity indices as measures of landscape texture. Landscape Ecology 8: 201–211.
Pykh Y.A. 2002. Lyapunov functions as a measure of biodiversity: Theoretical background. Ecological Indicators 2: 123–133.
Regan H.M., Colyvan M. and Burgman M.A. 2000. A proposal for fuzzy International Union for the Conservation of Nature (IUCN) categories and criteria. Biological Conservation 92: 101–108.
Riitters K.H., O'Neill R.V., Hunsaker C.T., Wickham J.D., Yankee D.H., Timmins S.P., Jones K.B. and Jackson B.L. 1995. A factor analysis of landscape pattern and strucutre metrics. Landscape Ecology 10: 23–39.
Simoyi R.H., Wolf A. and Swinney H.L. 1982. One-dimensional dynamics in a multicomponent chemical reaction. Physical Review Letters 49: 245–248.
Snacken F. and Antrop M. 1983. Structure and dynamics of landscape systems. In: Drdos J. (ed.). Landscape synthesis: geoecological foundations of the complex landscape management, pp. 10–30. Veda Publishing House of the Slovak Academy of Sciences, Bratislava.
Solé R.V., Alonso D. and McKane A. 2002. Self-organised instability in complex ecosystems. Philosophical Transactions of the Royal Society of London Series B-Biological Sciences 357: 667–681.
Solé R.V. and Manrubia S.C. 1995. Are rainforests self-organised critical? Journal of Theoretical Biology 173: 31–40.
Solé R.V., Manrubia S.C. and Benton M. 1999. Criticality and scaling in evolutionary ecology. Trends in Ecology and Evolution 156–160.

Solé R.V., Miramontes O. and Goodwin B.C. 1993. Oscillations and chaos in ant societies. Journal of Theoretical Biology 161: 343–357.

SPIP 2001. The scanning probe image processor. Image metrology APS. Lyngby, Denmark.

Sprott J.C. 2003. Chaos and time-series analysis. Oxford University Press, Oxford.

Sprott J.C., Bolliger J. and Mladenoff D.J. 2002. Self-organised criticality in forest-landscape evolution. Physics Letters A 297: 267–271.

Taylor P.D., Fahrig L., Henein K. and Merriam G. 1993. Connectivity is a vital element of landscape structure. Oikos 68: 571–573.

Thompson C.M. and McGarigal K. 2002. The influence of research scale on bald eagle habitat selection along the lower Hudson River, New York (U.S.A). Landscape Ecology 17: 569–586.

Turner M.G. 2005. Landscape ecology in North America: past, present, and future. Ecology 86: 1967–1974.

Turner M.G., Gardner R.H. and O'Neill R.V. 2001. Landscape ecology in theory and practice: pattern and process. Springer Verlag, New York, U.S.A.

Usher M.B. 1992. Statistical models of succession. In: Glenn-Lewin D.C., Peet R.K. and Veblen T.T. (eds.), Plant succession: theory and prediction, pp. 215–248. Chapman and Hall, London.

Wackernagel H. 1998. Multivariate Geostatistics. 2nd, completely revised. Springer, Berlin.

Wade T.G., Wickham J.D., Nash M.S., Neale A.C., Riitters K.H. and Jones K.B. 2003. A comparison of vector and raster GIS methods for calculating landscape metrics used in environmental assessments. Photogrammetric Engineering and Remote Sensing 69: 1399–1405.

Wagner H.H. 2003. Spatial covariance in plant communities: an integration of ordination, variogram modelling and the variance test of species richness. Ecology 84: 1045–1057.

Wagner H.H. 2004. Direct multiscale ordination with canonical correspondence analysis. Ecology 85: 342–351.

Wagner H.H. and Fortin M.J. 2005. Spatial analysis of landscapes: concepts and statistics. Ecology 86: 1975–1987.

Webb S.L. 1987. Beech range extension and vegetation history: pollen stratigraphy of two Wisconsin, U.S.A lakes. Ecology 68: 1993–2005.

Wickham J.D., Jones K.B., Riitters K.H., Wade T.G. and O'Neill R.V. 1999. Transitions in forest fragmentation: implications for restoration opportunities at regional scales. Landscape Ecology 14: 137–145.

Wiens J.A. 1989. Spatial scaling in ecology. Functional Ecology 3: 385–397.

Wiens J.A. 2000. Ecological heterogeneity: an ontogeny of concepts and approaches. In: Hutchings M.J., John E.A. and Stewart A.J.A. (eds.). The ecological consequences of heterogeneity, pp. 9–31.

With K.A. 1994. Using fractal analysis to assess how species perceive landscape structure. Landscape Ecology 9: 25–36.

With K.A. 1997. The application of neutral landscape models in conservation biology. Conserv Biol 11: 1069–1080.

With K.A. 1997. The theory of conservation biology. Conserv Biol 11: 1436–1440.

With K.A. 2002. Landscape connectivity and metapopulation dynamics. In: Gergel S.E. and Turner M.G. (eds.), Learning landscape ecology, pp. 208–227. Springer, New York.

Wu J. and Hobbs R. 2002. Key issues and research priorities in landscape ecology: An idiosyncratic synthesis. Landscape Ecology 17: 355–365.

Wu J. and Loucks O.L. 1995. From balance of nature to hierarchical patch dynamics: a paradigm shift in ecology. The Quarterly Review of Biology 70: 439–466.

Wu J., Shen W., Sun W. and Tueller P.T. 2002. Empirical patterns of the effects of changing scale on landscape metrics. Landscape Ecology 17: 761–782.

Wu X.B. and Sui D.Z. 2001. An initial exploration of a lacunarity-based segregation measure. Environment and Planning B: Planning and Design 28: 433–446.

Yemshanov D. and Perera A.H. 2002. A spatially explicit stochastic model to simulate boreal forest cover transitions: general structure and properties. Ecological Modelling 150: 189–209.

Zadeh L.A. 1965. Fuzzy sets. Information Control 8: 338–353.

Zimmermann N.E., Washington-Allen R.A., Ramsey R.D., Schaepman M.E., Mathys L., Kötz B., Kneubühler M., and Edwards T.C. 2007. Modern remote sensing for environmental monitoring landscape states and trajectories. In: Kienast F., Wildi O. and Ghosh S. (eds.). A Changing World. Challenges for Landscape Research, Vol. 8: 65–91. Springer Landscape Series, Dordrecht.

Zollner P.A. and Lima S.L. 1997. Landscape-level perceptual abilities in white-footed mice: perceptual range and the detection of forested habitat. Oikos 80: 51–60.

Essay on the Study of the Vegetation Process

Otto Wildi[1*] and László Orlóci[2]

[1] Swiss Federal Research Institute WSL, Zürcherstrasse 111, CH-8903 Birmensdorf, Switzerland
[2] University of Western Ontario, Dept. of Plant Sciences, NGA 5B7 London, Ontario, Canada
[*] Corresponding author: e-mail: otto.wildi@wsl.ch

Abstract
Vegetation is understood as the joint occurrence of plant species within any given site. Species combinations tend to reoccur at different points in space and time suggesting the existence of assembly rules. The issue of this essay is the vegetation process, reflecting the fact that vegetation changes permanently. Finding a universal prediction theory of vegetation dynamics is considered the ultimate goal in vegetation science. The essay follows a traditional scientific scheme discussing the questions of "what" can be assessed, "how" processes operate and whether there is an explanation to "why" the systems behave as observed. We argue that sampling, apart from statistical considerations, is very much a question of choice: choice of the variables, the temporal and spatial frame. For finding the appropriate sample size, however, there are methods available to find efficient solutions. Concerning the understanding of processes, a combination of determinism and randomness has to be taken into account. Some deterministic processes, however, show chaoticity. This is difficult to distinguish from stochasticity. In this context the role of linear processes is discussed as well. Identifying causes is often beyond the reach of ordinary investigations and one of the reasons is the requirement to access interacting scales. The essay underlines the relevance of flexibility and empiricism in a holistic approach, whereas a reductionist approach will only reveal isolated properties of the vegetation process.

Keywords: chaoticity, empiricism, flexibility, hierarchy, holistic approach, linearity, sampling stochasticity, uncertainty

F. Kienast, O. Wildi & S. Ghosh (eds.), A Changing World. Challenges for Landscape Research, 195–207.

Introduction

The term "vegetation" implies "species combination", that is, the joint occurrence of plant species within any given site. The fact that a species combination tends to reoccur at different points in space and time suggests the existence of assembly rules that emerge from underlying processes. Ever since Kerner von Marilaun's (1864) land-mark work "Das Pflanzenleben der Donauländer", the central paradigm of vegetation studies is process related. The existence of a process is implied when plant communities change. The change is continuous and it is inseparably linked with the Kernerian mechanism of "facilitation" caused via changes in the environment. Naturally, there is an apparent lack of order in the process and the elementary variables involved in change are stochastic in nature. The vegetation process is thus a convolution of determinism and randomness. The result is manifested in the existing spatio-temporal patterns.

The purpose of practice in vegetation science is to describe the processes in quantitative terms and to reveal the governing principles. Through this, scientific and practical objectives are served. By "practical" we address contributions to the solution of real-world problems among which global warming is presently the most intriguing (Orlóci 2001). The main scientific objective is the development of a general theory. Incredibly, this is a common objective in all vegetation studies at all times in post-Kernerian history, yet it proved to be elusive hitherto in all but the simplest anecdotal terms.

In search of this universal predictive theory of vegetation dynamics (Rees *et al.* 2001) we first point to complexity as the cause of the persisting deficit in the multitude of past and recently offered explanations (Anand and Orlóci 1996). Complexity has its origin in the nature of the ecological medium and in the conceptual and mechanistic limitations of dealing effectively with an unwinding sampling environment. In this environment, paraphrasing Orlóci (2001), heterogeneity rules, responses and interactions are likely to be non-linear; the sampling units are fuzzy, and neither "catalogued" nor is there "unhindered" access; dynamics is as given by nature, and pattern, random and deterministic, is a rule. This poses constraints on prediction and requires looking into nonlinear models, chaos theory, fuzzy probability and so on.

In order to describe the phenomena contributing to complexity in greater detail they shall be structured according to a traditional, scientific scheme, the "what", "how" and "why" questions (Goodall 1975; Orlóci *et al.* 2002). The question "what" concerns recognition. In this, the issue is pattern – and pattern recognition – and in turn, the processes generate the patterns. Knowing the processes answers the question "how" patterns evolve. Patchy vegetation, as an example, can have its origin in the heterogeneity of the substrate, in the behaviour of herbivores, or in the population dynamics of dominating species. Typical examples of causes, answering the "why" question, are competition or facilitation at small, and climate change at large scale (Connell and Slatyer 1977).

Whatever be the answer to the questions, it is restricted to a specific spatial, temporal and thematic frame of reference (Parker and Pikett 1998). By selecting the extent and resolution of any measurement, as well as its type (e.g., discrete vs. continuous), we define the scale within which the results are viable. Variable selection is an early step in this. Eventually the question of how to cope with results obtained at different hierarchical levels is posed, which is an issue in hierarchy theory (Allen and Starr 1982).

We begin our essay by presenting a brief history of vegetation science. This depicts the ongoing introduction of new conceptual tools, new theories, and the evolution of new paradigms. Generally, the "old" is not abandoned, but the "new" is added as alternatives or extensions. As an example, the detection of the chaotic nature of many processes may explain the lack of success with their predictability, but it does not imply that the notion of determinism in the manner of some directed mechanism, such as facilitation, should be denied (Anand 1997). However, the potential coexistence of process properties recognisable as scale-dependent mechanisms adds much to the complexity of the system.

Research in plant ecology is based on a number of seemingly arbitrary decisions, concerning variable selection, scaling and assumptions about the nature of processes. The subsequent section shows the possibility of countless alternatives in approach. Related to this, some fundamental principles are derived regarding the approach itself, such as the need for heuristicy, holisticity, and flexibility.

A Note on History

Understanding the true nature of species assemblages has been the main issue in the philosophical discussions and the ongoing methodological progress in the first half of the 20th century. Emphasis was put on the idea of systematically classifying vegetation pattern by defining re-occurring types of vegetation communities (Braun-Blanquet 1932) or reoccurring patterns of their successional process (Clements 1916). It seemed clear that strong forces would have to exist to generate the obvious order in space and also in time. Clements was convinced that a plant community has the rank of an organism. Tansley (1935) warned from this analogy, but still promoted that a plant community be a "quasi-organism", hence that classification would have a physical origin. When Braun-Blanquet (1932) developed his methods, known today as "Central European Phytosociology", CEPS (Ewald 2003), for which comprehensive review is found in Westhoff and van der Maarel (1978), he proposed a synsystematic system and a nomenclature for "communities" that adopted in form the taxonomic classification system. Braun-Blanquet assumed that, although a plant community may emerge and eventually get extinct, the "type" called association is constant or at least long lasting.

The standards established by Braun-Blanquet were widely accepted and they resulted in the collection and accumulation of gigantic quantities of commensurable information on the vegetation cover all over the globe. The classification rules and terminologies used in CEPS are nowadays seen as "philosophical and methodological anachronisms" (Ewald 2003), but they are not when viewed as a means for artificially cataloguing nature. Data collected for more than a century are in fact generating more attention than some had expected. The consensus that emerged holds that since plant communities continuously change in time and the changes may be quite rapid, the standards adopted for sampling document historic states and provide us with time series data of excellent quality. Furthermore, as will be discussed below, the current statistical tools and state of the art computing facilities allow fast access to archived information (Feoli and Orlóci 1991). Reconciling old and new standards in vegetation research therefore represents a most striking methodological trade-off.

In an opposing view, Gleason (1926, 1939) promoted his "individualistic concept of the plant association". The term "individualistic" is used to emphasise that each species exhibits its own unique pattern of dynamics. The Gleasonian view does in fact emphasize what separates communities rather than what makes them similar, not considering the importance that chaotic variation can play in making communities different in appearance. To those rejecting the idea of the existence of reoccurring species combinations, compositional patterns are merely seen as transient phenomena on a space-time continuum, an idea about which "Anglo-American plant ecology" revolved (AAPE, Ewald 2003). The continuum view promoted the development of ordination as the method of pattern recognition in vegetation (Bray and Curtis 1957). Hubbell (2001) pointed out that Gleason's view implies another frequently overlooked assumption, the so-called "neutrality". Although neutrality does not assume identical properties for all species nor equal population size, it assigns species populations equal chance for success and extinction. Hubbell (2001) argues that the dominance of the individualistic behaviour of species means that mechanisms of interaction like competition and facilitation, key issues in modern ecology, play a minor long-term role.

A new trend emerged in the second part of the 20th century, unifying classification and ordination: the ascent of statistical analysis. Goodall's (1954) application of principal components analysis (PCA) and Greigh-Smith's (1983) pattern analysis are first salvoes from a methodology that quickly grew to dominate ecological data analysis. In this, a set of vegetation relevés constitutes a statistical sample. The relevé is the record of a sampling unit encompassing a multitude of variables, the performance of plant species (Pielou 1984). The methods allow recognition of patterns through dimension reduction in multivariate data sets. Many methods for clustering, ordination, identification, ranking and discrimination rely on variance decomposition (Orlóci 1978). An alternative is the use of entropy and information, through partitioning of information (Kullback 1959; Shannon 1948). Among the specialised standard texts describing this we mention Orlóci (1978, 1991a), Pielou (1984), Jongmann *et al.* (1987) and Legendre and Legendre (1998). The application of these methods is neither restricted to a single type of data, nor to single specific purpose. Consequently, the focus may shift away from species as being the only possible variables. Plant functional types (PFT's, Box 1996; Smith *et al.* 1993) and character based community analysis (Pillar and Orlóci 1991) address the life strategy of plants at the global scale and in the latter case in the context of a powerful numerical methodology which offers an alternative to users of plant functional types or character set types. Character set types (CST's, Orlóci and Orlóci 1985; Orlóci 1991b) are defined flexibly through a character state based hierarchy, as the analysis progresses. The survey requires pre-selection of characters and character states. The approach yields alternatives to standard species based or serial character-based classifications (Orlóci 1991b). A character set type may unify common features of species of different families occurring in different continents. Analysing data of this type is considerably more complex than treating ordinary relevé data.

Another approach probes the survival and dispersal strategies of plants. Harper (1977) emphasised the importance of the breeding system, Grime (1977) proposed three fundamental survival strategies (stress-tolerant, ruderal, competitive), and Tilman (1988) investigated the importance of resource allocations into functionally distinct parts of the plant (roots, stem, leaves). More recently, molecular genetics brought the enquiry below the species level, where meta-populations are distinguished by their genome (Manel *et al.* 2003). Through this, very small evolutionary changes have become detectable and these may happen within only a few generations. Consequently, the species have lost their role as discrete, unequivocally distinguishable variables. This clearly encompasses the population level - from which some tried rather unsuccessfully to deduce the governing principles at the community process by the method of superposition of the low-level manifestations of the process.

Progress in other branches of science has fostered the theoretical basis of plant ecology as well (Rees *et al.* 2001). To mention a few, fuzzy-set theory in general (Zimmermann 1985) and applied to vegetation data (De Patta Pillar and Orlóci 1991; Feoli and Zuccarello 1991; Roberts 1986), the fractal view of nature including chaos theory (Mandelbrot 1977), the awareness of the scale problem (Podani 1984), space-time interactions (Legendre and Fortin 1989; Parker and Pikett 1998) and again molecular genetics (Manel *et al.* 2003) are examples. Like in many other fields of science, static and dynamic modelling (Brzeziecki *et al.* 1993; Guisan and Zimmermann 2000) have become routinely used approaches.

Basic Choices

It may seem strange to the reader that personal choices should be a major issue in methodological considerations in natural sciences. At first glance, the subject seems to be defined and well conditioned with no room for ambiguity: we are concerned with vegetation cover of the surface of the earth in four dimensions (including time). But there are practicalities, such as setting limits and other "bare necessities" as called in Orlóci (2001). One is the response type encountered in ecosystems. A general rule: in all realistic cases the response is non-linear (Fig. 1). In scientific papers and textbooks, however, graphs and functions show linear relationships or linear first differentials as a rule, not an exception. The motivation to describe functions linearly lies in the search for simplicity, the objective to find the most parsimonious solution to a problem.

As will be shown below, sampling offers a first example of a "what question". When choosing between random and preferential sampling design, a trade-off is encountered. Random sampling may lead to the collection of much useless material, while preferential may lead to irreparable bias:

1. *Sampling: random or preferential?* Sampling is needed because we cannot measure the state of the entire sampled environment. Much of what we know today on all scales from life zone to molecule as well as on temporal scales from population processes to the time scale of Ice Ages is coming from smart people looking at material considered suitable for their analysis. They were in fact looking at typical species, typical community stands, extreme situations, frequently occurring or strange phenomena, and the likes. As pointed out by Orlóci (1993, 2001) random sampling would often lead to a waste of effort, and as

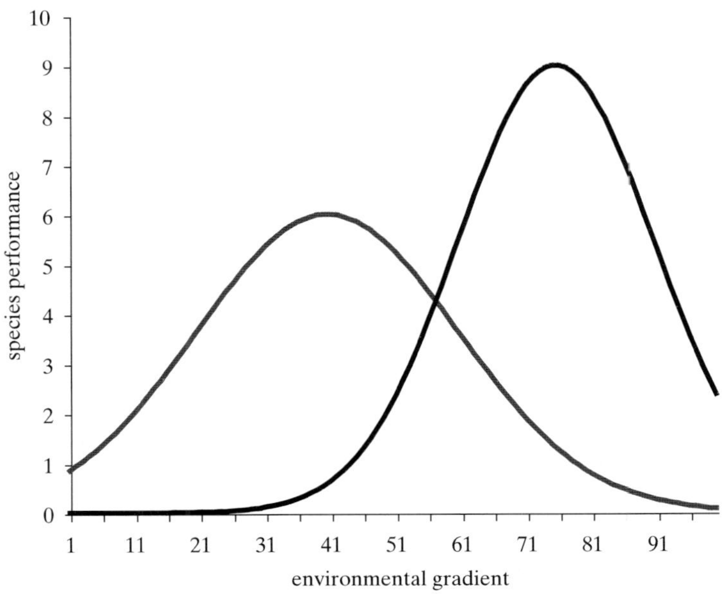

Fig. 1. An artificial series with nonlinear (Gaussian) change of performance of two species along a hypothetical environmental gradient of length 100.

a rule, is often not a rational option (Edgington 1987; Pielou 1975). On the other hand, the choice between random and preferential sampling is often scale determined. The selection and delimitation of the sampling environment is almost always preferential, while the estimation of statistical parameters such as mean, variance or spatial autocorrelation are bound to random sampling. In any case, the complication arising from the fact that there are no identifiable sample units remains. The user has to resolve the scale problem by choosing the size, the shape and the time interval in plot sampling.

2. *In focus: centroid or type?* Vegetation patterns can often be described as assemblages of types. It can be shown that centroids derived from sampling units are descriptions of types in environments where responses are linear (Orlóci 1993). In vegetation data a horse-shoe point cluster is most likely to occur when gradients are analysed. Figure 2 shows the response of species from Figure 1 in an ordination diagram. After performing a principal component analysis the centroid falls into the centre of the graph. This centre is a least likely state as it is located in the least dense region of the sample space. It is nonsensical and should not be used to represent the "type".
3. *Taxa: species-based or species-free?* At the outset of vegetation science species-based taxa proved to be most successful for the purpose of comparison of heterogeneous sampling sites (Braun-Blanquet 1932). This approach fails when comparison is needed between distant sites. While physiognomically very similar, the stands may be composed of different species only so that in the extreme none will be in common. The fact that species from different families tend to form similar life forms and life strategy under similar environmental conditions is known as convergent evolution (Lausi and Nimis 1985; Orians and Solbrig 1977). One of the best-known examples is the succulent life form of *Cactaceae* in the New World and its counterpart of Euphorbiaceae in the Old World. An example of

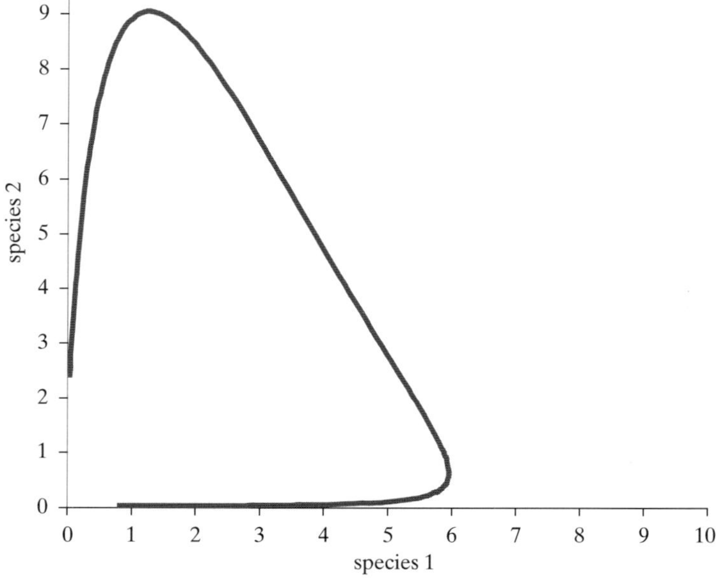

Fig. 2. Plot of relevés from fig. 1 in phase space. Species 1 and 2 now form axes x and y.

spectacular similarity is found in four remote locations, on Maui, Hawaii with *Argyroxiphium sandwichense* (Asteraceae), in Southern California with *Yucca whipplei* (Agavaceae), on dry sites in Chile with *Puya berteroniana* (Bromeliaceae), and on the Canary Islands with *Echium wildpretii* (Boraginaceae), all belonging to different families. There are three different approaches to handle this analytically. The plant species mentioned have the same life form (Raunkiaer 1934), but they also use the same strategy to respond to stress and therefore belong to the same plant functional type (PFT, Box 1996). Using character set types, as a third option, can be a generalisation of both (Orlóci 1991b; Pillar and Orlóci 1991). The latter involves scoring plants individually based on many properties that may include life form and the functional type. A unique difference of the Orlóci-Pillar system is its nested arrangement as compared to a sequential arrangement typically in Raunkiaer's case. The taxonomy is then a result of data analysis rather than some pre-established categories. The nested system for which Pillar and Orlóci (1991) present an efficient algorithm is extremely flexible and the set of descriptors can be optimised. As a result, individuals of the same species are allowed to belong to different character set types, should they differ in some significant respects. An example for the nested system, taken from Orlóci (2001), is shown in Table 1.

4. *Stopping rule in sampling: error variance or structure stability?* How much should be sampled? In the usual statistical context sampling variance is measured by SV=(V/n)(1-n/N) written by Sampford (1962) for practitioners in agriculture. When a stopping rule for sample size n is built, it is based on an assumed threshold for SV. This is arbitrary and so is the resulting sampling size to provide the desired accuracy. In ecology, stopping rules are best built around structures and their stability (Orlóci and Pillar 1989). Figure 3 clarifies the point. The issue is the stress function σ (D, Δ). Two distance configurations are compared, both $n \times n$ symmetric matrices. In a concrete case, D may depict the vegetation configuration and Δ the configuration of environmental factors. The optimal sample size is reached when σ (D, Δ) levels off. Thus, the criterion to stop is the stability of the relationship of the structure of environmental factors and the vegetation structure within the sample. The graph in Figure 3 indicates that this will no longer improve if the sample size is increased beyond n = 20 where any further investment in sampling effort would be wasted.

Other questions refer to processes ("how" questions) and they are subject to choices as well. In most cases, the process cannot be identified as such. Instead, hypotheses have to be considered and tested by data. As their number is steadily growing due to the ongoing progress in science, the choices have also increased in number and combination. Some examples:

Table 1. A typical data set to be used for character-set type analysis (from Orlóci 2001). In addition to standard multivariate analysis, the similarity between species is also considered, yielding an "alternative taxonomy".

Characters	Genus	*Styphelia*	*Vaccinium*	*Dubautia*	*Nephrolepis*	*Cibotium*	*Metrosideros*	*Andropogon*	*Psilotum*	*Myrica*
Life-form	(6 states)	2	2	3	3	1	1	4	3	2
Stem* tissue	(4 states)	3	3	3	3	1	3	4	3	3
Leaf texture	(4 states)	3	3	3	3	3	3	2	4	3
...										
Cover %		5	<1	10	<1	5	20	45	<1	<1

* or stem like organ

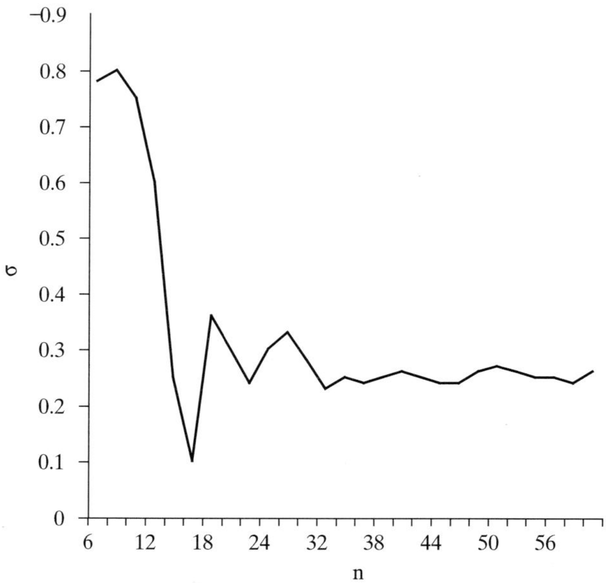

Fig. 3. Graph of stress function. Divergence σ between environmental data and vegetation data levels off around sample size n = 20 (from Orlóci 2001).

5. *Responses and interactions: linear or nonlinear?* As mentioned above, nonlinearity is the rule in the world of ecology. While linearity is often a good approximation for local description, it is the exception in a global view. The point is illustrated in Figures 1 and 2. The two species populations (X_1, X_2) are responding as a function of a hypothetical environmental factor, and since the gradient is long, the scatter diagram (Fig. 2) is a horseshoe in two or a spiral in three dimensions (Groenewoud 1965). This may have devastating effects on the reliability of standard statistical manipulations like the one explained under 2 (centroids). The same holds for density functions being far from the frequently assumed Gaussian type (Orlóci 1993). Consequently, central limit theorems may come into play.
6. *Unconventional phase space properties in focus:* The properties of phase space are not straightforward. Concentrating on the behaviour of moments and product moments is rather arbitrary as it implies the existence of a characteristic function. The Markov process as a promising approximation of linear vegetation change (Lippe *et al.* 1985, Orlóci *et al.* 1993) presents a case in point. Considering shape, the fractal nature of phase trajectories may be an appropriate scalar (Mandelbrot 1977). The fuzzyness of systems may be applied (Feoli and Zuccarello 1991) in other cases, and entropy (Margalef 1958), information (Rényi 1961) and probability (Goodall 1964) are important concepts.
7. *Coping with uncertainty: chaos or stochasticity?* It has been shown by Anand (1997) that trajectories of vegetation succession are characterised by chaotic behaviour. This is an alternative to the standard statistical view in which regularity is given by a deterministic component and uncertainty comes from a stochastic source. What has to be taken into account is the possibility that chaotic behaviour can spring from determinism and it may convolute with an additional stochastic component (Anand 1997). The issue is illustrated

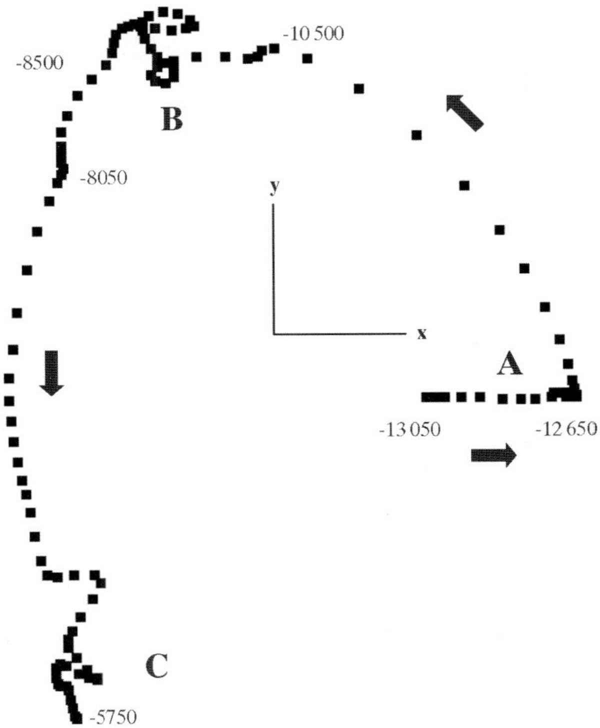

Fig. 4. Trajectory of pollen records of tree species. Two-dimensional space (x, y) obtained by principal coordinates analysis. The numbers are years before present. A, B and C are attractor areas. Data are from Lischke *et al.* (2002).

in Figure 4, displaying an ordination of a time series from a pollen diagram from Soppensee, Switzerland (Lischke *et al.* 2002). The trajectory is characterised by the occurrence of several strongly directed phases, separated by time segments of apparently random fluctuation. This kind of pattern can be observed in almost all long time series of vegetation data (Orlóci *et al.* 2002).

At first glance the "why" question appears to be out of reach for personal choices. However, scale effects restrict the type of causes that can be identified. And scale is set by the design of the investigation.

8. *Coping with scale and hierarchy: what level?* The scale of measurement has two parts: grain (resolution) and extent (overall size) (Wildi *et al.* 2004). Scale effects exist in space and time, but there is also a thematic resolution, i.e. the number and type of variables considered. The effect of scale is quite obvious as a spatial reference. For example, competition and facilitation among plant individuals, at one extreme, can be observed within the range of a few meters. The effect of climate, however, may best be seen on the scale of an entire region. As Parker and Pickett (1998) pointed out, the scales above and below the one chosen is relevant too: "The middle level represents the scale of the investigation, and processes of slower rate act as the context and processes of faster rates reflect other mechanisms, initial conditions or variance."

It can be seen that how to choose among options is far from being trivial as it will affect the results. It is in fact a most crucial task in ecological research and in the forging of a general agreement about coordinated research strategies.

General Principles

It turns out that any new insight into the vegetation process has been reached when the study approach respected general principles of design, well known to ecologists. Orlóci (2001) sees these principles as flexibility, empiricism, local relevance and proper selection of lines of approach.

Flexibility: Flexibility is very much the hallmark of what Poore (1955, 1962) and Orlóci 1993) described as successive approximation and Wildi and Orlóci (1991) as flexible analysis. In these, refinements and reasoned change are allowed in problem direction and technique at any stage. Therefore, a conclusion is not a "one shot" undertaking, but a process. It matures by refinements. When no further refinement can be achieved with the expenditure of reasonable effort, the conclusion becomes an inference.

Empiricism: Empiricism implies that the basic propositions about nature are derived from observation and experiment. This empiricism goes hand in hand with abstract reasoning to fill gaps in knowledge, left blank by things intangible in nature whose existence is suspected from conditions found elsewhere. This is like seeing an apple falling down and not up, and inferring the existence of a force, gravitation. It is also like observing populations do not stay put, but migrate and thereby avoid destruction; hence the Cainean principle of organic evolution being slower than environmental change (Cain 1944). Furthermore, it is like inferring the existence of a process, community development, from observation of the existence of the connection between sediment age and concordant differences in community composition, which Kerner has done.

Local relevance: When the conclusion about nature is reached by a method that is logically connected to the actual conditions of the sampling environment, the conclusion is "relevant". But is it generally relevant or locally relevant? General relevance is not the opposite of local relevance, provided that generality comes about not as a consequence of unreasonable assumptions made about the sampling environment.

Lines of approach: The study of community dynamics traditionally involves a choice between fundamentally different lines of approach:

Reductionist line: Elementary manifestations of community dynamics are emphasised, which appear most likely at the site (stand, patch, gap) level and at short period length, easily linked to population processes, such as tolerance and inhibition (Connell and Slatyer 1977), life-history and competitive relations (Grime 1977; Peet and Christensen 1982), cohort senescence (Mueller-Dombois 1992), reproductive strategies (Harper 1977), and propagule bank composition (Egler 1954).

Holistic line: Community level manifestations are emphasised, such as the process trajectory attributes, phase structure (linear, non-linear), attractor migration (direction of process), directedness (determinism), periodicity (velocity oscillations), dimensionality (trajectory shape complexity), parallelism or global process concordance (Orlóci 2000; Orlóci *et al.* 2002).

Implicit in the choice of reductionism as ones guiding line of philosophy is the assumption, that the governing principles of vegetation dynamics are derivable from manifestations of the process, at the individual and population levels. The history of the last three decades or so is ample proof that this assumption is specious. This conclusion is repeated in experience that Gleick (1987) and Çambel (1993) narrate in a science-wide sense.

The holistic mode of search for governing principles does not imply that the search stops at the community level. Rather the search is hierarchical in that it fragments into finer levels down through scales where the search absorbs increasing amounts of the information revealed about the process, ultimately at the population and individual levels. In this way the search for governing principles is indeed an exercise in successive approximation through scales, and possibly involving the joint long-term efforts of researchers at different places of the globe.

References

Allen T.F.H. and Starr T.B. 1982. Hierarchy: perspectives for ecological complexity. The University of Chicago Press. 310 p.

Anand M. and Orlóci L. 1996. Complexity in plant communities: the notion and quantification. J. theor. Biol. 179: 179–186.

Anand M. 1997. The fundamental nature of vegetation dynamics – a chaotic synthesis. COENOSES 12 (2–3): 55–62.

Box E.O. 1996. Plant functional types and climate at the global scale. Journal of Vegetation Science 7: 309–320.

Braun-Blanquet J. 1932. Plant sociology. The study of plant communities. (Translated by G.D. Fuller and H.S. Conard.) McGraw-Hill, New York and London.

Bray J.R. and Curtis J.T. 1957. An ordination of the Upland Forest communities of southern Wisconsin. Ecological Monographs 27: 325–349.

Brzeziecki B., Kienast F. and Wildi O. 1993. A simulated map of the potential natural forest vegetation of Switzerland. Journal of Vegetation Science 4: 499–508.

Cain S.A. 1944. Foundations of Plant Geography. Harper, New York.

Çambel A.B. 1993. Applied Chaos Theory: a Paradigm for Complexity. Academic Press, New York.

Clements F.E. 1916. Plant succession. An analysis of the development of vegetation. Carnegie Institute, Washington, Publication 242, Washington D.C.

Connell H.J. and Slatyer R.O. 1977. Mechanisms of succession in natural communities and their role in community stability and organisation. American Naturalist 111: 1119–1144.

De Patta Pillar V. and Orlóci L. 1991. Fuzzy components in Community Level Comparisions. In: Feoli E. and Orlóci L. Computer assisted vegetation analysis. Kluwer Academic Publishers, Dortrecht: 87–93.

Edgington E.S. 1987. Randomization Tests. 2nd ed. Marcel Dekker, New York.

Egler, F.E. 1954. Vegetation science concepts. I. Initial floristic composition – a factor in old-field vegetation development. Vegetatio 4: 412–417.

Ewald J. 2003. A critique for phytosociology. Journal of Vegetation Science 14: 291–296.

Feoli C. and Zuccarello V. 1991. Syntaxonomy: a source for useful fuzzy sets for environmental analysis? In: Feoli E. and Orlóci L. (eds). Computer assisted vegetation analysis, pp. 265–271. Kluwer, Dortrecht.

Feoli E. and Orlóci L. 1991. Computer assisted vegetation analysis. Handbook of vegetation science 11. Kluwer Academic Publishers, Dortrecht, Boston, London.

Gleason H.A. 1926. The individualistic concept of the plant association. Bulletin of the Torrey Botanical Club 53: 7–26.

Gleason H.A. 1939. The individualistic concept of the plant association. Amer. Midl. Nat. 21: 92–110.

Gleick J. 1987. Chaos. Making a New Science. Penguin Books, New York.

Goodall D.W. 1954. Objective methods for the classification of vegetation. III. An essay in the use of factor analysis. Aust. J. Bot. 2: 304–324.

Goodall D.W. 1964. A probabilistic similarity index. Nature 203: 1098.

Goodall D.W. 1975. Setting objectives for ecological research. Bull. Ecol. Soc. Aust. 5: 3–8.

Green R.H. 1979. Sampling design and statistical methods for environmental biologists. Wiley-Interscience, New York, Chichester, Brisbane, Toronto. 257 p.

Greig-Smith P. 1983. Quantitative Plant Ecology. 3rd. Ed. Blackwell Scientific Publications, Oxford. 359 pp.
Grime J.P. 1977. Evidence for the existence of three primary strategies in plant and its relevance to ecological and evolutionary theory. Am. Nat. 111: 1169–1194.
Groenewoud van H. 1965. Ordination and classification of Swiss and Canadian coniferous forests by various biometric and other methods. Diss. ETH Nr. 3700. Berichthaus Zürich.
Guisan A. and Zimmermann N.E. 2000. Predictive habitat distribution models in ecology. Ecological Modelling 135: 147–186.
Harper J.L. 1977. The population biology of plants. Academic Press, London.
Hubbell S.P. 2001. The unified theory of biodiversity and biogeography: a synopsis of the theory and some challenges ahead. In: Silvertown J. and Antonovics J. (eds.). Integrating ecology and evolution in a spatial context. Blackwell, pp. 393–411.
Jongman R.H.G., ter Braak C.J.F. and van Tongeren O.F.R. 1987. Data analysis in community and landscape ecology. Pudoc Wageningen.
Kenkel N.C. and Orlóci L. 1986. Applying metric and nonmetric multidimensional scaling to ecological studies: some new results. Ecology 67: 919–928.
Kerner A. von Marilaun 1864. Das Pflanzenleben der Donauländer. 2nd ed. by F. Vierhapper, 1929. Univ.-Verlag Wagner, Innsbruck.
Kullback S. 1959. Information theory and statistics. New York, Wiley.
Landolt E. 1977. Oekologische Zeigerwerte zur Schweizer Flora. Veröff. Geobot. Inst. ETH, Stiftung Rübel 64. 208 p.
Lausi D. and Nimis P.L. 1985. The study of convergent evolution in plants and plant communities. A quantitative approach. Abstracta Botanica 9: 67–77.
Legendre P. and Legendre L. 1998. Numerical Ecology. 2nd ed. Elsevier, Amsterdam.
Legendre P. and Fortin M.-J. 1989. Spatial analysis and ecological modelling. Vegetatio 80: 107–138.
Lippe E., De Smitt J.T. and Glenn-Lewin D.C. (1985): Markov models and succession: a test from a heathland in the Netherlands. Journal of Ecology 73: 775–791.
Lischke H., Lotter A.F. and Fischlin A. 2002. Untangling a Holocene pollen record with forest model simulations and independent climate data. Ecological Modelling 150: 1–21.
Mandelbrot B.B. 1977. Fractals: form, chance and dimension. Freeman, San Francisco.
Manel S., Schwartz M.K., Luikart G. and Taberlet P. 2003. Landscape genetics: combining landscape ecology and population genetics. Trends in Ecology and Evolution 18: 189–197.
Margalef R. 1958. Information theory in ecology. General Systems 3: 36–71.
Mueller-Dombois D. and Ellenberg H. 1974. Aims and Methods of Vegetation Ecology. John Wiley and Sons, New York Chichester Brisbane Toronto. 547 p.
Mueller-Dombois D. 1992. Vegetation of the Wet Windward Slope of Haleakala, Maui, Hawaii. Pacific Science 46, 2: 197–220.
Orians G.H. and Solbrig O.T. 1977. Convergent Evolution in Warm Deserts. Dowden, Hutchinson Ross, Stroutdsburg, Pennsylvania.
Orlóci L. 1978. Multivariate analysis in vegetation research. 2nd ed. Dr. W. Junk, The Hague. 451 p.
Orlóci L. 1991a. CONAPACK: A program for Canonical Analysis of Classification Tables. Ecological Computations Series: Vol. 4. SPB Academic Publishing, The Hague.
Orlóci L. 1991b. On character-based plant community analysis: choice, arrangement, comparison. Coenoses 6: 103–107.
Orlóci L. 1993. The complexities and scenarios of ecosystem analysis. In: G. P. Patil and C. R. Rao (eds.). Multivariate Environmental Statistics, Elsevier Scientific, New York. pp. 423–432.
Orlóci L. 2000. From Order to Causes. A personal view, concerning the principles of syndynamics. http://sites.netscape.net/lorloci
Orlóci L. 2001. Prospects and expectations: reflections on a science in change. Community Ecology 2 (2): 187–196.
Orlóci L and Orlóci M. 1985. Comparison of communities without the use of species: model and example. Ann. Bot. (Roma) 43: 275–285.
Orlóci L. and Pillar de Patta, V. 1989. On sample size optimality in ecosystems survey. Biometrie-Praximetrie 29: 173–184.

Orlóci L., Anand M. and He X. 1993. Markov chain: a realistic model for temporal coenosere? Biom. Praxim. 33: 7–26.

Orlóci L., Pillar V.D., Anand M. and Behling H. 2002. Some interesting characteristics of the vegetation process. Community Ecology 3(2): 125–146.

Parker V.T. and Pickett S.T.A. 1998. Historical contingency and multiple scales of dynamics within plant communities. In: Peterson D.L. and Parker V.T. (eds.). Ecological Scale: Theory and Applications. Columbia University Press. pp. 171–191.

Peet R.K. and Christensen, N.L. 1980. Succession: a population process. Vegetatio 43: 131–140.

Pielou E.C. 1975. Ecological Diversity. Wiley, New York.

Pielou E.C. 1984. The interpretation of ecological data. John Wiley and Sons. New York, London, Sidney, Toronto. 263 p.

Pillar V. De Patta and Orlóci L. 1991. Fuzzy components in community level comparisions. In: Feoli E. and Orlóci L. Computer assisted vegetation analysis. Kluwer Academic Publishers, Dortrecht: 87–93.

Pillar V. De Patta 1993. Character-based community analysis: the theory and an application program. Ecological Computations Series (ECS). Vol. 5. SPB Academic Publishing bv. The Hague, the Netherlands.

Podani J. 1984. Spatial processes in the analysis of vegetation: theory and review. Acta Botanica Hungarica 30: 75–118.

Poore M.E.D. 1955. The use of phytosociological methods in ecological investigations. I–III. J. Ecol. 43: 226–244, 245–269, 606–651.

Poore M.E.D. 1962. The method of successive approximation in descriptive ecology. Advances in Ecological Research 1: 35–68.

Raunkiaer C. 1934. The life forms of plants. Oxford University Press, Oxford. (Translated from the original published in Danish, 1907)

Rees M., Condit R, Crawley M., Pacala S. and Tilman D. 2001. Long-Term Studies of Vegetation Dynamics. Science 293: 650–655.

Rényi A. 1961. On measureso fo entropy and information. In: J. Neyman (ed.), Proceedings of the4th Berkeley Symposium on Mathematical Statistics and Probability, pp. 547–561. Univeristy of California Press, Berkeley.

Roberts D.W. 1986. Ordination on the basis of fuzzy set theory. Vegetatio 66: 123–131.

Sampford M.R. 1962. An Introcuction to Sampling Theory with Applications to Agriculture. Oliver and Boyd, Endinburgh.

Shannon C.E. 1948. A mathematical theory of communication. Bell System Tech. J. 27: 379–423.

Smith T.M., Shugart H.H., Woodward F.I. and Burton, P.J. 1993. Plant Functional Types. In: Solomon A.M. and Shugart H.H. (eds.). Vegetation Dynamics and Global Change, pp. 272–292. Chapman and Hall, London.

Tansley A.G. 1920. The classification of vegetation and the concept of development. J. Ecol. 8: 118–148.

Tansley A.G. 1935: The use and abuse of vegetational concepts and terms. Ecology 16, 3: 284–307.

Tilman D. 1988. Dynamics and Structure of Plant Communities. Princeton, New Jersey.

Warming E. 1909. Oecology of plants. An introduction to the study of plant communities. Oxford University Press, London. 422 p.

Westhoff V. and van der Maarel E. (1978): The Braun-Blanquet approach. In: Whittaker, R.H. (ed.). Classification of plant communities, pp. 289–399, W. Junk, The Hague.

Whittaker R.H. 1978. Direct gradient analysis. In: Whittaker R.H. (ed.). Ordination of Plant Communities. Junk, The Hague: 7–50.

Wildi O. and Orlóci L. 1991. Flexible gradient analysis: a note on ideas and an application. In: Feoli E. and Orlóci L. (eds.). Computer assisted vegetation analysis, pp. 265–271. Kluwer, Dortrecht.

Wildi O., Feldmeyer-Christe E., Ghosh S. and Zimmermann N.E. 2004. Comments on vegetation monitoring approaches. Community Ecology 5: 1–5.

Zimmermann, H.J. 1985. Fuzzy set Theory – and its Applications. Kluwer-Nijmhoff Publishing. Dortrecht.

Statistical Analysis of Landscape Data: Space-for-time, Probability Surfaces and Discovering Species

Sucharita Ghosh* and Otto Wildi

Swiss Federal Research Institute WSL, Zürcherstrasse 111, CH-8903 Birmensdorf. Switzerland
* Corresponding author: e-mail: rita.ghosh@wsl.ch

Abstract
In statistical terms, a landscape may be perceived as the realization of a space-time stochastic process. In this treatise, we focus mainly on space and address three very specific problems namely, space-for-time substitution, construction of probability or quantile surfaces and prediction of a species-area curve.

Keywords: chronosequence, curve and surface estimation, self-similarity, space-for-time substitution, species-area relations

F. Kienast, O. Wildi & S. Ghosh (eds.), A Changing World. Challenges for Landscape Research, 209–221.

Introduction

Space-time stochastic processes play important roles in many areas of landscape research. One example is stochastic population dynamics which includes predator-prey interactions, competition, spread of diseases, dispersal, birth and death, migration etc. (Bartlett 1960). In some instances, not the magnitude of an observation, but rather the points of occurrence of events in space and/or in time are of interest. This leads to point processes or point patterns (Cressie 1993; Cox and Isham 1980; Diggle 2003). In other examples, average behavior of a stochastic process is of interest (Cressie 1993).

The classical approach to model such processes is to use parametric likelihood based methods (maximum likelihood estimation; Cox and Hinkley 2000) for estimating the model parameters. Nonparametric models and optimal smoothing methods are likely to extend the scope of applicability. These methods are useful when there is little reason to believe in a completely specified form of the underlying probability distribution function (Eubank 1988; Fan and Gijbels 1996; Silverman 1986).

Developing statistical methods to update physical or ecological models using sampled observations, to improve upon or even to simplify the so far known models and to make realistic estimation and prediction are also of interest. In some situations, efficient and parsimonious statistical methods are needed to avoid lengthy simulations in numerical models. Two possible approaches to achieve these are (a) semiparametric methods and (b) Bayesian inference. In a Bayesian approach (Berger 1999), parametric prior density functions are used and posterior inference is carried out using observed data that enter through the likelihood function; for an example see Wikle *et al.* (2001). Semiparametric methods combine parametric models that use specific forms of probability distributions with flexible nonparametric components. An example is a partial linear model that is a regression type relationship between two or more processes, where the functional form of the regression function is not fully specified. Beran and Ghosh (1998) discuss such a model for comparing temperature trends in the northern and the southern hemispheres. Drawing statistical inference for changes in landscapes and to find probability models for uncertainties in model-outputs are also important areas of future research (also see Nychka 2000).

Statistical analysis of landscape data will thus typically involve analysis of space-time data, although in specific cases, one may have either spatial or time series data. What distinguishes spatial data from time series is that generalization of temporal ordering is not obvious in space. What one needs in space are the concepts of *neighborhoods* and *directions* (Mardia 1972). Analysis of environmental time series data is treated in Ghosh *et al.* (2007). In this paper we bring in space and discuss three examples from landscape research. We consider space-for-time substitution for understanding long-term changes in vegetation, construction of probability and quantile surfaces as an alternative to fitting mean surfaces as for instance considered in geostatistics, and the problem of extrapolating a species-area curve. The space-for-time substitution described here is based on Euclidian distances between vectors of observations from different plots. The method described here is fully data-driven. In future research, spatial (space-time) probability models will have to be considered for generating the chronosequences. The problem of constructing probability and quantile surfaces is motivated by the fact that species may react more to major environmental variability and extreme events than to gradual changes in means. High environmental fluctuations can cause species extinction and invasion. Therefore, quantile surfaces for environmental variables are of particular interest. Species-area relations occur in the context of biodiversity research. The key problem here is to estimate the number of species in the landscape by extrapolating the information obtained in a sample. It turns out that extrapolation of a species-area curve can be affected by a self-similarity type property. This is discussed in the context of estimating the number of species in an infinitely large landscape.

Space-for-time Substitution

As far as analyses of space-time data are concerned, important issues are the length of the time series at each location and the total number of spatial locations where data are available. For instance when the length of the time series at a given location is small, one may combine information from 'nearby' spatial locations to achieve precision. Repeated measures type ideas (Diggle *et al.* 1994) are useful in these contexts where observations from nearby spatial locations serve as replicates (Wildi 2002; Wildi and Schütz 2000). Ghosh (2001) uses a Bayesian type argument in a time series context to show that pooling information from repeated measures helps reduce some of the uncertainty. Generalization of this theory to space will be taken up in future research.

This discussion takes us to another very different concept, namely the hypothesis of space-for-time substitution also known as a chronosequence approach. Filed data of succession over long time periods are rare (Foster and Tilman 2000; Wildi 2002). For the investigation of long-term change, the chronosequence approach sometimes offers the only alternative to excessively long observation time. In a chronosequence, trends observed in plots from different locations are averaged. It is therefore also known as space-for-time substitution (Pickett 1989). The approach involves several well-known risks. The most obvious is that the ecological conditions in two different plots are often close, but never identical.

Despite many potential pitfalls in the interpretation of results, various investigations have shown the fact that rates of change in succession as well as the direction are directly related to the initial conditions (Myster and Pickett 1994), a prerequisite of space-for-time substitution. This is shown schematically in Figure 1 explaining the univariate case. Three time series are recorded simultaneously at different locations (heavy lines in Fig. 1). A comparison of all measurements, after shifting the series (light line in Fig. 1) shows that they all belong to

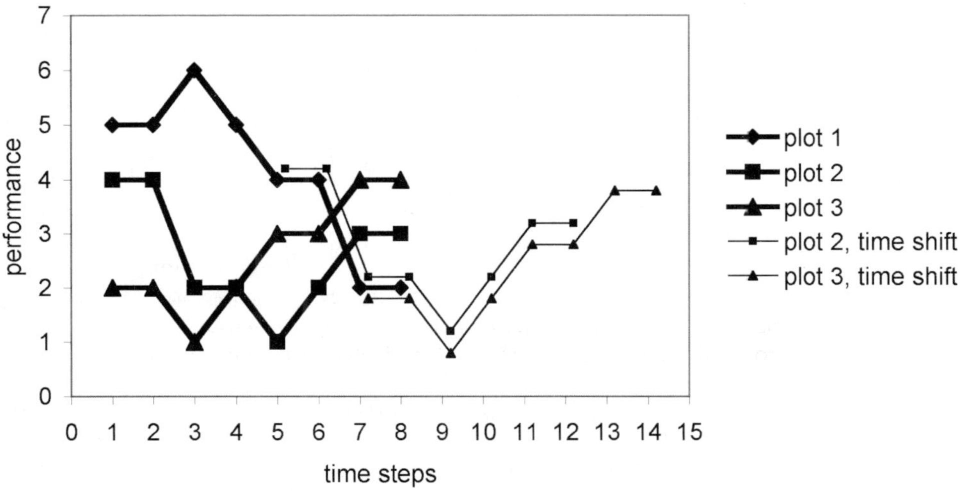

Fig. 1. Schematic representation of the one dimensional case. When the measurements simultaneously taken at different locations (heavy line) are properly shifted (light line) a unique chronosequence emerges.

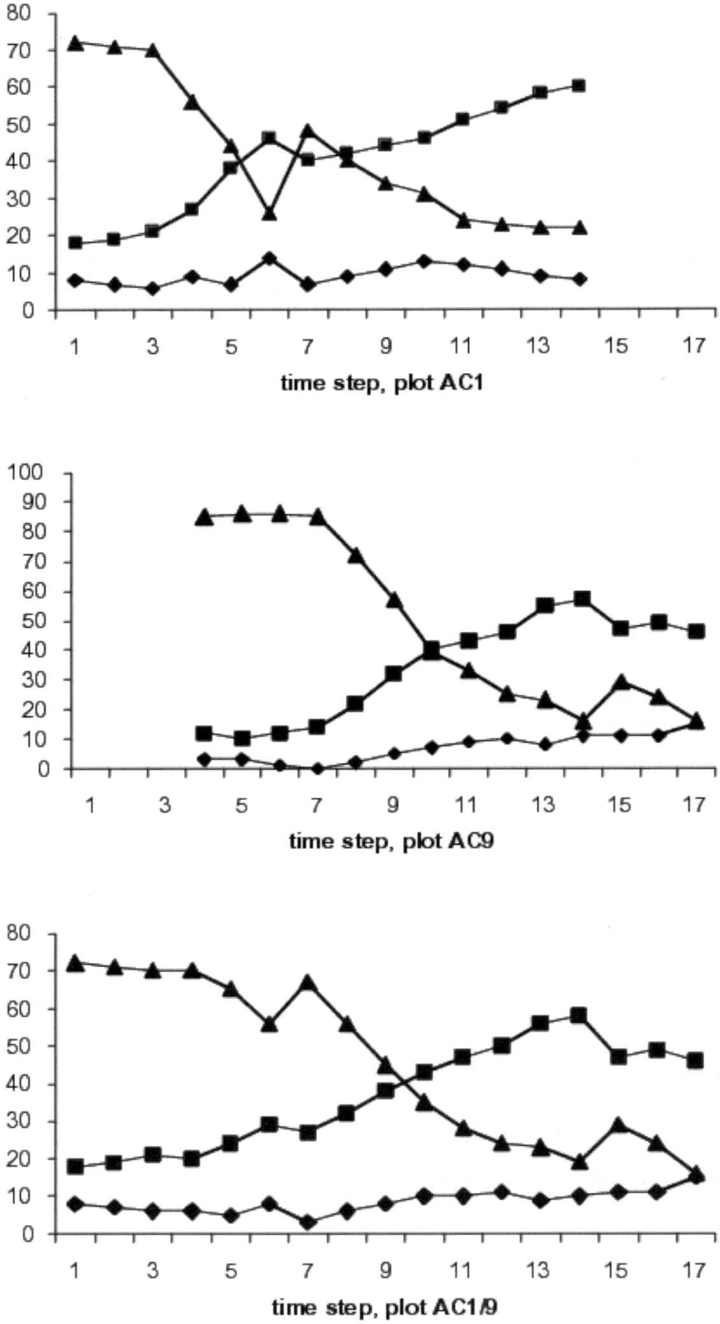

Fig. 2. Space-for-time substitution for three selected plant functional types from the Swiss National Park (Wildi and Schütz 2000). A new time series AC1/9 (time steps 1 through 17) is generated by superimposing data from plots AC1 (time step 1 through 14) and AC9 (time step 4 through 17). The plant functional types are *Aconitum* (triangles), *Deschampsia* (squares) and *Trisetum* (rhomboids).

the same chronosequence. In the univariate case, Euclidean distance e_{ij} is used for comparison. This is defined as

$$e_{ij} = \left[\sum_{k=1}^{p} (X_{ik} - X_{jk})^2 \right]^{1/2}$$

for any relevé i and j with p species. An example is given in Wildi and Schütz (2000). They take sample data from 59 plots in the Swiss National Park, where the earliest observations started in 1917 until present with time intervals of 5 years. These plots were established to document secondary succession from fertilized pastures poor in nutrients towards early forest stages. Hence, each plot describes a short but a different fraction of succession. Using space-for-time substitution (Fig. 1) they obtain succession series of 415 years (Figs. 2 and 3).

Unlike e.g. in dendrochronology, where the age of individual records can be measured by ^{14}C dating, a succession series cannot be verified in practice or even observed directly. Under these circumstances, developing a model generating the observed temporal pattern offers a chance to verify the results. Wildi (2002) concludes that it was relatively easy to find appropriate functions generating the pattern. However, to include realistic assumptions, spatial processes need to be considered simultaneously.

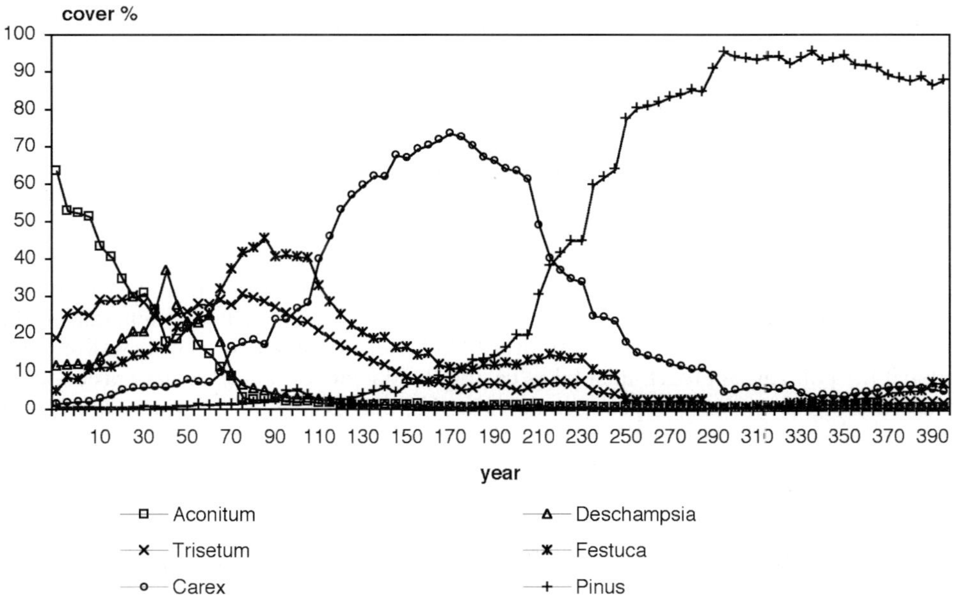

Fig. 3. Artificial time series (chronosequence) composed of 59 measured series described by six plant functional types (Wildi and Schütz 2000). The data document succession in the Swiss National Park.

Probability and Quantile Surfaces

In many instances of landscape research, the central issue is the recognition of the underlying probability distribution. Using precipitation as an example, the mean yields an estimation of water availability *on the average*. But the lower end of the probability distribution is an expression of the risk of drought, whereas the upper end may be investigated while considering the risk of a flood. Statistical inference concerning the whole probability distribution or its quantiles provides a comprehensive view of the specific landscape process. The graphical representation of the high or low quantiles or the extreme ends of the underlying probability distribution can be used as maps for indicating potential areas of risks. A spatio-temporal representation of such maps would then provide the space-time development of these risks. In the time series context, these ideas have been introduced in Ghosh *et al.* (2007).

To extend these concepts to space, suppose that we have an observation Y or more precisely $Y(t,s)$ recorded at a spatial location s, being followed up in time which we denote by t. Typically, s is a two dimensional vector, but it need not be; s can also have one dimension. The observations themselves can be vectors including physical, biological and other descriptions. In the purely spatial context, we suppress t and write $Y(s)$. We now proceed with the idea that the magnitude of Y is our main concern and the problem is to quantify in what way the values of Y differ in space and time.

When only the average behavior or the expected value of Y namely $\mu(t,s) = E[Y(t,s)]$ is of interest, one estimates $\mu(t,s)$ directly from the data. This for instance is the concern of a geostatistical analysis (Matheron, 1963) in the purely spatial context. In this approach, weighted averages of spatially located observations are taken and the weights are chosen to satisfy some criterion involving the variogram. For geostatistics in particular, the concepts of stationarity and isotropy are crucial (ibid.), so that instead of using raw data where such assumptions need not apply, one may use the residuals obtained from a regression fit for a subsequent geostatistical analysis (Cliff and Ord 1981; for an application in forestry see Ghosh *et al.* 1997b). Other methods for fitting smooth surfaces also exist. For an overview of statistical inference for spatial processes, see for instance, Cressie (1993) and Ripley (1988).

In landscape research, one important aim is to quantify and predict changes. One proposal for doing so is via a direct study of the underlying probability distribution function (pdf) F. This results in a very general approach because average, variance, skewness or any other parameter say θ can be mathematically expressed as a functional $\theta(F)$. Thus if we start with the axiom that variability or changes in the landscape process are in fact inherent in the changes in the underlying probability distribution function, then we can as well directly model the pdf. This approach then takes us into the domain of non-stationary processes in the sense that the pdf of Y depends on its space-time coordinates. As for which parameter θ is to be investigated will depend on the particular question in mind. In general, in the light of global changes and environmental risks, a description of the average or the expected value of Y need not suffice and some assessment of high variability and other extreme behavior may need to be taken into account. An example of this is given in section 5.1 in the temporal context and similar examples can also be found in space (Draghicescu 2002). The marginal pdf of $Y(t,s)$, at a given point of time and a spatial location, is defined as

$$F_y(t,s) = P(Y(t,s) \leq y), y \in R$$

where R refers to the real line $(-\infty,\infty)$. A mathematically equivalent quantity is the space-time quantile, and it is defined as the inverted F function. Thus, an a quantile is a number $y_a(t,s)$ such that

$$1-\alpha = P(Y(t,s) \leq y_\alpha(s,t)), 0 < \alpha < 1$$

and the inversion formula is

$$y_\alpha(s,t) = \inf\{y \mid F_y(s,t) \geq 1-\alpha\}.$$

By fixing time, and for a given cut-off value (threshold) $y \in R$ and probability is α, one can compute the probability surface $F_y(t,s)$ or the quantile surface $y_\alpha(t,s)$ Although, quantiles and probability distributions are mathematically equivalent, in applications, quantiles may be easier to interpret as these have the same units of measurement as the raw observations. Large and small values of α correspond to extreme or near-extreme events.

If there is reason to believe in a parametric form of the underlying probability distribution, then the method of maximum likelihood can be used to estimate the parameters. On the other hand, if there is no prior knowledge concerning the shape of the probability distribution of Y, nonparametric smoothing methods can be applied using moving windows. In space, a window refers to the concept of a neighborhood or a collection of points that are near a specified location in space or in space-time. One may optimize the size of the window either by the size of the space-time interval or by the number of data points that are in it. Mathematical analysis can demonstrate that the size of the window will have opposite effects on bias and variance of the estimator. In particular, a larger window size would increase bias but reduce variance. A compromise is reached by choosing that particular window size that minimizes the mean squared error which is equal to the sum of the variance and the square of the bias. This is known as optimal bandwidth selection.

Nonparametric estimation of space-time quantiles can be carried out either by inverting an estimate of the probability distribution function or by direct calculation of local quantiles. For instance, suppose that long-term time series data are available from k spatial locations. Then a two-stage smoothed estimate of the $(1-\alpha)$ quantile can be calculated as follows. At the first stage, let $\hat{\theta}_\alpha(t,s_j)$ be the estimated $(1-\alpha)$ quantile at time obtained from time series of observations at location s_j. The method to compute $\hat{\theta}_\alpha(t,s_j)$ is described in chapter 5.1. In the first stage, $\hat{\theta}_\alpha(t,s_j)$ is computed at every fixed spatial location. At the second stage, $\theta_\alpha(t,s_j)$ is smoothed over all spatial locations using the following formula for interpolation:

$$\bar{\theta}_\alpha(t,s) = \frac{1}{kh}\sum_{j=1}^{k} K(s_j,s,h)\hat{\theta}_\alpha(t,s_j), 0 < \alpha < 1.$$

Figure 4 shows 0.95 quantile surfaces for summer precipitation in Switzerland (*reproduced from* Draghicescu 2002). In this formula, $K(s_j,s,h)$ is a symmetric kernel with window-width h centered at the spatial location s. The idea is that as a function of s_j the value of the kernel function goes down with increasing distance of s_j from s. Thus the above sum is a weighted average of estimates $\hat{\theta}_\alpha(t,s_j)$ from locations s_j, $j=1,2,\ldots,k$. In this weighted average, the estimates from locations that are further away from the target location get lower weight. Specifically, suppose that the spatial locations are two-dimensional and of the type $s_j = (s_{1,j},s_{2,j})$, $j=1,2,\ldots,k$.

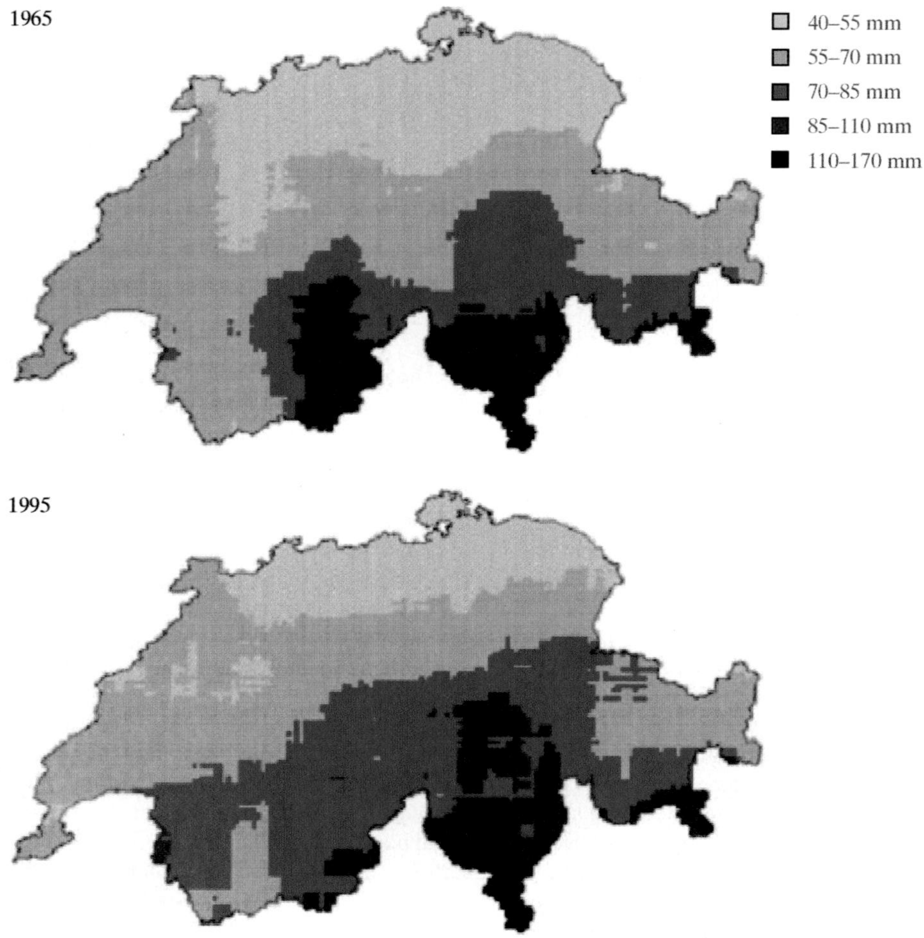

Fig. 4. 0.95 Quantile surfaces for summer maxima of daily precipitation in Switzerland showing climate variations. Darker shade shades represent higher values. The years 1965 and 1995 were chosen for the purpose of illustration. (From Draghicescu (2002); reproduced with permission of the author).

A possible kernel in this case is the product Gaussian kernel. Thus if we consider rectangular windows with area $h=b_1 \times b_2$, the corresponding kernel is a product

$$K(s_j, s, h) = K(s_{1,j}, s, b_1) \times K(s_{2,j}, s, b_2)$$

where

$$K(s_{i,j}, s, b_i) = \frac{1}{\sqrt{2\pi b_i^2}} \exp\{-(s_{i,j} - s)^2 / 2b_i^2\}, i = 1,2.$$

Square windows, i.e. $b_1=b_2$ are often used. For examples of other kernels such as the truncated Gaussian kernel, see Ghosh and Draghicescu (2002).

Self-similarity and Species-area Curves

Self-similarity can be understood as repetition of the same structure at different scales. This concept is related to the idea of fractals. For examples of such processes in nature, see Mandelbrot and Wallis (1968). Beran (1994), Hampel (1987) and Künsch (1986) discuss the mathematical background and statistical aspects of self-similarity in the time series context. Self-similarity is not just a temporal phenomenon; it also occurs in space (Matheron 1962, 1973); for examples, see Vicsek (1989; growth phenomena), Bramson *et al.* (1996; spatial models for species-area curves) and Ghosh (2005; random fields for explaining power-laws in species-area curves). The estimated self-similarity parameter (Ghosh *et al.* 1997a) can be used to give estimate and confidence intervals for the expected total number of species in the landscape.

In what follows, we elaborate on the method of Ghosh *et al.* (1997a). These authors analyze presence-absence data on epiphytic lichens from forest plots in Switzerland (Dietrich 1997) and estimate the mean species-area curves (forest plots) by a bootstrap method combined with a log-linear regression model. Bootstrap approach to species-area curve has been discussed by other authors as well (Smith and van Belle 1984). However, Ghosh *et al.* (1997a) resample *paths* rather than plots, and fit a log-linear model to the *increments* (first difference) of the estimated species-area curve. Parametric estimation of increments has appeared elsewhere in the literature, e.g. Mingoti and Meeden (1992), who assume a finite total number of species and independence between plots. The method presented in this article is most applicable when independence of the plots need not hold and the number of species need not converge to a finite value in an infinitely large landscape. Ghosh (2005) postulates a random field model that explains power-law species-area relationships in a large landscape.

Suppose that the set of all observations in the survey corresponds to a random sample ω of randomly selected sites from the population of all ecologically similar sites. Suppose that the randomly selected sites are indexed by the integers $1,2,...,k$. During field work, the observer chooses the plots in a certain way, so that the specific sequence of the plots visited is simply a permutation of the integers $1,2,...,k$. We call this sequence a *path* denoted by π so that $\pi=\{i_1,i_2,...,i_k\}$. Here $i_1,i_2,...,i_k$ are permutations of the integers $1,2,...,k$. The set of all such paths or the set of all possible permutations of these integers is simply the all possible ways the observer can visit the plots. To define a species-area curve in terms of these paths consider the following. For every randomly sampled path π corresponding to the random sample of plots we define a species-area curve as the total number of species in the *first m* plots on this path. We denote this number by $Y_{\pi,\omega}(m)$. Thus, for every π and ω, $Y_{\pi,\omega}(m)$ is a function of m. This function is non-negative (being the number of species) and non-decreasing, i.e. given π and ω, $Y_{\pi,\omega}(m_1)$ will be larger or equal to $Y_{\pi,\omega}(m_2)$ if $m_1 \geq m_2$. However, for different paths π the species-area curves may intersect each other. The *mean species-area curve* is the function $\mu(m)$ such that,

$$Y_{\pi,\omega}(m) = \mu(m) + \varepsilon_{\pi,\omega}(m),\ m \geq 1$$

$$\mu(m) = E_{\omega} E_{\pi|\omega} Y_{\pi,\omega}(m).$$

In these equations, $\mu(m)$ denotes the mean species-area curve whereas $\varepsilon_{\pi,\omega}(m)$ is the difference between the mean species-area curve and the single species-area curve. The $\varepsilon_{\pi,\omega}(m)$ have zero mean but they are *not uncorrelated.* We would like to draw statistical inference for $\mu(m)$ and for simplicity of notations, we suppress ω.

Suppose that the total number of plots in the survey is k. To resample species-area curves, resample paths by simple random sampling with replacements from the collection of all possible paths which is the collection of all possible permutations of the digits 1,2,...,k. This is repeated B times which is typically a large number such as 1000. Let π_j denote the simulated path during the jth simulation, $j=1,2,...,B$. For every m=1,2,..., the sample mean of $Y_{\pi_1}(m), Y_{\pi_2}(m),..., Y_{\pi_B}$(m) is the bootstrap estimate of $\mu(m)$. This is given by the formula

$$\hat{\mu}(m) = \frac{1}{B}\sum_{i=1}^{B} Y_{\pi_i}(m),\ m \geq 1.$$

For extrapolation of the $\hat{\mu}(m)$ curve, various approaches can be taken. For nonparametric methods, see Beran and Ocker (1999) and Ghosh and Draghicescu (2002). However, if a parametric model can be found that fits the data well, then further insight into the spatial distribution of the species can be obtained. In the species-area literature log-linear models are popular as many datasets show straight line features in a log-log scale. Exactly what is to be log-transformed however, will depend on the specific spatial distribution of the species. For the epiphytic lichen dataset log-log model for the increments or the first differences of $\hat{\mu}(m)$ fitted the data with over 99% value for the R^2 coefficient. We discuss the consequences of this fit. Define increments $D(j)=\hat{\mu}(j)-\hat{\mu}(j-1)$ and consider the log-linear model

$$\log(D(j)) = \alpha + \beta \log(j) + e(j),\ j = 1,2,..$$

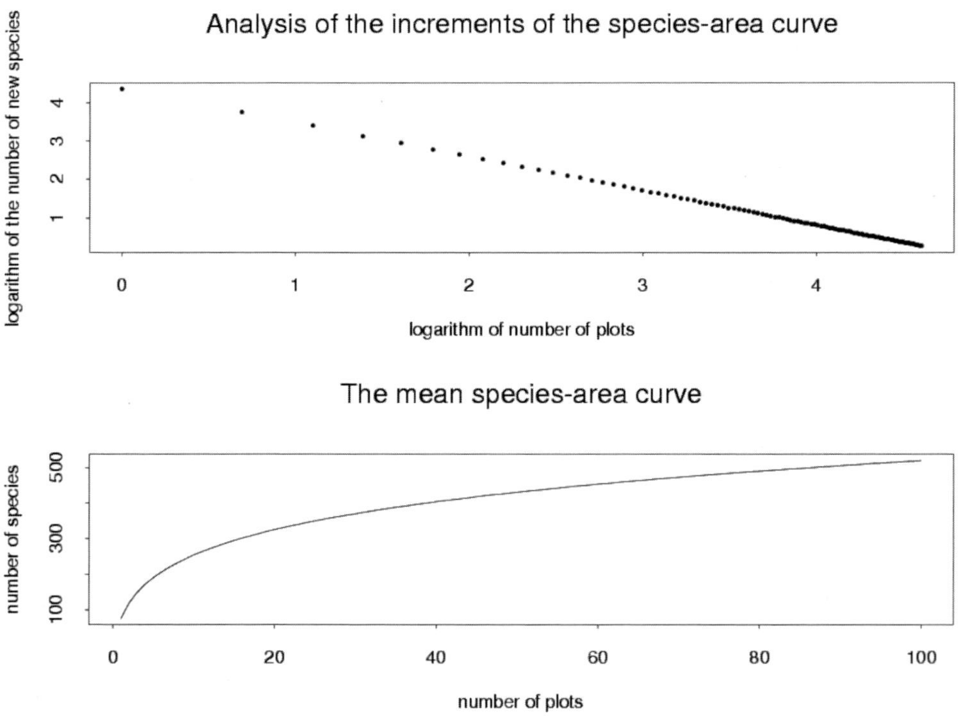

Fig. 5. An example of a species-area relation. Above: number of new species vs. number of plots in log-log coordinates. Below: cumulative number of species vs. number of plots.

The $e(j)$ are assumed to be uncorrelated zero mean normal variables with variance σ^2. For this model, the expected value of the increments is

$$E[D(j)] = \exp\{\alpha + \frac{1}{2}\sigma^2\} j^{\beta}, \; j = 1,2,...$$

Sum of the $D(j)$ values over $j = 1,2,...,m$ is the mean number of species when m sites have been visited. The theoretical expected value of this quantity is,

$$E\sum_{j=1}^{m} D(j) = \exp\{\alpha + \frac{1}{2}\sigma^2\}\sum_{j=1}^{m} j^{\beta}.$$

This is the expected total number of species in sites. In this formula, the parameter β plays an important role. By examining the sum on the right hand side of this equation, we see that, when $\beta<0$, $E[D(j)]\rightarrow\infty$ as $m\rightarrow\infty$. In other words, on the average with increasing number of sites visited, the rate at which new species are encountered diminishes. This is not contrary to intuition. In addition, if β is sufficiently small i.e. $\beta<-1$, then the expected total number of species in the infinitely large landscape (i.e. as $m\rightarrow\infty$) will be finite; this means the species-area curve will reach an asymptote after a moderate number of plots have been visited, even if the landscape is very large. However, if $\beta<0$ but not small enough such that $-1<\beta<0$, then although the rate at which new species are encountered will diminish, the decay will not be fast enough but *hyperbolic*. In this case, as the number of plots visited tends to infinity, the expected cumulative number of species will theoretically diverge! The implication is that, if β is negative but not small enough, one will continue to encounter new species, although at a slower rate, and the cumulative number of species will continue to increase as more plots are visited. An example of such a situation is shown in Figure 5. In this figure, a log-linear species-area relation is shown using $\alpha = 4.351$, $\beta = -0.882$ and $\sigma = 0.04$. These are the regression parameter estimates for the fitted epiphytic lichen species-area curve from forest areas in the Swiss pilot project (Dietrich 1997). The negative value of the slope parameter in the log-linear model ensures that the increment curve has a downward tendency. However since $-1<\beta<0$, the cumulative species-area curve does not converge to a finite value in a moderate number of steps. Clearly, β needs to be estimated well, in order to estimate the total number of species in the landscape. For this, well designed pilot studies are necessary. It is important to note that, extrapolation should be done only for ecologically similar areas where the same laws can be expected to prevail. Otherwise, heterogeneity would have to be taken into account, for instance by stratification of the landscape.

Acknowledgements

The authors would like to thank Dana Draghicescu, CUNY, New York for giving us the permission to use Figure 4 for illustration. Detailed remarks by two referees on an earlier version of the manuscript are also gratefully acknowledged.

References

Bartlett M.S. 1960. Stochastic population models. Methuen and Co., London.

Beran J. 1994. Statistics for Long-Memory Processes. Chapman and Hall, New York.

Beran J. and Ghosh S. 1998. Root-n-consistent estimation in partial linear models with long-memory errors. Scandinavian Journal of Statistics 25: 345–357.

Beran J. and Ocker D. 1999. SEMIFAR forecasts, with applications to foreign exchange rates. Journal of Statistical Planning and Inference 80: 137–153.

Berger J.O. 1999. Statistical decision theory and Bayesian statistics. Springer Verlag, New York.

Bramson M., Cox J.T. and Durrett R. 1996. Spatial models for species area curves. Annals of Probability 24: 1727–1751.

Cliff A.D. and Ord J.K. 1981. Spatial processes: Models and applications. Pion Corp., London.

Cox D.R. and Hinkley D.V. 2000. Theoretical statistics. Chapman and Hall, New York.

Cox D.R. and Isham V. 1988. Point processes, Chapman and Hall, New York.

Cressie N.A.C. 1993. Statistics for spatial data. Wiley, New York.

Diggle P.J. 2003. Statistical analysis of spatial point patterns. Arnold, London.

Diggle P.J., Liang K.Y. and Zeger S.L. 1994. Analysis of longitudinal data. Oxford University Press, Oxford.

Dietrich M. 1997. Häufigkeit, Diversität, Verbreitung und Dynamik von epiphytischen Flechten in Schweizerischen Mittelland und den Voralpen. Inauguraldissertation der Philosophisch-naturwissenschaftlichen Fakultät der Universität Bern.

Draghicescu D. 2002. Nonparametric quantile estimation for dependent data. Ph.D. thesis no. 2592, EPFL.

Eubank R. 1988. Spline smoothing and nonparametric regression. Marcel Dekker, New York.

Fan J. and Gijbels I. 1996. Local polynomial modeling and its applications. Chapman and Hall, London.

Foster B.L. and Tilman D. 2000. Dynamic and static views of succession: testing the descriptive power of the chronosequence approach. Plant Ecology 146: 1–10.

Ghosh S. 2001. Nonparametric trend estimation in replicated time series. Journal of Statistical Planning and Inference 97: 263–274.

Ghosh S. 2005. Random fields based explanation for power-laws in species-area curves. (Preprint).

Ghosh S., Dietrich M. and Scheidegger C. 1997a. Bootstrap based species-area curve estimation for epiphytic lichens in Switzerland. (Preprint; appears in the PhD thesis of the second author, University of Bern).

Ghosh S. and Draghicescu D. 2002. Predicting the distribution function for long-memory processes. International Journal of Forecasting 18: 283–290.

Ghosh S., Landmann G., Pierrat J.C. and Müller-Edzards C. 1997b. Spatio-temporal variation in defoliation. In: Müller-Edzards C., De Vries W. and Erisman J.W. (eds.). 10 years forest condition monitoring in Europe. Studies on temporal development, spatial distribution, and impacts of natural and anthropogenic stress factors. pp. 35–50. Geneva and Brussels, United Nations Economic Commission for Europe / European Commission.

Ghosh S., Beran J., Heiler S., Percival D. and Tinner W. 2007. Memory, non-stationarity and trend: analysis of environmental time series. In: Kienast F., Wildi O. and Ghosh S. (eds.). A Changing World. Challenges for Landscape Research, Vol. 8: 223–247. Springer Landscape Series, Dordrecht.

Hampel F.R. 1987. Data analysis and self-similar processes. Proceedings of the 42nd session of the ISI, Tokyo, Book 4, 235–254.

Künsch H. 1986. Statistical aspects of self-similar processes. Invited paper, Proceedings of the first world congress of the Bernoulli society, Tashkent, vol. 1, Prohorov Yu. and Sazonov V.V. (eds). VNU Science press, Utrecht, 67–74.

Mandelbrot B.B. and Wallis J.R. 1968. Noah, Joseph and operational hydrology. Water Resources Res. 4: 909–918.

Mardia K.V. 1972. Statistics of directional data. Academic Press, London.

Matheron G. 1962. Traité de Géostatistique Appliquée. Cambridge Philosophical Society, Tome 1, Technip, Paris.
Matheron G. 1973. The intrinsic random functions and their applications. Advances in Applied Probability 5: 439–468.
Mingoti S.A. and Meeden G. 1992. Estimating the total number of distinct species using presence and absence data. Biometrics 48: 863–875.
Nychka D. 2000. Challenges in understanding the atmosphere. Journal of the American Statistical Association 95: 972–975.
Pickett T.A. 1989. Space-for-time substitution as an alternative to long-term studies. In: Likens E. (ed.). Long-Term Studies in Ecology: approaches and alternatives. Springer, New York. pp. 110–135.
Ripley B.D. 1988. Statistical inference for spatial processes. Cambridge University Press, Cambridge, UK.
Silverman B.W. 1986. Density estimation. Chapman a and Hall, London.
Smith E.P. and van Belle G. 1984. Nonparametric estimation of species richness. Biometrics 40: 119–129.
Vicsek T. 1992. Fractal growth phenomena. World Scientific cop., Singapore.
Wikle C.K., Milliff R.F., Nychka D. and Berliner L.M. 2001. Spatiotemporal hierarchical Bayesian modeling: tropical ocean surface winds. Journal of the American Statistical Association 96: 382–397.
Wildi O. 2002. Modelling succession from pasture to forest in time and space. Community Ecology 3: 181–189.
Wildi O. and Schütz M. 2000. Reconstruction of a long-term recovery process from pasture to forest. Community Ecology 1: 25–32.

Memory, Non-stationarity and Trend: Analysis of Environmental Time Series

Sucharita Ghosh[1*], Jan Beran[2], Siegfried Heiler[2], Donald Percival[3] and Willy Tinner[4]

[1] Swiss Federal Research Institute WSL, Zürcherstrasse 111, CH-8903 Birmensdorf, Switzerland
[2] Department of Mathematics and Statistics, University of Konstanz, D-78457 Konstanz, Germany
[3] Applied Physics Laboratory, University of Washington, Seattle, WA 98195-5640. U.S.A.
[4] Institute for Plant Sciences, University of Bern, CH-3013 Bern, Switzerland
[*] Corresponding author: e-mail: rita.ghosh@wsl.ch

Abstract
This chapter is a collection of short contributions on environmental time series analysis. The focus is on nonparametric trend estimation, role of nonstationarity vs. long-memory or slowly decaying correlations, wavelets and extreme quantiles. Data examples illustrate the methods.

Keywords: long-range dependence, nonparametric curve estimation, nonstationarity, quantiles, time series, wavelets

F. Kienast, O. Wildi & S. Ghosh (eds.), A Changing World. Challenges for Landscape Research, 223–247.

Introduction

In landscape research, the most important questions concern changes over time, or trend. Trend is defined as the expected value of the stochastic process as a function of time. We will say that there is trend if this function is not identically equal to a constant everywhere. By definition thus, trend is of first order importance. It turns out however, that to model this function, a key issue is the rate at which the auto-correlations in the time series data decay with increasing distance in time. Thus modeling of the auto-correlations is an additional important step towards modeling and prediction of the trend function. For instance, if the auto-correlations decay slowly to zero or have long-memory, then one may see trend-like behavior in the data where there may be no trend at all.

In classical statistical analysis of data, the role of variability is typically understated. Estimates of the variance or the standard deviation are used in calculations of the standard error of the mean, or variances are compared in an ANOVA where the main intention is to compare means. In time series analysis however, variability has a front seat. The reason is that changes in variability indicate a certain lack of stability in the process and in some cases, strong variations can even cause spurious trend-like behavior. For instance, in random walk, the variance explodes over time. This forces the process to wander off, possibly for a long time, although theory tells the process to stay on the average, exactly where it had started. Thus one of the important tasks in time series analysis is to distinguish between deterministic trend and trend-like behavior. Variability can have differing patterns at different scales such as weeks, months and years. Analysis of wavelets addresses this behavior of variability in the process.

Often time series analysis is based on the assumption of stationarity. But this assumption, however convenient, may not be practical at all. A number of things may change over time; in particular, the underlying probability distribution function can have different shapes at different points of time. In questions related to risk, this becomes the most basic issue. Analysis of the probability distribution function, especially its behavior in the tails, can answer a number of questions; situations where this becomes important include species *extinction, strong winter storms, droughts, heavy precipitation,* and so on. A branch of statistics devoted to studies of such events is the *theory of extreme values.* A complementary approach is via a study of the *quantile processes.* A related issue is that of *changing seasonality* where changing amplitudes and phases are of concern in processes which are otherwise periodic. For instance, while winter storms occur every year, the severity of the storm may change over time, having serious implications for wind throw and other damages.

Occasionally, prior experience may suggest that the trend curve can be expressed as a parametric model; e.g. a growth curve may follow a quadratic pattern within a specific range. In natural processes however, the trend curve may be too complex to be formulated in terms of a finite number of parameters and one may have to be content with only a set of weak assumptions. Thus, one may assume that long-term change is a slow process or that *changes occur smoothly.* A standard approach in this case is to use *nonparametric smoothing* or *nonparametric regression* methods which involve taking weighted averages of observations in narrow moving windows. It then turns out that a low resolution estimation procedure results in lower variance but larger bias. This is an important issue in nonparametric smoothing techniques for trend estimation. Thus a trade-off must be allowed for an optimal assessment of the trend curve.

This chapter is divided into four parts. The first part: *trends and seasonal variations in time series,* contributed by Siegfried Heiler, introduces nonparametric curve estimation methods and modeling of seasonal variations in time. The second part: *systematic vs. random development, long-range dependence and nonstationarity,* contributed by Jan Beran, addresses the basic issue of distinguishing between deterministic trend and stochastic trend-like behavior. The third part: *wavelets,* contributed by Donald B. Percival, introduces wavelet analysis of environmental time series data and the fourth part: *smoothing quantiles,* contributed by Sucharita Ghosh and Willy Tinner, discusses the use of extreme quantiles to assess extreme events. The equation numbers refer to equations within each of these short contributions.

Trends and Seasonal Variations in Time Series (Siegfried Heiler)

A first step in the analysis of environmental time series is, like in most of other fields, plotting the measurements against time (Heiler and Michels 1994). Very often the plot will reveal some trends in the course of time, which also in many cases cannot be modeled by a simple analytic function as, e.g. a low degree polynomial, over the whole observation period.

A first step towards grasping such phenomena is nonparametric regression. The idea, which was first put forward by Nadaraya and Watson in two independent papers in 1964, is based on nonparametric density estimation, where in contrast to the classical histogram with fixed binwidths, the estimates are evaluated by a moving average over the data frame. These weights are taken from a suitable kernel function K and the width of the local neighborhood is controlled by fixing a bandwidth h. This gives

$$f_n(x) = \frac{1}{nh}\sum_{i=1}^{n} K\left(\frac{x_i - x}{h}\right). \tag{1}$$

Some customary choices of kernels are exhibited in Figure 1. The normal density is also a popular choice. For multivariate data $x \in \Re^d$, the univariate kernels in (1) may be replaced by a *product kernel* (a d-dimensional product of univariate kernels with bandwidths h_j, $j = 1, \ldots, d$ or by a *norm kernel*, where a univariate kernel is applied to a suitable norm in $\Re^d$. If in the formula for the conditional expectation

$$\begin{aligned} \mu(x) &= \int y f(y|x) dy \\ &= \frac{1}{f(x)} \int y f(x, y) dy \end{aligned} \tag{2}$$

the unknown densities $f(x)$ and $f(x, y)$ are replaced by kernel density estimates of the above form, then we arrive at the *Nadaraya-Watson nonparametric regression estimator*

$$m(x) = \frac{\sum_{i=1}^{n} K\left(\frac{x_i - x}{n}\right) y_i}{\sum_{j=1}^{n} K\left(\frac{x_j - x}{h}\right)} \tag{3}$$

which may be written as a weighted mean

$$m(x) = \sum_{i=1}^{n} w_h(x_i, x) y_i$$

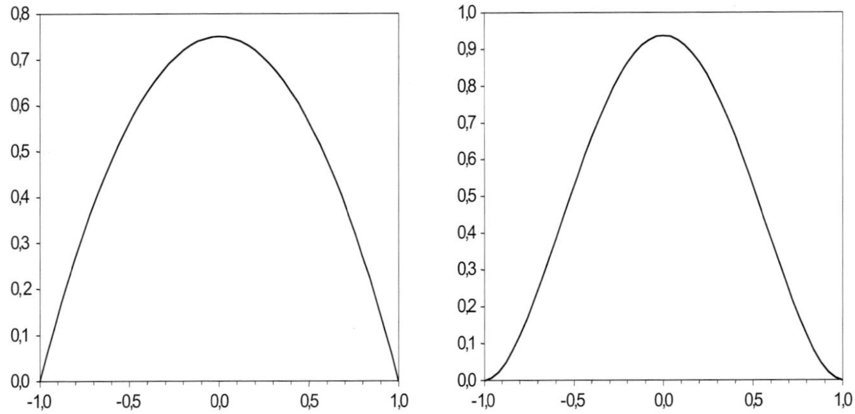

Fig. 1. Two popular kernels for density estimation and nonparametric regression: Epanechnikow (left) and Bisquare (right).

with the neighborhood weights

$$w_h(x_i, x) = \frac{K(\frac{x_i - x}{h})}{\sum_{j=1}^{n} K(\frac{x_j - x}{h})}.$$

The corresponding regression model is

$$y_i = \mu(x_i) + \varepsilon_i \tag{4}$$

with zero mean residuals ε_i which are, in the simplest case independent. These weights sum up to 1. The most critical point in all kinds of nonparametric regression is the choice of the bandwidth h. If it is too small, then a wiggly and noisy estimate for m will result. If it is too large, then interesting details in the course of μ will be flattened out. What one needs for good estimation is consistency of the estimate m. This property ensures precision of the estimate in large samples. For this,

$$h \rightarrow 0 \; and \; nh \rightarrow \infty \quad for \; n \rightarrow \infty$$

is a necessary condition. But this condition is not very helpful in practice. An optimal choice for a constant bandwidth (i.e. one which minimizes the mean squared error between m and μ) for simple nonparametric regression is given by

$$h_{opt} = c_k * n^{-1/5}.$$

However, this is not very helpful either, since the constant does not only depend on the chosen kernel, but also on the derivatives of the μ–function and the variances of the underlying process, which are all unknown. For density estimation a rule of thumb bandwidth $h = 1.06sn^{-1/5}$ was derived by Silverman (1986), assuming normal data, where s is the standard deviation of the x's. A flood of papers with the concern of bandwidth choice has appeared in the meantime. A detailed review may be found e.g. in Feng (2004). A first step for a practitioner may be to compare the graphs of various choices of bandwidths. We see, the freedom for the practitioner of not having to choose a particular form for the trend function goes with a price. He has to choose bandwidths.

Another disadvantage of Nadaraya-Watson type regression is an increasing bias in areas of highly unbalanced design and at the boundaries of the x-space. In time series applications the x-variables are the time indices or a transformed version of time. Hence for regularly arising measurements, unbalancedness is not a problem. Also, typically the observations are equidistant in time. What however remains is the problem at the boundaries. As one possible remedy, special boundary kernels have been developed. Another way to cope with this problem is to use locally weighted regression (Fan and Gijbels 1996). The motivation for locally weighted regression arises from the following argument. First of all, when the conditional expectation is a smooth function of time, Taylor series expansion gives

$$\mu(t) = \mu(t_0) + (t - t_0)\mu'(t_0) + \frac{1}{2}(t - t_0)^2\mu''(t_0) + \ldots. + \frac{1}{r!}(t - t_0)^r\mu^{(r)}(t_0) + e_r(t*) \tag{5}$$

where $\mu'(t)$, $\mu''(t)$, .. are the derivatives of $\mu(t)$ and the rest, namely the remainder term $e_r(t*)$ can be assumed to become small with increasing r. Neglecting the remainder and reformulating

$$\alpha_j(t_0) = \mu^j(t_0)/j!$$

we arrive at a local polynomial representation for $\mu(t)$ as,

$$\mu(t) \sim \sum_{j=0}^{r} \alpha_j(t_0)(t - t_0)^j \tag{6}$$

where t is time, standardized to the interval [0,1]. This approach motivates the nonparametric estimation of μ as a local polynomial by solving the (local) least squares problem:

$$minimize \quad \sum_{i=1}^{n}[y_i - \sum_{j=0}^{r}\alpha_j(t_i - t)^j]^2 K(\frac{t_i - t}{h}). \tag{7}$$

Hence the intercept

$$\hat{\alpha}_0(t_0) = m(t_0) \tag{8}$$

is a local estimator for μ at t_0 and with the 'slope'

$$\hat{\alpha}_j(t_0)j! = m^{(j)}(t_0),$$

an estimator of the jth derivative of μ is given. With $r = 0$ (local constant estimation), the solution corresponds to Nadaraya-Watson estimation. The big advantage of local polynomial regression is in the automatic adaptation property in the boundary area. Therefore the bias of estimation in boundary parts is of the same order of magnitude as in the central part of the time series. For many applications, a local linear approach ($r = 1$) will already be sufficient for estimating μ and $r = j + 1$ for estimating $\mu^{(j)}$. In general $r - j$ should be odd for estimating $\mu^{(j)}$.

Finding a suitable bandwidth which leads to an optimal balance between bias and variance is again a difficult job. A relatively simple procedure that may be helpful is cross validation. Calculate

$$CV(h) = \frac{1}{n}\sum_{i=1}^{n}[y(t_i) - m_{h,i}(t_i)]^2,$$

where $m_{h,i}(t_i)$ is the so-called leave-one-out estimator of μ at time t_i, where the observation $y(t_i)$ is not being used in the estimation procedure. Then choose $h_{cv} = argminCV(h)$, i.e. the value of h that minimizes $CV(h)$ on a grid of h values. Again, quite a few sophisticated procedures are proposed in the literature (e.g. Feng 2004).

In Figure 2, the monthly time series of global temperature from January 1856 to December 1996 is exhibited. The measurements (in centigrade) are differences from an average over a period from 1961 to 1990. In order to estimate a trend in the temperature, a Nadaraya-Watson estimator and a local linear estimator, both with a bandwidth of $h = 150$ and the Epanechnikov-kernel were evaluated. They mainly differ in the boundary parts, where the local linear estimator is less biased.

Like in other areas of statistics, also environmental time series may exhibit some rhythmic pattern, the so-called seasonal variations, Heiler and Feng (2000). These may be due to daily changes in temperature, weekly or monthly production or consumption patterns or caused by climate changes in the course of a year. The typical characteristic of these seasonal variations is that they have a known, fixed periodicity P, but the pattern may change slowly with time. This again suggests a type of local modeling. Since each periodic function – measured only at discrete, equidistant time points – can be represented by a finite Fourier series, seasonal variations may locally be approximated by a trigonometric series with angular frequencies pertaining to the known periodicity P of the season,

$$S(s) = \sum^{q}[\beta_j(t_0)cos\lambda_j(s - t_0) + \gamma_j(t_0)sin\lambda_j(s - t_0)] \tag{9}$$

where $\lambda_1 = 2\pi/P$ is the frequency of the season and $\lambda_j = j\lambda_1$ are the corresponding harmonics for $j = 2, ..., q$ such that $q \leq P/2$. For monthly time series we would have $P = 12$, for weekly series $P = 52$ etc.

If the plot of the time series reveals a local trend and seasonal variation, then combining models (4) and (9) for estimation leads to the local least squares problem

$$min\sum_{t=1}^{n}[y_t - \mu(t) - S(t)]^2K(\frac{t - t_0}{h}).$$

Evaluating $y_t - \hat{S}(t)$ is called seasonal adjustment in the time series literature.

In the upper part of Figure 3 monthly measurements of the content of phosphorus on the surface of lake Constance for a period of almost 25 years are exhibited. The time series shows a strong seasonal variation with high values during winter season and an overall declining trend, where the decrease is obviously non-linear. The decreasing trend goes along

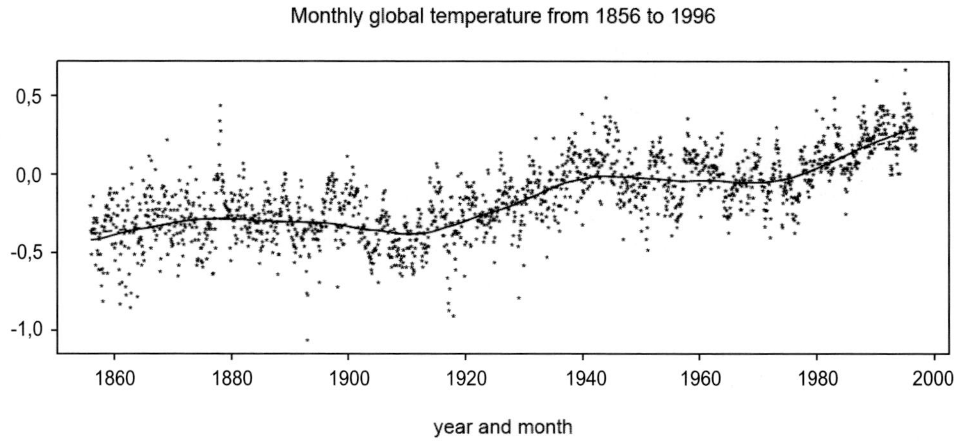

Fig. 2. Monthly global temperature from 1856 to 1996. The plotted values are deviations from a mean.

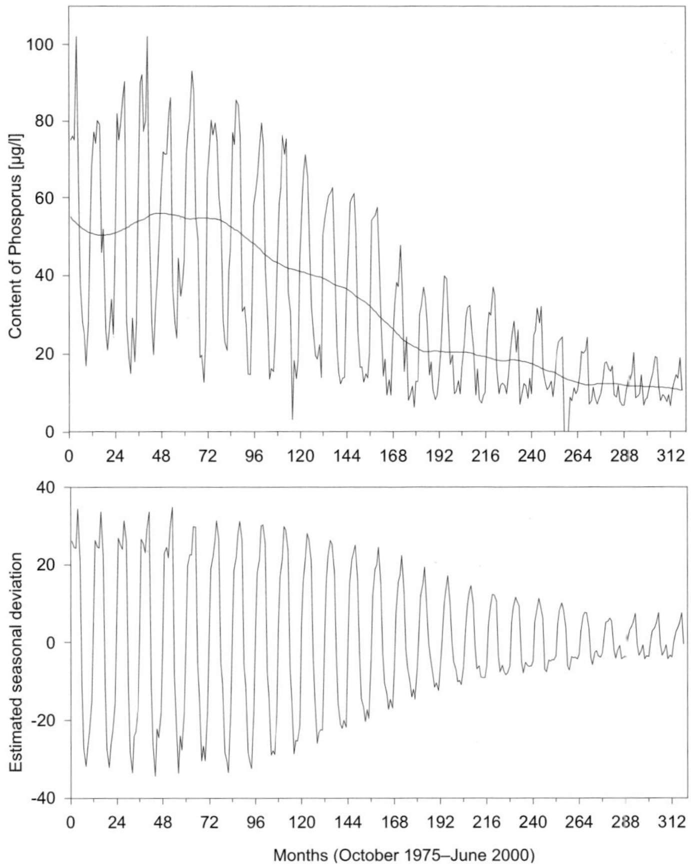

Fig. 3. Content of phosphorus on the surface of Lake Constance. Above: time series plot and trend component. Below: seasonal component.

with a decrease in the strength of the seasonal variations. This highly non-stationary time series was analysed using nonparametric local estimation procedure with bandwidth $h = 30$, local polynomial degree equal to three and periodicity of the season $P = 12$. One can see that due to the local nature of the procedure it works in a flexible way and can cope well with the non-stationary behaviour of the series. The estimated trend reflects the non-linear overall movement, which after some decrease over one and a half years increases again for about two years before it starts decreasing for a period of roughly eleven years. During the last eleven years the decrease is slowly levelling out. This overall movement is accompanied by a steady decrease in the strength of the seasonal deviations with a reduction of the amplitude by a factor of about six.

Systematic vs. Random Development, Long-range Dependence and Nonstationarity (Jan Beran)

Some of the main features one is looking for in a time series are trends, periodicities and dependence structure. In particular, for environmental time series it is important to tell systematic from random developments. Figure 4a shows a typical series where this question is essential. The series consists of rainfall measurements for the rainfall season (June to September) in the Sahelian zone of northern Africa from 1900 to 1995. The values are percentage departures from the 1961–90 mean for the months of June to September. The data were obtained by averaging station data for the area 10–20N and 15W–30E (see Hulme 1996; Hulme and Kelly 1993; Rowell *et al.* 1995). Drought in the Sahelian zone is a serious problem, as it has led to famines in the region in recent decades. The question is whether this may be attributed to a random occurence of reduced rainfall or whether there has been a systematic change in the pattern. Figure 4a seems to suggest the latter, however, in addition to the visual impression conclusive statistical evidence is required. The difficulty is that purely random processes with constant expected value can generate very similar fluctuations. For instance, Figure 4b shows a simulated "rainfall" series obtained from the random walk model (Spitzer 1964) $Y_t = \mu + X_t$ with

$$X_t = X_{t-1} + \epsilon_t \ (t \geq 0), \tag{1}$$

where ε_t are independent zero mean normal random variables, and $X_o = 0$. For this model, the expected value is constant. However, due to the increasing variance and autocorrelation, trend like behaviour is observed. The decrease in Figure 4b is even much more pronounced than for the observed Sahel rainfall record. Nevertheless, the process Y_t is recurrrent, i.e. for any constant $c > 0$, the probability that Y_t is in the interval $[\mu - c, \mu + c]$ infinitely often is equal to one. For rainfall this would imply that, even if there had been droughts for a number of consecutive years, precipitation will come back to its 'normal' (average) level if one waits long enough. Since random walk is not stationary, this may take quite long though. The sample path of a process where recurrence to the average level happens much faster is displayed in Figure 4c. The series was generated by a so-called fractional ARIMA (FARIMA) model with long-memory parameter $\delta = 0.4$. This process is stationary with constant mean. The parameter δ is closely related to the fractional dimension of sample paths of corresponding continous-time modifications of the process (Mandelbrot 1983). The autocorrelations $\rho(k) = \mathrm{corr}(X_t, X_{t+k})$ decay very slowly, namely like a constant times $k^{2\delta-1}$ as the k tends to infinity. This stochastic long-range dependence generates local spurious trends. Again, if the rainfall

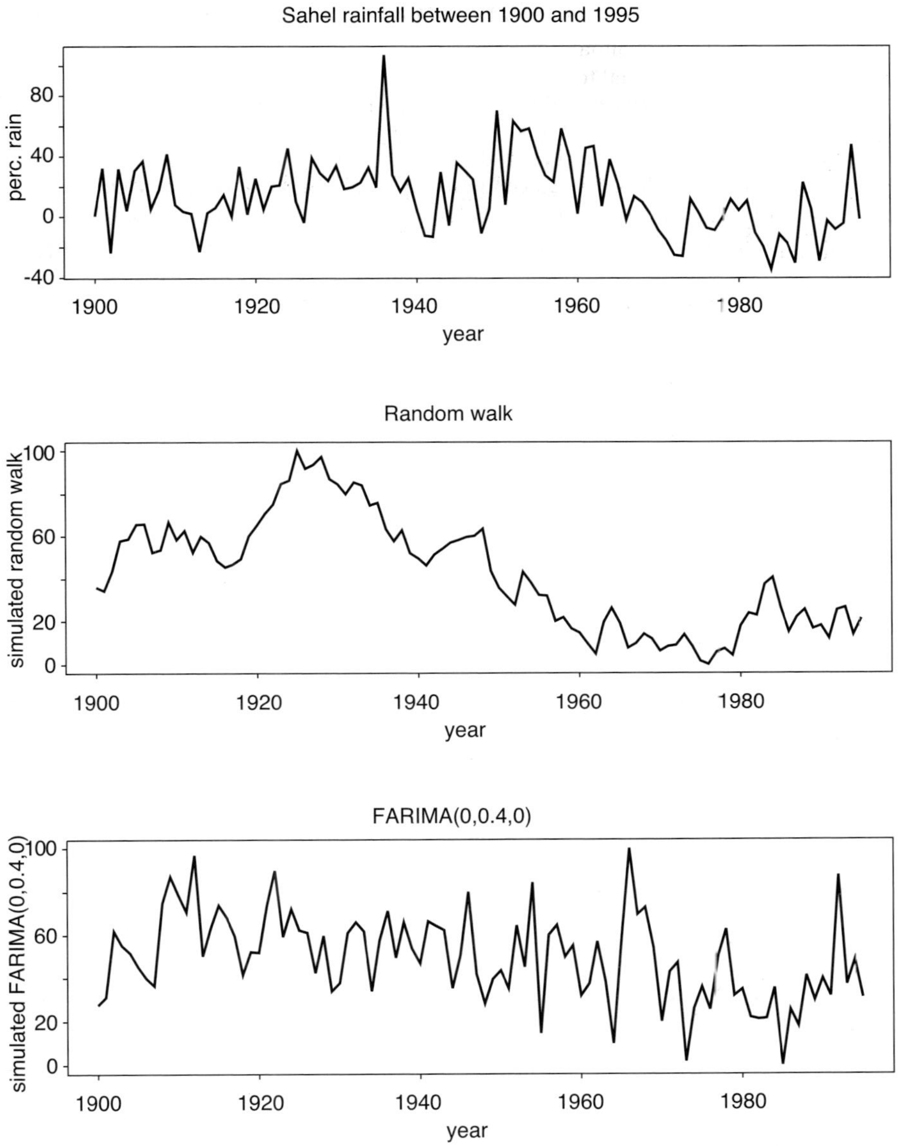

Fig. 4. a) Sahel rainfall between 1900 and 1995 – the measurements consist of percentage departures from the 1961–90 mean for the months of June to September; b) simulated random walk; c) simulated FARIMA(0,0.4,0) process.

were generated by this process, then the process would return to its average level sometimes in the near future.

The question therefore arises whether the drop in rainfall in Figure 4a may indeed be generated by one of the purely random models above (or a similar recurrent process), or whether there has been a systematic change so that a return to a normal level is less likely. To investigate this, alternative models with a systematic change of the expected value need to be considered for comparison. The simplest way of comparing models is to define a general class of processes that includes all models under consideration. Statistical testing together with suitable model choice criteria can then be used to find the statistically most plausible model. A class of models that includes stationary long- and short-range dependence, as well as deterministic trends and random walk type nonstationarity, are so-called SEMIFAR-models (Beran and Feng 2002a,b). A SEMIFAR(p,d) model is defined as the solution of

$$\phi(B)(1-B)^{\delta}\{(1-B)^{m}Y_i - g(t_i)\} = \epsilon_i \tag{2}$$

where $m = 0$ or 1, $\delta \in (-0.5, 0.5)$, ε_i are iid $N(0, \sigma_\varepsilon^2)$–variables, B is the backshift operator with $BY_i = Y_{i-1}$, $\phi(x) = 1 - \sum_{j=1}^{p} \phi_j x^j$ is a polynomial with roots outside the unit circle, and $(1 - B)$ is the fractional differencing operator defined by

$$(1-B)^{\delta} = \sum_{k=0}^{\infty} b_k(\delta)B^k \tag{3}$$

with

$$b_k(\delta) = (-1)^k \frac{\Gamma(\delta+1)}{\Gamma(k+1)\Gamma(\delta-k+1)} \tag{4}$$

(see Granger and Joyeux 1980; Hosking 1981; Beran 1994, 1995). Moreover, g is a general deterministic trend function of rescaled time $t_i = i/n$. In particular, the random model (1) corresponds to (2) with $g \equiv 0$, $\phi(x) = 1$, $\delta = 0$ and $m = 1$. The model generating figure 1c corresponds to $g \equiv$ constant, $\phi(x) = 1$, $m = 0$ and $\delta = 0.4$.

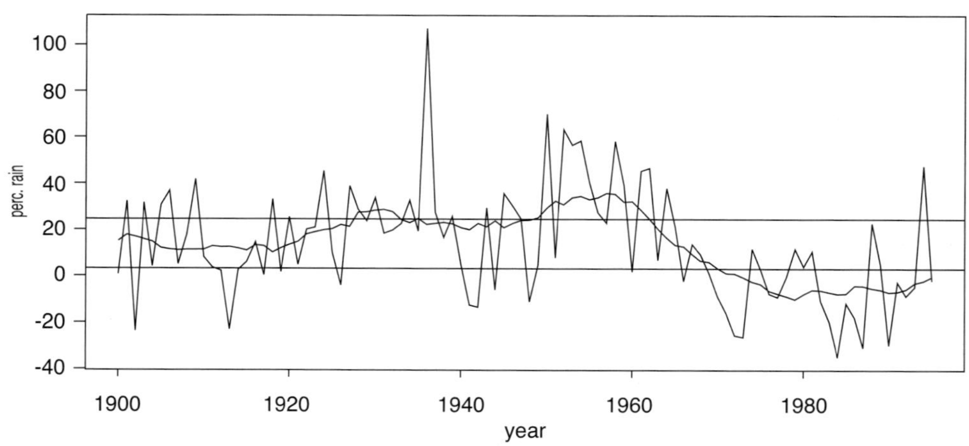

Fig. 5. Sahel rainfall – estimated trend and critical limits for the hypothesis of no trend.

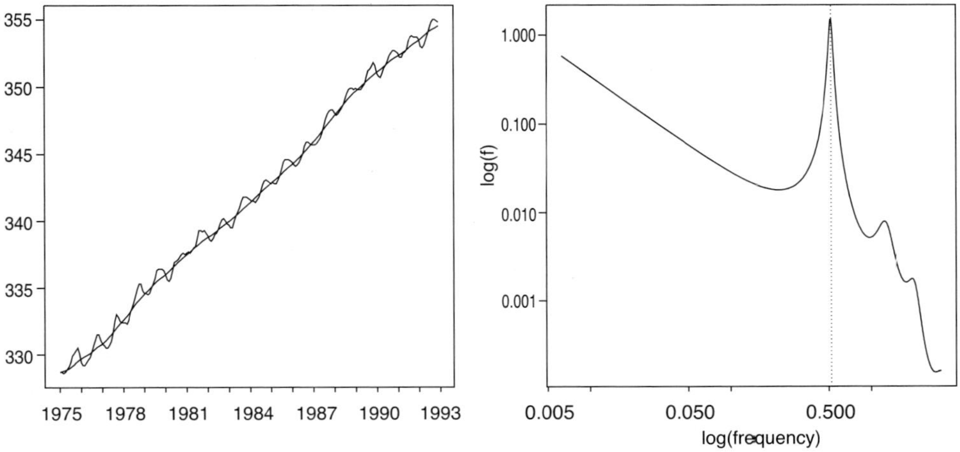

Fig. 6. The annual atmospheric CO_2 mixing ratio at the South Pole (left) and spectral density of detrended series (right).

The parameters of the model can be estimated by an iterative semiparametric algorithm. Confidence intervals and tests can then be calculated to identify the statistically most plausible model. Application to the Sahel rainfall data yields the following conclusions: There is a significant deterministic trend function g. The detrended process is stationary, and almost uncorrelated. Of particular interest is the time after 1960. The trend g is indeed below the critical line (lower bound of rejection region) since the late 1960s (Fig. 5). This confirms the conjectures of Hulme (1996) and other climatologists, and refutes, within the given framework, the claim that the drop in rainfall may be a purely random phenomenon.

Another example illustrates additional features that may be detected by SEMIFAR models. Figure 6 (left) displays monthly atmospheric CO_2 mixing ratios (in ppm) at a South Pole site (Ice and snow-covered plateau 89°59' S, 24°48' W; Conway *et al.* 1994). In this case, there is a very clear trend function and a seasonal pattern. Figure 6 (right) shows the spectral density of the detrended series obtained from the SEMIFAR-fit (in log-log-coordinates). The peaks in the spectrum indicate periodicities. The main peak (marked by a vertical line) corresponds almost exactly to the seasonal period of 12 months, but there are also two other distinct peaks indicating faster oscillations. Moreover, the residual process appears to be a nostationary fractional random walk type process, the estimated fractional differencing parameter being $\hat{d} = \hat{m} + \hat{\delta} = 0.56$ (95%-confidence interval [0.40, 0.73]) and ϕ a polynomial of order 7.

Acknowledgement

The Sahel rainfall record is maintained by Mike Hulme, Climatic Research Unit, University of East Anglia. The source for the CO_2-data is T.J. Conway, P. Tans and L.S. Waterman, National Oceanic and Atmospheric Administration, Climate Monitoring and Diagnostics Laboratory, Boulder, Colorado.

Wavelets (Donald B. Percival)

Wavelets have become increasingly popular over the last decade as a means for analyzing environmental time series (see, for example, Bradshaw and Spies 1992; Brillinger 1994; Brunet and Collineau 1994; Craigmile *et al.* 2004, Csillag and Kabos 2002; Dale and Mah 1998; Lark and Webster 1999, 2001; Percival *et al.* 2004; Whitcher *et al.* 2002). Wavelet analysis allows us to look for patterns in variability in a time series at different times and over different scales (i.e., intervals of time). This type of analysis is particularly well suited for time series that are either (i) well modelled as long memory processes, (ii) exhibit certain fractal-like properties or (iii) have nonstationary characteristics. Here we present some of the key ideas behind wavelet analysis. Our presentation is a condensation of a somewhat longer introduction to wavelets given in Percival *et al.* (2004), but, for extensive details, the reader is referred to Percival and Walden (2000).

The word 'wavelet' means a 'small wave,' so we start by considering what is meant by a wave and in exactly what sense a wavelet is a small version of a wave. We take a wave to be a real-valued function of time t that oscillates back and forth about zero, with the amplitude of its oscillations being relatively the same everywhere. The top plot of Figure 7 shows part of a typical wave, namely, a function that is defined by $\sin(t)$ for $-\infty < t < \infty$. By contrast, the bottom two plots show two different wavelets. The top one is the Haar wavelet $\psi^{(H)}(\cdot)$, which is zero everywhere except when t in the interval [–1, 1]. The value that the Haar wavelet has over the first half of this interval has the opposite sign from what it has over the second half. The Haar wavelet is balanced about zero because it is above and below zero by an equal amount over an equal amount of time. The bottom wavelet is proportional to the first derivative of the probability density function for a normal (Gaussian) random variable. The plots of these two wavelets make it clear in what sense a wavelet is a small version of a wave: whereas a wave has substantial fluctuations over the entire real axis, a wavelet is zero (or nearly so) outside of some finite interval ([–1, 1] for the Haar wavelet and approximately [–1.5, 1.5] for the other wavelet).

We can use a wavelet to tell us something interesting about a time series. To see this, suppose we let xt represent the tth value of a time series, where t is an integer. Let us regard x_t as a sample from a real-valued function $x(\cdot)$; i.e., we have $x_t = x(t\Delta)$, where Δ is the spacing between adjacent observations. Let us now consider what happens if we multiply $x(\cdot)$ and the Haar wavelet $\psi^{(H)}(\cdot)$ together and integrate the resulting function; i.e., we form

$$W^{(H)} = \int_{-\infty}^{\infty} \psi^{(H)}(t')x(t')\,dt'.$$

The variable $W^{(H)}$ is called a Haar wavelet coefficient. From the shape of the Haar wavelet in Figure 7, we can argue that $W^{(H)}$ is proportional to

$$\int_0^1 x(t')\,dt' - \int_{-1}^0 x(t')\,dt'.$$

Texts on elementary calculus tell us that an integral of the form

$$\frac{1}{b-a}\int_a^b x(t')\,dt'$$

defines the average value of the function $x(\cdot)$ over the interval $[a, b]$. The integral $\int_0^1 x(t')\,dt'$ can be regarded as the average value of $x(\cdot)$ over the interval [0, 1], and similarly $\int_{-1}^0 x(t')\,dt'$ is its average value over [–1, 0]. The width of these intervals is commonly referred to as the scale associated with $W^{(H)}$. Thus the Haar wavelet coefficient is proportional to a difference between two averages, each of unit scale. If $|W^{(H)}|$ is large, these unit scale averages of $x(\cdot)$ differ substantially; on the other hand, if $W^{(H)}$ is close to zero, there is not much difference between the two averages.

While the Haar wavelet in Figure 7 tells us about the difference in averages over intervals of unit scale centered about the origin, it would be nice to obtain similar information about averages over other scales centered about arbitrary time points. We can do so by adjusting the scale of the wavelet and then relocating it. This process is illustrated in Figure 8, which shows three Haar wavelets of scale two centered at times t = –3, 0 and 3 (top to bottom plots). Let $\psi^{(H)}_{\lambda,t}(\cdot)$ stand for the Haar wavelet of scale $\lambda > 0$ that is centered at time t. If we multiply $\psi^{(H)}_{\lambda,t}(\cdot)$ and $x(\cdot)$ together, we obtain a wavelet coefficient that depends on λ and t:

$$W^{(H)}(\lambda, t) \equiv \int_{-\infty}^{\infty} \psi^{(H)}_{\lambda,t}(t')x(t')\,dt'. \tag{1}$$

Note that $W^{(H)}(1, 0)$ and $W^{(H)}$ are identical. This function of scale and location is known as the Haar wavelet transform of $x(\cdot)$. The particular value $W^{(H)}(\lambda, t)$ is called the Haar wavelet coefficient at λ and t. The coefficient $W^{(H)}(\lambda, t)$ is proportional to a difference in scale λ averages located just before and after time t. If the difference between these averages is large (small), then $|W^{(H)}(\lambda, t)|$ will be large (small).

In a similar manner we can define wavelet transforms other than the Haar by starting with a basic wavelet $\psi(\cdot)$ that has unit scale and is centered at the origin. We can define $\psi_{\lambda,t}(\cdot)$ by suitably stretching or shrinking $\psi(\cdot)$ to scale λ and then relocating it to time t. The

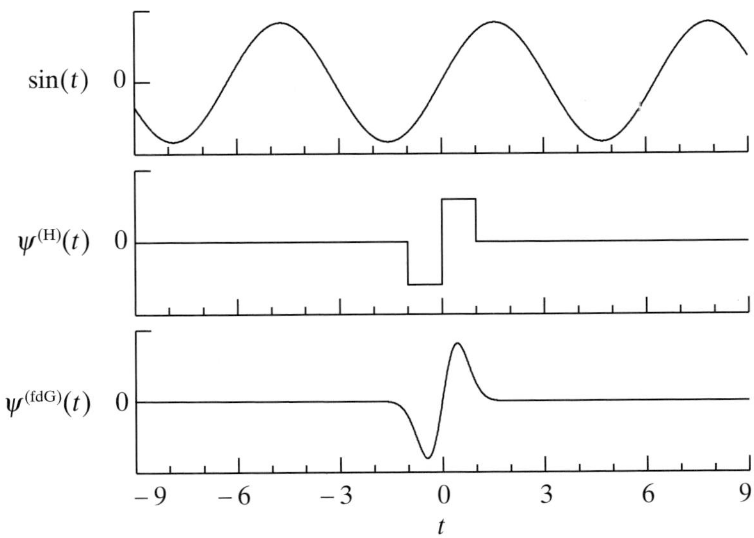

Fig. 7. A wave (top plot) and two wavelets (bottom two plots). The wave is just a sine function with a period of 2π. The wavelets are the Haar wavelet $\Psi^{(H)}(\cdot)$ (middle plot) and a wavelet $\Psi^{(fdG)}(\cdot)$ (bottom) based upon the first derivative of the standard Gaussian (normal) probability density function.

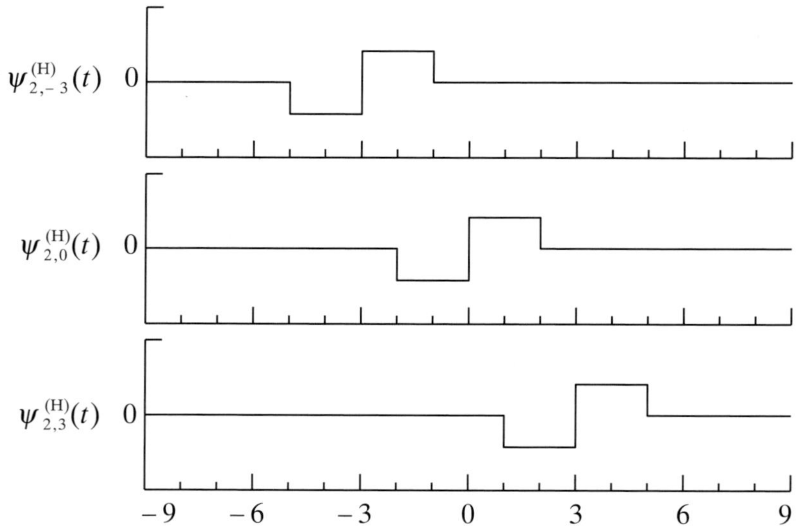

Fig. 8. Three Haar wavelets of scale $\lambda = 2$ centered at times $t = -3, 0$ and 3 (top to bottom).

wavelet transform itself is given by the obvious analog of Equation (1). The interpretation of the resulting wavelet coefficients $W(\lambda, t)$ can be deduced from a study of the shape of the basic wavelet $\psi(\cdot)$. For example, while the Haar wavelet $\psi^{(H)}(\cdot)$ depicted in Figure 7 leads to coefficients that are proportional to the difference between adjacent averages, the plot of $\psi^{(fdG)}(\cdot)$ in this figure suggests that this wavelet yields coefficients that are proportional to a difference between adjacent weighted averages.

The interpretation of wavelet coefficients as the difference between averages is fundamental for transforms based upon wavelets similar to those shown in Figure 7. The fact that we can make this physical interpretation explains why the wavelet transform has proven to be useful for many different types of time series. While average values of time series are important, what is more often of scientific interest is how stable such averages are over time. To cite a well-known example, the study of a possible change in the climate leads us to consider how much change there has been over time in averages of various climate indicators. Since an effective way to determine this would be to see how much change there has been in averages associated with different scales (decades, centuries, etc.), the wavelet transform emerges as a important tool because of its physical interpretation.

In addition to its appealing physical interpretation, let us note two other key properties of the wavelet transform. The first says that the continuous time series $x(\cdot)$ and its wavelet transform $W(\cdot,\cdot)$ are equivalent in the sense that, given either $x(\cdot)$ or $W(\cdot,\cdot)$, we can recover the other. The 'forward' wavelet transform, i.e., going from $x(\cdot)$ to $W(\cdot,\cdot)$, is simply a matter of definition (e.g., Equation (1)). The inverse wavelet transform, i.e., going from $W(\cdot,\cdot)$ to $x(\cdot)$, is not immediately obvious, but in fact can be done as follows:

$$x(t) = \frac{1}{C_\psi}\int_0^\infty \left[\int_{-\infty}^{\infty} W(\lambda, t')\frac{1}{\sqrt{\lambda}}\psi\left(\frac{t-t'}{\lambda}\right)dt'\right]\frac{d\lambda}{\lambda^2}, \qquad (2)$$

where C_ψ is a constant depending on just $\psi(\cdot)$. The fact that we can recover $x(\cdot)$ from its wavelet transform is important because this means that all the information about the time series must also be contained in $W(\cdot,\cdot)$ in some manner. We can consider $x(\cdot)$ and $W(\cdot,\cdot)$ to be two representations for a single mathematical entity, with $W(\cdot,\cdot)$ offering a reexpression of the time domain formulation $x(\cdot)$ in a manner that can bring out certain key information more succinctly. Another way of stating this equivalence is that knowledge of how $x(\cdot)$ changes at all scales λ and all time locations t is equivalent to knowing $x(\cdot)$ itself.

The second key property is that the wavelet transform can be used as the basis for an 'energy' decomposition of $x(\cdot)$. By definition, the term 'energy' is just short-hand for the quantity $\int_{-\infty}^{\infty} x^2(t)\, dt$, which we assume to be finite (for certain $x(\cdot)$, this definition would be consistent with the physical notion of energy). We can then write

$$\int_{-\infty}^{\infty} x^2(t)\, dt = \int_0^{\infty} \int_{-\infty}^{\infty} \frac{W^2(\lambda, t)}{C_\psi \lambda^2}\, dt\, d\lambda. \tag{3}$$

Thus, whereas a plot of $x^2(t)$ versus t tells us how the energy in $x(\cdot)$ is distributed with respect to time, the right-hand integrand yields an energy distribution across both scale and time.

The transform $W(\cdot,\cdot)$ is called a continuous wavelet transform (CWT) since it is applied to a function $x(\cdot)$ defined over the entire real axis. In practical applications, we only have a finite number N of sampled values. If we only have these samples, it is not possible to compute the CWT exactly, but we can resort to approximations; however, rather than approximating the CWT, an alternative approach is to define a wavelet transform specifically tailored for sampled values. This approach leads to the discrete wavelet transform (DWT), but it does require that the samples be of the form $x_t = x(t\Delta)$, where for convenience we take the integer t to range from 0 up to $N-1$; i.e., we need the data to be collected at equal intervals (if this is not the case, we can resort to an interpolation method to create a regularly sampled time series). To some degree of approximation, we can regard the DWT as 'slices' through a corresponding CWT, but it is important to note that the DWT can stand on its own, independent of any connection to a CWT. The slices are restricted to scales λ that assume the dyadic values $2^{j-1}\Delta$, where j = 1, 2, . . . , J_0. Here J_0 is the total number of scales used and is usually dictated by the sample size N and/or by the application at hand. This particular discretization of λ proves to be sufficient for most applications. Typically J_0 is set to a value no greater than $\log_2(N)$, so the number of scales that we can meaningfully analyze is restricted by the amount of data available. It is convenient to define $\tau_j = 2^{j-1}$ as a standardized dyadic scale, with the actual physical scale then being given by $\tau_j\,\Delta$.

The DWT also requires a discretization of the continuous time variable t. One choice gives us N wavelet coefficients on each scale, yielding a version of the DWT called the maximal overlap DWT (MODWT). Let X be an N dimensional vector whose t^{th} element is x_t. For a given J_0, the MODWT transforms X into J_0+1 new vectors, each of dimension N. The first J_0 of these are denoted by $\widetilde{\mathrm{W}}_1, \ldots, \widetilde{\mathrm{W}}_{J_0}$ and constitute the MODWT wavelet coefficients associated with (standardized) scales τ_j, $j = 1, \ldots, J_0$. To some degree of approximation, each element in $\widetilde{\mathrm{W}}_j$ can be associated with a CWT wavelet coefficient on scale $\tau_j\Delta$, so the MODWT wavelet coefficients have the same physical interpretation as the CWT coefficients. The final vector is denoted by $\widetilde{\mathrm{V}}_{J_0}$ and contains the so-called MODWT scaling coefficients. Whereas the wavelet coefficients $\widetilde{\mathrm{W}}_j$ are proportional to changes in averages over a scale of τ_j, these scaling coefficients are proportional to averages over a scale of τ_{J_0+1}. Roughly speaking, the scaling coefficients are summaries of the information in the CWT at all physical scales λ greater than or equal to $\tau_{J_0+1}\Delta$.

The MODWT is equivalent to the original time series in the sense that, given the MODWT coefficients, we can reconstruct X (cf. Equation (2) for the CWT). This leads to the following additive decomposition, which is known as a multiresolution analysis (MRA):

$$X = \sum_{j=1}^{J_0} \widetilde{D}_j + \widetilde{S}_{J_0}. \tag{4}$$

In the above, $\widetilde{D}_j$ is an N dimensional vector that depends upon just $\widetilde{\mathrm{W}}_j$ and hence is constructed using just those MODWT wavelet coefficients that are associated with changes of averages on a scale of τ_j. This vector is called a 'detail series' and is the part of the MRA of X that can be attributed to variations on a scale of τ_j. The final term in the MRA is $\widetilde{S}_{J_0}$, which again is an N dimensional vector, but this depends just on the scaling coefficients $\widetilde{\mathrm{V}}_{J_0}$. The vector $\widetilde{S}_{J_0}$ is called the 'smooth series' because it is associated with averages over scales $2\tau_{J_0}$ and longer and hence captures the slowly varying portion of X. Thus an MRA is an additive decomposition that expresses a time series as the sum of several new series, each of which can be associated with variations on a particular scale.

The MODWT also offers a scale-based decomposition of the energy in X. Recalling that we took $\int_{-\infty}^{\infty} x^2(t)\, dt$ to be the energy in $x(\cdot)$, let us define $\|\mathrm{X}\|^2 = \sum_{t=0}^{N-1} x_t^2$ to be the energy in X. The MODWT-based energy decomposition says that

$$\|\mathrm{X}\|^2 = \sum_{j=1}^{J_0} \|\widetilde{\mathrm{W}}_j\|^2 + \|\widetilde{\mathrm{V}}_{J_0}\|^2 \tag{5}$$

(cf. Equation (3) for the CWT). Thus the energy in X is preserved in its MODWT coefficients, and $\|\widetilde{\mathrm{W}}_j\|^2$ represents the contribution to the energy that is attributable to variations on the scale of τ_j. This energy decomposition in turn allows us to decompose a well-known measure of the variability in a time series, namely, the sample variance, defined as

$$\hat{\sigma}_X^2 = \frac{1}{N}\sum_{t=0}^{N-1}(x_t - \bar{x})^2 = \frac{1}{N}\sum_{t=0}^{N-1} x_t^2 - \bar{x}^2 = \frac{1}{N}\|\mathrm{X}\|^2 - \bar{x}^2,$$

where $\bar{x} = \frac{1}{N}\sum_t x_t$ is the sample mean of the time series. If we use Equation (5) to substitute for $\|\mathrm{X}\|^2$ in the above, we obtain the following scale-based decomposition of the sample variance:

$$\hat{\sigma}_X^2 = \frac{1}{N}\sum_{j=1}^{J_0} \|\widetilde{\mathrm{W}}_j\|^2 + \frac{1}{N}\|\widetilde{\mathrm{V}}_{J_0}\|^2 - \bar{x}^2. \tag{6}$$

We can use this expression to argue that the contribution to the sample variance due to variations on a scale of τ_j is given by $\frac{1}{N}\|\widetilde{\mathrm{W}}_j\|^2$. We can take this quantity to be an estimator for a corresponding theoretical quantity called the wavelet variance, which gives us a scale-based decomposition of a theoretical variance associated with a population that x_t is assumed to be drawn from.

Let us now illustrate the above ideas by considering a time series that is a surrogate for March–September coastal land temperatures along the Gulf of Alaska (Wiles *et al.* 1999). This surrogate series is based upon tree ring records, and its formation is discussed in detail in Wiles *et al.* (1998). The series is plotted at the bottom of Figure 9 and consists of one value (measured in degrees celsius) for each year from 1600 to 1990 (there are N = 391 values in

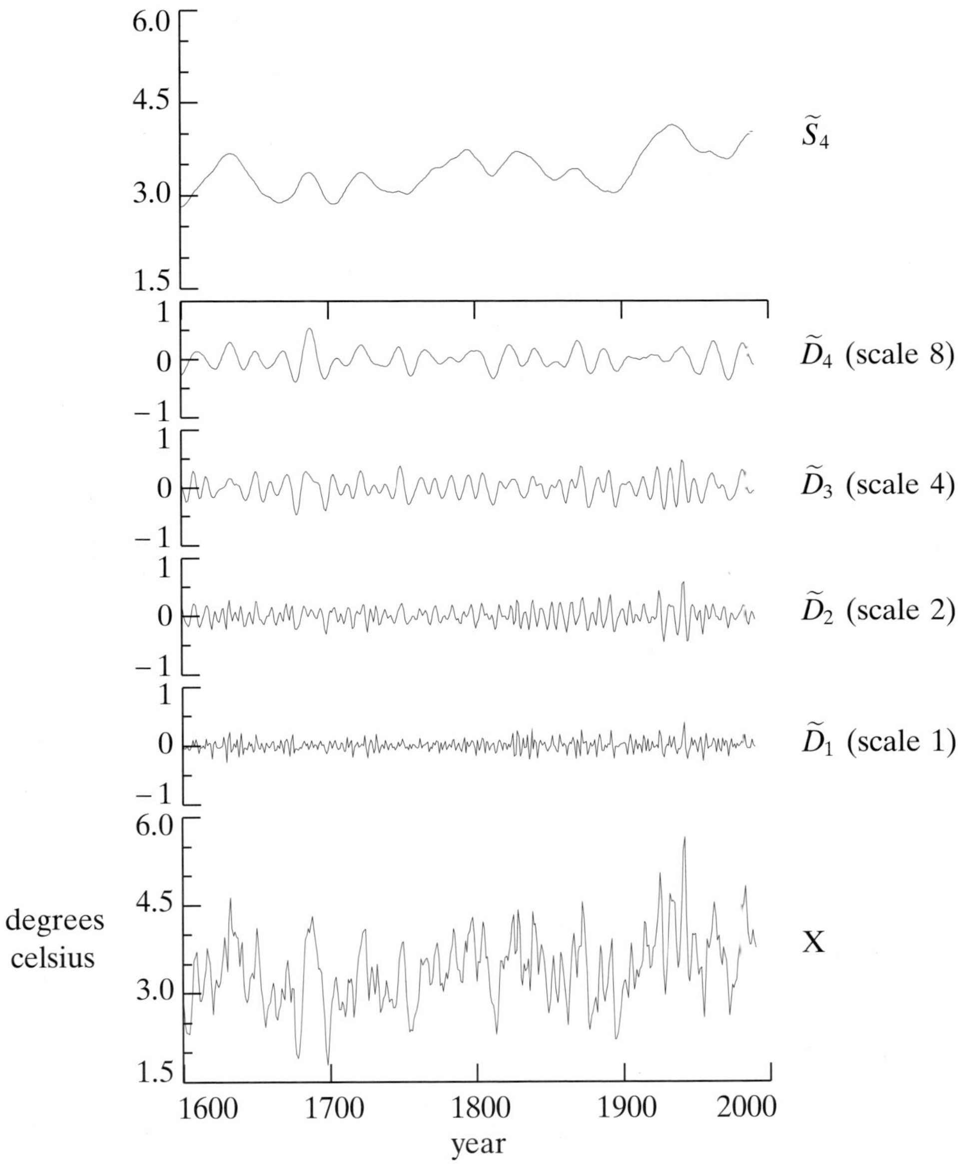

Fig. 9. Multiresolution analysis (MRA) for Gulf of Alaska temperature time series based upon the Haar MODWT wavelet transform. The bottom-most curve shows the time series X itself, while the other five curves show the components of the MRA.

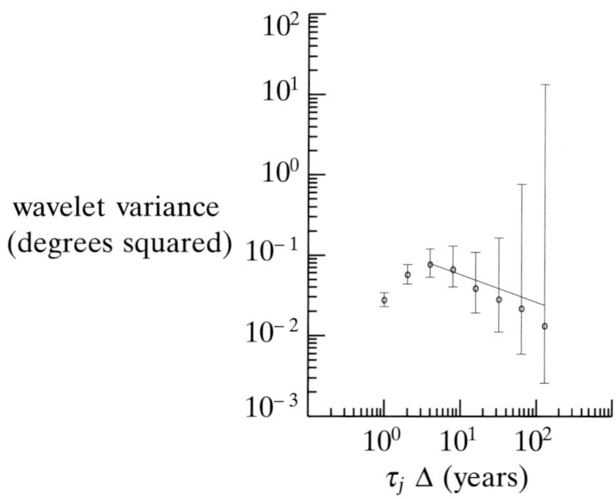

Fig. 10. Haar MODWT wavelet variance estimates. The o's indicate MODWT-based unbiased estimates of the Haar wavelet variance for scales $\tau_j\Delta$ equal to 1, 2, 4, 8, 16, 32, 64 and 128 years. The vertical lines emanating from the o's give 95% confidence intervals for a corresponding theoretical wavelet variance. The thin line through the o's depicts a weighted regression fit over the largest six scales of the log of the wavelet variance estimates versus the log of the scale.

all). Above this series we have plotted an MRA based upon a MODWT using the Haar wavelet. Here we have set $J_0 = 4$ so that we get four detail series $\tilde{D}_1, \ldots \tilde{D}_4$ (associated with changes in averages on scales of 1, 2, 4 and 8 years) and the smooth series $\tilde{S}_4$ (associated with averages over all scales greater than 8 years).

A study of this MRA indicates that the variations at any given scale appear to be homogeneous across time for the most part; i.e., there is little evidence that the statistical characteristics of the series have changed much over the last four centuries (there is, however, a suggestion of increased variability in the last halves of $\tilde{D}_1$ and $\tilde{D}_2$ as compared to their first halves). The smooth series $\tilde{S}_4$ reflects large scale averages of the series and captures what is often called the long-term trend. This part of the MRA seems to have increased somewhat in the 1900s compared to its average over the previous three centuries.

We can quantify how much each scale contributes to the overall variability in this temperature series via the wavelet variance, which is a population version of the variance decomposition described by Equation (6). Figure 10 shows a plot of unbiased MODWT-based wavelet variance estimates versus scales ranging from $\tau_1\Delta = 1$ year up to $\tau_8\Delta = 128$ years (details about this estimator are given in Percival and Walden 2000, Chapter 8). Each wavelet variance estimate is denoted by an 'o', out of which vertical lines emanate above and below. These lines represent 95% confidence intervals for the theoretical wavelet variance. Note that the confidence intervals associated with the largest scale of $\tau_8\Delta = 128$ years encompass nearly four orders of magnitude, a reflection of the fact that we really cannot expect to say much about changes on a scale of 128 years with only 391 years of data.

The wavelet variance estimates in Figure 10 indicate that the dominant contribution to the sample variance is due to variations at a scale of four years. The smaller variability at the two smallest scales might be due to the way in which this surrogate series was constructed and hence might not be indicative of the true small scale temperature variations. The wavelet variance estimates over the six largest scales seems to decay roughly linearly on the log/log plot. A linear rate of decay with a slope of, say, β is consistent with what we can expect for a time series that can be modelled by a fractionally differenced (FD) process with a parameter $\delta = (\beta + 1)/2$. The thin line through the last six wavelet variance estimates is the result of a weighted regression fit that takes into account the sampling variability in the individual estimates (Percival and Walden 2000, §9.5). Here the estimated slope $\hat{\beta}$ yields an estimate for the FD parameter of $\hat{\delta} \doteq 0.3$, with an estimated standard error of 0.1. In general, a study of the rate of decay of the wavelet variance versus scale can give us considerable insight into the correlation structure of a time series.

Acknowledgement

This work was supported by the US. National Science Foundation under Grant No. 0222115.

Smoothing Quantiles (Sucharita Ghosh and Willy Tinner)

Often one is faced with the problem of quantifying extremal events and their future development. Extremal events are catastrophes like flood, hurricanes, tornadoes, dam bursts, earthquakes, forest fires, avalanches, snowstorms and so on. Sigma (1996) lists 30 worst catastrophes and 30 most expensive insurance claims in the world during a 25-year period between 1970 and 1995 (also see Embrechts *et al.* 2003). Typical questions concern estimation of the mean waiting time between extreme events, frequencies of extreme events in past records, evaluating the probability distribution function (pdf) of extremes and their high quantiles (e.g. to assess, how high can the highest observation be, say in questions of engineering constructions to prevent disastrous events), distribution and in particular high quantiles of random sums (e.g. in insurance problems), analysis of clustering of extremal events etc. Typically, such analyses involve dealing with the order statistics of the set of observations. While dealing with extreme observations, the concept of robustness also comes into play; for instance, estimation of the the mean may be replaced by a median (e.g. Beran *et al.* 2002) that is less affected by the outliers.

Plenty of research has been done about analyzing extreme value data that have arisen from stationary processes or are independently and identically distributed (iid) sequences. In other words, if it is known that the underlying marginal probability distribution of the process (including all joint distributions) remains the same when time is shifted, then much work has been done for developing methods to estimate probability distributions of extreme values. In this chapter, we present nonparametric methods for estimation of quantiles that is particularly suitable for handling non-stationary pdf as well as various types of auto-correlation structures in the data (e.g. Ghosh *et al.* 1997).

Quantiles are inverted distribution functions (Fig. 11). This means, to get smooth estimates of quantiles, one approach would be to first smooth the empirical distribution function (edf) to get an estimate of the underlying pdf. As a function of threshold values, an edf is simply a (non-)exceedance probability function i.e. the probability that the process will not exceed a certain threshold. To describe the estimation method formally, let us first introduce the notations. Let $Y_1, Y_2, \ldots.., Y_n$ denote n observations from the stochastic process of interest. Suppose that i denotes time. We will also often be speaking in terms of re-scaled times $x_i = \frac{i}{n}$ in stead of the real time i.

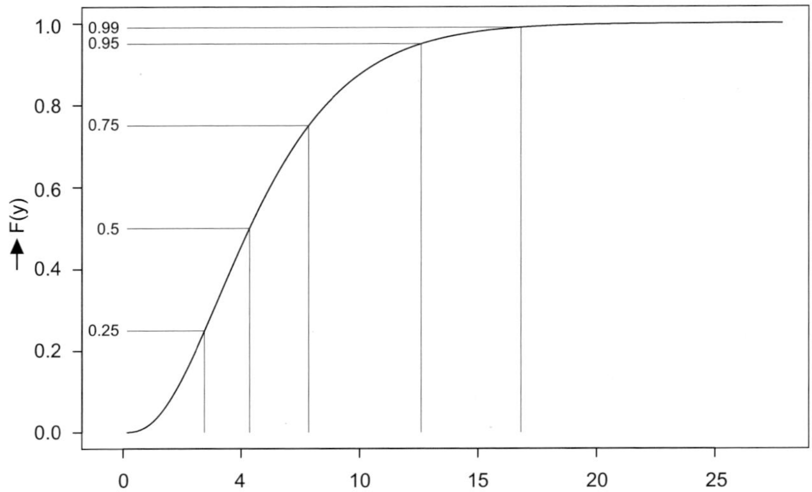

Fig. 11. A pdf and its quantiles.

Let y be a given threshold and we are concerned about estimating the non-exceedance probability $F_{x_i}(y) = P(Y_i \le y)$ at time i or at rescaled time $x_i = i/n$. More generally, if x denotes a number between 0 and 1, then, we want to estimate the probability $F_{x_i}(y)$ that is the probability of not exceeding the threshold y at re-scaled time x.

A nonparametric estimator of this probability can be immediately constructed (also see Eubank 1988, Gasser and Müller 1984) by averaging the indicator variable f_i defined as follows:

$$f_i = 1 \text{ if } Y_i \le y \quad (1)$$

$$= 0 \text{ if } Y_i > y. \quad (2)$$

Making use of a suitable kernel function K that is symmetric around zero and has other suitable mathematical properties, our estimator for F is

$$\hat{F}_x(y) = \frac{1}{nh} \sum_{i=1}^{n} K\left(\frac{x_i - x}{h}\right) f_i \quad (3)$$

In this weighted average, the observations that are located far from the rescaled time x get lower weight. The letter h stands for the bandwidth or the window-width.

For a given data set, one can plot this estimate of the (non-)exceedance probability against time for various different threshold values y. If the probability distribution of Y is stationary, then ideally, the curves would be horizontal straight lines.

In computations, this smoothing method is repeated for a number of threshold values y. The inversion formula to obtain the estimated quantile at rescaled time x is:

$$Q_x(\alpha) = inf\,\{y | \hat{F}_x(y) \ge 1 - \alpha\},\ 0 < \alpha < 1. \quad (4)$$

Figures 12 and 13 show an analysis of Northern and Southern hemisphere land and sea temperature records for 1854–1989. The original observations were monthly data, that were deviations from the monthly averages in 1950–1979. In this analysis, these monthly records were averaged within years. The figures show the data with their 75th and 25th percentiles (top) as well as the interquartile range (bottom). For the sake of illustration, the calculations were done for 21-year and 31-year windows for the Northern and the Southern data respectively, although in detailed analysis, optimal bandwidths will have to be used. The figures show very different patterns in the interquartile range, showing changes in the pattern of the variability over time.

A second example of changing pattern of variability even where no trend can be seen is in Figure 14. Here, estimated quantiles are drawn for a number of trees *(Abies alba)*, Abetone, Italy. The data are the so-called *tree ring indices*, typically used to remove *age-trend* from ring-width data. Any deviation from the horizontal line at ordinate = 1 is to be interpreted as climatic or other environmental influences (see Cook and Kairiukstis 1990).

When one uses a box kernel, where all observations in a window get equal weight, this method can be easily implemented using a standard statistical software that has a macro for computing sample quantiles directly for a set of observations. One simply needs to apply this macro on a moving window. For other kernels, the two-step procedure can be used.

Beran and Ocker (1999) propose a nonparametric prediction method by making use of a Taylor series expansion of the trend function. To predict probabilities, one can adopt the same idea for the log odds-ratio. For instance, if

$$U_x = log \frac{F_x(y)}{1 - F_x(y)} \tag{5}$$

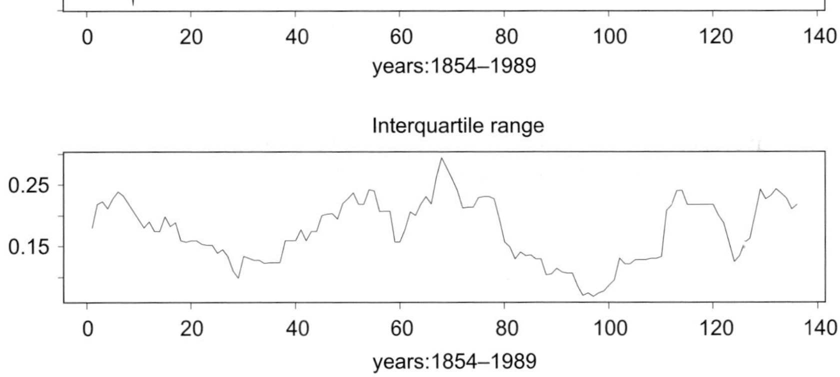

Fig. 12. Analysis of temperature data from the Northern hemisphere.

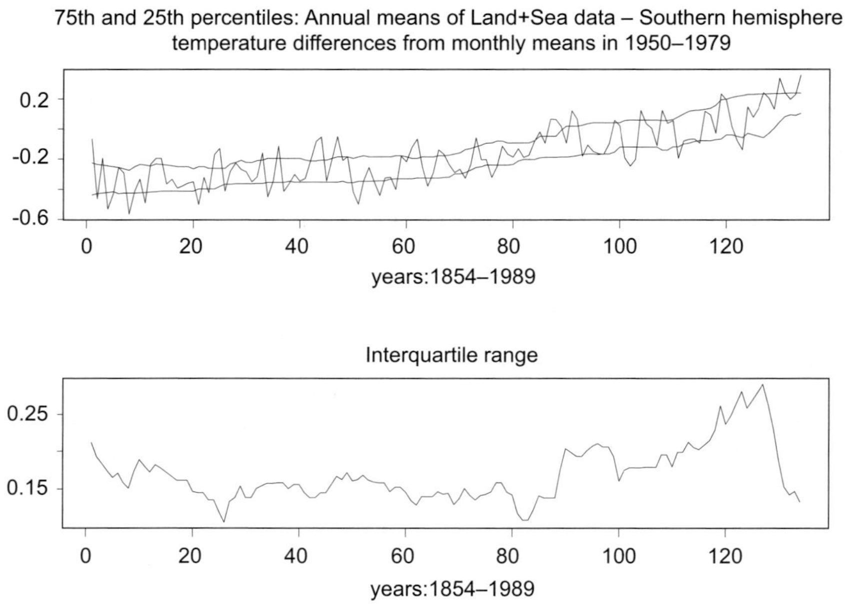

Fig. 13. Analysis of temperature data from the Southern hemisphere.

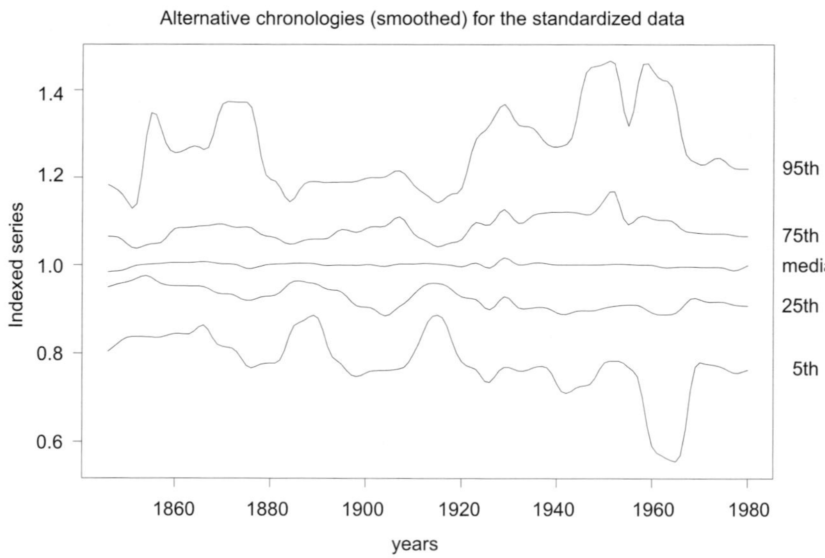

Fig. 14. Alternative chronologies or time dependent quantiles for *Abies alba*, Abetone, Italy.

and $U_{x+\delta}$ denotes the predicted value at rescaled time $x + \delta$ then the predicted value of F at rescaled time $x + \delta$ would be

$$F_{x+\delta} = \frac{exp(U_{x+\delta})}{1 + exp(U_{x+\delta})}. \tag{6}$$

Detailed formulas for the prediction limits are in Ghosh and Draghicescu (2002a).

An algorithm for the optimal choice of the bandwidths derived for a locally-stationary model is in Ghosh and Draghicescu (2002b). Here, local-stationarity roughly means that when the observations are not too far apart in time, the auto-covariances are approximately stationary; i.e. they depend only on the lags. This model has been discussed in Beran (1994) and Ghosh *et al.* (1997); also see Taqqu (1975). The convenience of this model in environmental research is that, it is broad enough to encompass non-Gaussian pdfs, which may even change with time (non-stationary). Details are in Ghosh *et al.* (1997) and Ghosh and Draghicescu (2002b); also see Engel *et al.* (1994).

Concluding Remarks

Analysis of quantiles is of special relevance in many fields of environmental research. For instance, vegetation may react more to extreme environmental conditions or catastrophic events than to the means, leading to abundance or extinction of species. An example is discussed in Tinner *et al.* (1999) concerning the regional extinction of important tree species (e.g. *Abies alba*) because of fire disturbance; also see Carraro *et al.* (1999). Estimated quantiles can be used to identify any 'instabilities' in vegetation records, with the aim to link these with evidences of external disturbances in a regression modeling approach involving high or low quantiles (Koenker and Bassett 1978). In paleo-ecology for instance, very little is known about the effect of Late Glacial and Holocene climate and disturbance variability on long-term vegetation dynamics. Investigation of such questions are important since – driven by global change – ecosystems in Europe are likely to be exposed to extinction and immigrations during the next decades and centuries. Quantitative analysis of environmental data with these methods can provide new information and address key questions with major scientific and societal relevance.

Acknowledgement

The tree ring data were collected by Fritz Schweingruber, WSL, Birmensdorf. The source of the global temperature data is Climate Research Unit, University of East Anglia, UK. The precipitation data source is Meteoswiss, Switzerland.

References

Beran J. 1994. Statistics for long-memory processes. Chapman and Hall, New York.

Beran J. 1995. Maximum likelihood estimation of the differencing parameter for invertible short- and long-memory ARIMA models. Journal of the Royal Statistical Society, Series B 57: 659–672.

Beran J. and Feng Y. 2002a. SEMIFAR models – a semiparametric framework for modelling trends, long-range dependence and nonstationarity. Computational Statistics and Data Analysis 40: 393–419.

Beran J. and Feng Y. 2002b. Data driven bandwidth choice for SEMIFAR models. Journal of Computational and Graphical Statistics 11: 690– 713.

Beran J., Feng Y., Ghosh S. and Sibbertsen P. 2002. On robust local polynomial estimation with long-memory errors. International Journal of Forecasting 18: 227–241

Beran J. and Ocker D. 1999. SEMIFAR forecasts, with applications to foreign exchange rates. Journal of Statistical Planning and Inference 80: 137–153.

Bradshaw G.A. and Spies T.A. 1992.. Characterizing canopy gap structure in forests using wavelet analysis. Journal of Ecology 80: 205–215.

Brillinger D.R. 1994. Some river wavelets. Environmetrics 5: 221–220.

Brunet Y. and Collineau S. 1994. Wavelet analysis of diurnal and nocturnal turbulence above a maize crop. Foufoula-Georgiou E. and Kumar P. (eds.). Wavelets in Geophysics. Academic Press, San Diego, CA, pp. 129–150.

Carraro G., Klötzli F., Walther G.-R., Gianoni P. and Mossi R. 1999. Observed changes in vegetation in relation to climate warming. Zürich: vdf, Hochschulverlag AG an der ETH Zürich, 87 pp.

Conway T.J., Tans P.P. and Waterman L.S. 1994. Atmospheric CO_2 records from sites in the NOAA/CMDL air sampling network. In T.A. Boden, D.P. Kaiser, R.J. Sepanski, and F.W. Stoss (eds.). Trends '93: A Compendium of Data on Global Change. ORNL/CDIAC-65. Carbon Dioxide Information Analysis Center, Oak Ridge National Laboratory, U.S. Department of Energy, Oak Ridge, Tenn., U.S.A.

Cook E.R. and Kairiukstis L.A. (eds.) 1990. Methods of Dendrochronology – Applications in the environmental sciences. Kluwer Academic Publishers, Dordrecht.

Craigmile P.F., Guttorp P. and Percival D.B. 2004. Trend assessment in a long memory dependence model using the discrete wavelet transform. Environmetrics 15: 313–335.

Csillag F. and Kabos S. 2002. Wavelets, boundaries, and the spatial analysis of landscape pattern. Ecoscience 9:177–190.

Dale M.R.T. and Mah M. 1998. The use of wavelets for spatial pattern analysis in ecology. Journal of Vegetation Science 9: 805–814.

Embrechts P., Klüppelberg C. and Mikosch T. 2003. Modelling extremal events for insurance and finance. Springer Verlag, New York.

Engel J., Herrmann E., Gasser T. 1994. An iterative bandwidth selector for kernel estimation of densities and their derivatives. Journal of Nonparametric Statistics 4: 21–34.

Eubank R. 1988. Spline smoothing and nonparametric regression, Marcel Dekker, New York.

Fan J. and Gijbels I. 1996. Local Polynomial Modeling and its Applications, Chapman and Hall, London.

Feng Y. 2004. Non- and Semiparametric Regression with Fractional Time Series Errors-Theory and Applications to Financial Data. Habilitation Thesis, Konstanz.

Gasser T. and Müller H. 1984. Estimating Regression functions and their derivatives by the kernel method, Scandinavian Journal of Statistics 11: 171–185.

Ghosh S., Beran J. and Innes J. 1997. Nonparametric conditional quantile estimation in the presence of long memory. Student 2: 109–117.

Ghosh S. and Draghicescu D. 2002a. Predicting the distribution function for long-memory processes. International Journal of Forecasting 18: 283–290.

Ghosh S. and Draghicescu D. 2002b. An algorithm for optimal bandwidth selection for smooth nonparametric quantiles and distribution functions. In, Statistics in Industry and Technology: Statistical Data Analysis based on the L1-norm and related methods. Birkhäuser Verlag, Basel, Switzerland, pp. 161–168.

Granger C.W.J. and Joyeux R. 1980. An introduction to long-range time series models and fractional differencing. Journal of Time Series Analysis 1: 15–30.

Heiler S. and Feng Y. 2000. Data-driven Decomposition of Seasonal Time Series. Journal of Statitsical Planning and Inference 91: 351–363.

Heiler S. and Michels P. 1994. Deskriptive und Explorative Datenanalyse. R. Oldenbourg, Mnchen.

Hosking J.R.M. 1981. Fractional differencing. Biometrika 68: 165–176.

Hulme M. 1996. Recent climatic change in the world's drylands. Geophysical Research Letters 23: 61–64.

Hulme M. and Kelly M. 1993. Exploring the links between desertification and climate change. Environment, 35: 39–45.

Koenker R.W. and Bassett G.W. 1978. Regression quantiles. Econometrica 46: 33–50.

Lark R.M. and Webster R. 1999. Analysis and elucidation of soil variation using wavelets. European Journal of Soil Science 50: 185–206.

Lark R.M. and Webster R. 2001. Changes in variance and correlation of soil properties with scale and location: analysis using an adapted maximal overlap discrete wavelet transform. European Journal of Soil Science 52: 547–562.

Mandelbrot B.B. 1983. The fractal geometry of nature. New York: Freeman.

Percival D.B. and Walden A.T. 2000. Wavelet methods for time series analysis. Cambridge University Press, Cambridge, UK.

Percival D.B., Wang M. and Overland J.O. 2004. An introduction to wavelet analysis with application to vegetation time series. Community Ecology 5: 19–30.

Rowell D.P., Folland C.K., Maskell K. and Ward M.N. 1995. Variability of summer rainfall over tropical North Africa (1906–1992) – observations and modeling. Quarterly Journal of the Royal Meteorological Society 121: 669–704.

Schmatz D.R., Ghosh S. and Heller I. 2001. Tree Ring Web and alternative chronologies. Abstracts book. International Conference Tree Rings and People. Davos, 22–26 September 2001, p. 120.

Sigma 1996. Natural catastrophes and major losses in 1995: decrease compound to previous year but continually high level of losses since 1989. Sigma publication no. 2, Swiss Re, Zürich.

Silverman B.W. 1986. Density Estimation. Chapman & Hall, London.

Spitzer F. 1964. Principles of Random Walk. Van Nostrand, Princeton.

Taqqu M. 1975. Weak convergence to fractional Brownian motion and to the Rosenblatt process. Zeitschrift für Wahrscheinlichkeitstheorie und Verwandte Gebiete 31: 287–302.

Tinner W., Hubschmid P., Wehrli M., Ammann B., and Conedera M. 1999. Long-term forest fire ecology and dynamics in southern Switzerland. Journal of Ecology 87: 273–289.

Whitcher B.J, Byers S.D., Guttorp P. and Percival D.B. 2002. Testing for homogeneity of variance in time series: long memory, wavelets and the Nile river. Water Resources Research, 38:10.1029/2001WR000509.

Wiles G.C., D'Arrigo R.D. and Jacoby G.C. 1998. Gulf of Alaska atmosphereocean variability over recent centuries inferred from coastal tree-ring records. Climatic Change 38: 289–306.

Wiles G.C., D'Arrigo R.D. and Jacoby G.C. 1999. Spring/Summer Gulf of Alaska temperature reconstructions. International Tree-Ring Data Bank, IGBP PAGES/World Data Center–A for Paleoclimatology, Data Contribution Series no. 1999–045, NOAA/NGDC Paleoclimatology Program, Boulder CO, U.S.A. Available at ftp://ftp.ngdc.noaa.gov/paleo/treering/reconstructions/gulf_of_alaska.

Model Up-scaling in Landscape Research

Heike Lischke[1*], Thomas J. Löffler[2], Peter E. Thornton[3] and Niklaus E. Zimmermann[1]

[1] Swiss Federal Research Institute WSL, Zürcherstrasse 111, CH-8903 Birmensdorf, Switzerland
[2] Geological Institute, Swiss Federal Institute of Technology, Zürich, Switzerland
[3] Climate and Global Dynamics Division, National Center of Atmospheric Research (NCAR), Boulder, CO 80305, U.S.A
[*] Corresponding author: e-mail: heike.lischke@wsl.ch

Abstract
Up-scaling of ecological models, i.e. translating a model from a source to a larger target scale is a topic of growing importance in landscape research. It is required for adapting the scales of applications, data and models and for simplifying models in a controlled way, in order to decrease simulation times and model uncertainties. Up-scaling from lower scales can corroborate existing models, yield reliable forecast models and enhance ecological understanding by identifying relevant source scale processes and explaining how target scale phenomena are constituted from source scale mechanisms.

This chapter is an overview of up-scaling techniques based on hierarchy theory as an ideal framework for a successful up-scaling. Hierarchy theory leads to a general formulation of the up-scaling process, which consists of (a) aggregating source scale variables to target scale variables and (b) deriving the associated target scale model functions.

Different up-scaling approaches are presented and categorized according to model nonlinearity, variable heterogeneity and the used approximations. The methods range from direct evaluation, to implicit up-scaling, heuristic up-scaling, meta-modeling, mean-field approaches, scale transition theory, moment methods, and to scale separation and adaptive mesh approaches.

Examples include the scaling of a) two logistically growing subpopulations to their metapopulation b) photosynthesis from chloroplasts to continental vegetation canopies, c) individual tree dynamics to the stand level, and d) small-scale active seed dispersal behavior of animals to a general seed dispersal kernel and plant migration model.

Keywords: up-scaling, ecological models, landscape models, hierarchy theory, aggregation, mean-field, scale transition, scale separation, moment equations

F. Kienast, O. Wildi & S. Ghosh (eds.), A Changing World. Challenges for Landscape Research, 249–272.

Introduction

Observations, analyses, or model predictions are all related to a certain scale, scale being defined by grain (or resolution) and extent (or size), measured in absolute (meters, seconds) or in relative (ratio) terms (Wu and Qi 2000). While the concept of scale is relatively old, its importance in (landscape) ecology has considerably grown since the 1990s (Levin 1992; Nikora *et al.* 1999; Peterson and Parker 1998; Schneider 2001; Wiens 1989; Wu 2004). In landscape ecology and modeling, scale primarily addresses space and time, but also embraces thematic categorizations, from biomes (Holdridge 1947; Kirilenko and Solomon 1998; Prentice *et al.* 1992; Thornton *et al.* 2002), to ecosystems (Beerling and Woodward 1996), vegetation types (Brown 1994; Zimmermann and Kienast 1999), individual species (Bolliger *et al.* 2000; Iverson and Prasad 1998), structured populations (Lischke *et al.* 1998), or to individuals (Pacala *et al.* 1993).

Scaling addresses the process of translating information from one scale to another (Auger and Lett 2003; Bugmann *et al.* 2000; King 1991; Melbourne and Chesson 2005; Wu 1999; 2004). Scaling requires the definition of a source and a target scale including grain and extent. Scaling from small to large scale is called **up-scaling**, in the other direction **downscaling**.

Scaling is needed for a variety of reasons. One is incompatibility of ecological data gathered at different scales. To make such data compatible, they must be transformed to a common target scale. Scaling is also necessary to translate model processes from one to another scale. For example, the lack of small scale data or restricted computing resources can impede large scale model simulations. Then the model's process functions or the data should be adapted to the coarser scale. For instance, when attempting to predict photosynthesis at regional, continental or global extents, simulations are not feasible for single leaves, i.e. at a temporal grain of minutes and a spatial grain of centimeters. Then, up-scaling of the leaves' photosynthesis to canopy processes at the level of large biomes or landscape units is required. Scaling is also used to make predictions on scales which are not accessible for direct observation, e.g. of the overall carbon sequestration and pools of an entire biome (Schulze *et al.* 1994; Thornton 1998). Additionally, if model parts act at different grains, they can be adjusted by scaling to a common grain.

The term scaling refers to data, processes, or models. However, in most cases scaling is related to models. This is because scaling is not a property inherent in the observed system, but a change of view of the observer, who has a model about how processes work or how the raw numbers of data have to be interpreted. Such models by the observer are often only implicit (e.g. "0°C means a temperature where water freezes") and thus not necessarily obvious. In the following we focus on the scaling of ecological models.

Hierarchy Theory and Scaling

A prerequisite for scaling models is to develop a concept of how the involved processes and components interrelate and how they can be ordered. Hierarchy theory provides a conceptual framework for this ordering (Allen and Hoekstra 1992; O'Neill *et al.* 1986; Pattee 1973; Peterson and Parker 1998; Salthe 1985; Schneider 2001). It deals with a ranking of levels, each consisting of interacting entities with their own dynamics (Fig. 1). From a landscape ecological view, a hierarchy consists of e.g. the levels cell, individual, population, community, ecosystem, catchment, biome, and biosphere. Each hierarchical level determines the grain of the associated scale. The extent is the section of the hierarchical level that will be looked at. For example, on the hierarchical level of individual trees, a reasonable grain of the spatial scale is several meters. If the extent is chosen as e.g. one ha, the hierarchical level of a forest stand is still covered, but not that of a catchment.

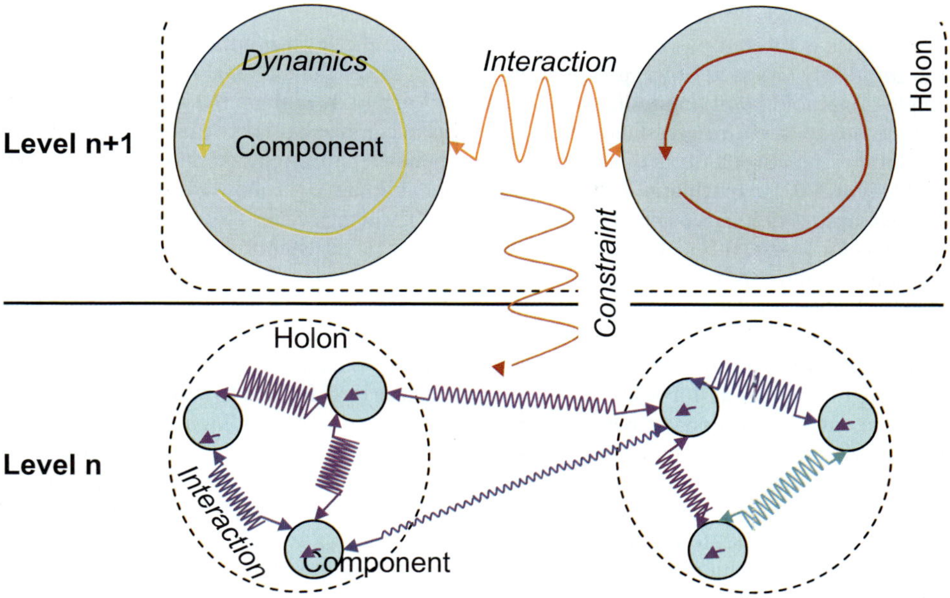

Fig. 1. Hierarchy defined by process frequency and interaction strength. Two hierarchical levels of a system are exemplified. Levels are defined by the time scales (frequencies) of the processes (dynamics and interactions). The lower the level the faster are the processes. Entities of level n can be grouped to holons according to the interaction strength and frequency. Holons of level n are the entities of level *n*+1, i.e. the functioning of level *n*+1 is founded at level *n*. The entities of level n are constrained by level *n*+1.

Ecological hierarchy and associated scales do not exist per se; rather they represent an instrument constructed by the observer or modeler to deal with ecological complexity and technical constraints (Wu and David 2002). An obvious criterion to group entities (organisms, communities etc.) into hierarchical levels is the entities' size. But for scaling issues the strength of the interactions among the entities, expressed by the size of process rates (King 1991), is more practical. The entities of each level *n* (source level) within a hierarchy have their own dynamics and interactions (Fig. 1). Entities with strong or fast interactions can be grouped into "holons". Consequently, the entities of the different holons are only weakly coupled among each other. The holons at level *n* form the entities at level *n*+1.

This implies that the dynamics of one entity at (target) level *n*+1 originates from the aggregated dynamics of the entities of the corresponding holon at level *n*. The interactions among two entities at level *n*+1 originate from the aggregated interactions of the corresponding entities at level *n*. For example, the leaves of the trees can be viewed as entities on the level "tree". They burst from the buds, develop, sequester carbon and shed at slightly different times and speeds. Leaves of the same tree interact more strongly with each other, e.g. by shading or by nutrient use, than with those of other trees. That means the leaves of a tree can be regarded as one holon, and consequently the tree's canopy as an entity of the stand level. The aggregated dynamics is the development of each tree's canopy during one season, the aggregated interaction between two trees is the shading of a tree by a taller one.

Thus, the behavior of the system at level *n*+1 is a consequence of the behavior at level *n*. However, how much of the behavior of a single level-*n*-entity eventually makes a way into level n+1 depends on its share in the aggregated dynamics. Usually, most of the single level-n-dynamics is filtered out, e.g. by smoothing between the level-n-entities. This implies that the frequency or speed of the dynamics at level n is lower than at level n+1. For example, a single graminoid plant accumulates nutrients and carbon over years but may die off after a few decades. In entire grassland ecosystems, this accumulation takes place over centuries. Sometimes, however, effects of lower levels can penetrate through several levels above (Wu and David 2002), particularly if they are strongly nonlinear, e.g. follow a threshold-function. For example, a single tree-seed (level "individuals"), which reaches a site in an old-field, can found a new forest, although the average seed number (level "stand") is far below the critical value for establishment.

The higher level forms the context of the lower level i.e. it modifies the lower level environment (Allen and Hoekstra 1992; King 1991) by low frequency constraints. These constraints are determined by the n+1-level-dynamics, which are themselves constrained by even higher levels and consist partly of the aggregated level-n-dynamics (=feedback). Consequently, any stresses or perturbations of level n+1 (e.g. by even higher levels) will have implications for level n, but a single level-n-entitiy has no or only little impact on level n+1. For example, the accelerated growth or death of a single tree does not have a huge influence on the forest dynamics, however the shadowing of the canopy, which is maybe reinforced by increased cloudiness, affects each single tree.

In the context of hierarchy theory, the key challenge of up-scaling is to derive aggregated variables and dynamics from the lower level, which is the focus of the subsequent sections.

Up-scaling Approaches

In the following, we primarily focus on **up-scaling**. The term **"source"** indicates level n from which we start, and the term **"target"** refers to level n+1 (Fig. 1). These levels can differ so strongly e.g. single cells and entire biomes, that several up-scaling steps are required ("scaling-ladder", Wu 1999).

In the literature, up-scaling as a technique refers traditionally to two possible actions; namely (1) increasing (spatial or temporal) extent only or (2) increasing extent and grain. Increasing the extent with an identical grain is merely a problem of computational resources. Increasing extent and grain often involves changing the hierarchical level and requires an understanding of interactions within and between the scales.

Principles

Prior to focusing on up-scaling techniques, we propose several principles for up-scaling that arise from hierarchy theory:

- Principle 1: Identify the goal of the up-scaling.
- Principle 2: Understand the source scale: entities, grain, and extent. Identify and formulate the dynamics, interactions and constraints of the entities in the source level.
- Principle 3: Describe the entities and their dynamics, interactions and constraints mathematically.
- Principle 4: According to the goal of the up-scaling, identify target scale or hierarchical level. If required, determine potential entities and processes of the target scale.
- Principle 5: Group the source scale entities to holons, according to the interaction strengths on that scale. The holons should be consistent as far as possible with the predefined target scale entities.

- Principle 6: Determine how the holons of the source scale can be transformed (aggregated) mathematically into suitable entities of the target scale.
- Principle 7: Given the transformation rules of principle 6, aggregate the process functions of the source to the target scale.
- Principle 8: Check whether the aggregated model reproduces the results of the original model adequately. If the check fails, start again with principle 4 and adjust.

Techniques

Developing an appropriate approach to derive the aggregated variables and functions is **the** challenge in the up-scaling process. Various categorizations have been used for up-scaling approaches (e.g. Auger and Lett 2003; Bugmann *et al.* 2000; Dieckmann *et al.* 2000; King 1991; Rastetter *et al.* 1992; Urban 2005). Here, we categorize the approaches by ordering them according to intricacy of the process functions and variables.

$$y_s(\xi) = m(y_s(x), y_d(x)), y_s = (y_{s,k}), y_d = (y_{d,l}), m = (m_k) \qquad (1)$$

represents the **source scale model**. The entities are the state variables $y_s(x)$ and driving variables $y_d(x)$. x is the scaling variable (e.g. location, time point or biotic organization), and ξ is one specific realization of it. The model function m describes how the state variables $y_s(\xi)$ at this specific realization depend on the state variables $y_s(x)$ and driving variables $y_d(x)$. A static model depends only on driving variables whereas a dynamic model additionally depends on state variables changing over time. The indices k and l describe the different elements of the state and driving variables and model functions.

The **aggregated variables**

$$Y_v(X) = T(y_v(x)), v = s, d \qquad (2)$$

on the target scale with scaling variable X are obtained by first grouping the holons on the source scale (principle 5) and by then applying transformation T (principle 6). In the following, we often abbreviate the vector of driving and state variables by $y(x)=(y_s(x),y_d(x))$ and $Y(x)=(Y_s(x), Y_d(x))$. The type of the aggregated variables Y_v depends on the goal of the up-scaling. If this is merely to increase the spatial or temporal grain, then Y_v is not predefined. In this case T can be chosen freely, usually as simple as possible, e.g. as a sum or average. In other cases, Y_v is predefined more specifically by the goal. Then the grouping and the transformation Y_v must be chosen accordingly. Usually, there are several possibilities for the Y_v and consequently for T. For example, when up-scaling an individual-based forest stand model intended to study species migrations over large regions and long periods, the aggregated variables might be pre-defined as spatio-temporal distributions of tree species populations, or also species specific migration rates.

Strictly applying the transformation T also to the model functions m yields the ideal **aggregated functions**

$$M^{ideal}(Y_s(X), Y_d(X)) = T(m(y_s(x), y_d(x)) \qquad (3)$$

which describe the ideal **up-scaled model**

$$Y_s(\Xi) = M^{ideal}(Y_s(X), Y_d(X)) \tag{4}$$

with Ξ representing one specific realization of X. However, the exact derivation of M^{ideal} is usually impossible, and M^{ideal} has to be approximated by M, which involves an approximation error Δ. As a consequence, $M^{ideal} = M+\Delta$ and the aggregated model equates to:

$$Y_s(\Xi) = M(Y_s(X), Y_d(X)) + \Delta \tag{5}$$

The up-scaling techniques differ mainly in how transformation T is derived and applied. Generally, T should be chosen in such a way that it results in a minimal approximation error Δ (Auger and Lett 2003) and is mathematically tractable. In the development of many ecological models, the process functions are formulated on the target scale alone, i.e. neither the source scale nor T are considered at all. The target scale process formulation in such models is based on qualitative reasoning and empirical knowledge on that scale only, instead of being up-scaled from the level below. The lower level processes are then only implicitly contained in the target scale functions and parameters (implicit up-scaling, Bugmann *et al.* 2000). In the often used **heuristic up-scaling** the source scale process functions exist, but the target scale model is not derived explicitly from these. Instead, it is chosen from experience or expert knowledge on the target scale, and checked against target scale data or source scale simulations.

In other cases, T is used explicitly in different ways: either by freely choosing it or by constructing it to suit predefined types of target scale variables Y_s. Although T is a general function, a commonly used transformation is the sum or average of the single variables of a holon, e.g. a population, large scale grid cell or time period. Below, we consider averages of continuous state variables $y_s(\xi)$ (6), but the reasoning also holds for sums of discrete variables. Here, h_X is the grain of the target scale.

$$Y_s(\Xi) = \frac{1}{h_X} \int_{x=\Xi}^{\Xi+h_X} y_s(x)dx \tag{6}$$

Then the aggregated function (7) is the average of the source scale functions. The central problem of up-scaling is to evaluate or appropriately approximate this integral, and ideally to express the solution as a function of the target scale variables $Y_s(\Xi)$.

$$M^{ideal}(Y_s(\Xi)) = \frac{1}{h_X} \int_{x=\Xi}^{\Xi+h_X} m(y_s(x), y_d(x))dx \tag{7}$$

Table 1 gives an overview over different methods to solve this problem. The example "Two logistic populations" (see below) demonstrates some of the methods. A straightforward approach is the evaluation of the integral (7) by numeric integration, e.g. by computing the function $m(y_s(x), y_d(x))$ for all x at the source scale and averaging over each Ξ. This so-called **direct evaluation** (Tab. 1c) by computing (direct extrapolation of King 1991) avoids the difficulties of mathematically solving (7), i.e. up-scaling the functions m. For example, one can simulate the dynamics of populations on each 100 m^2 sub-area of an area of 1 km^2 and then average it to predict the average population dynamics on 1 km^2. Since the variables and the process functions remain the same, this approach does not involve the change of hierarchical levels.

Tab. 1. Up-scaling approaches for different types of pattern variability (homogenous (b), variable (a, c, d) and large scale heterogenous (e)), model functions (linear or nonlinear). Each dot is a source scale entity x, its size corresponds to the value of one variable y, each of the four squares a holon and consequently target scale entity. The short lines in (d) and (e) indicate interactions between the entities.

	Variability	(Non-)linearity of model functions	Suitable Approaches
a	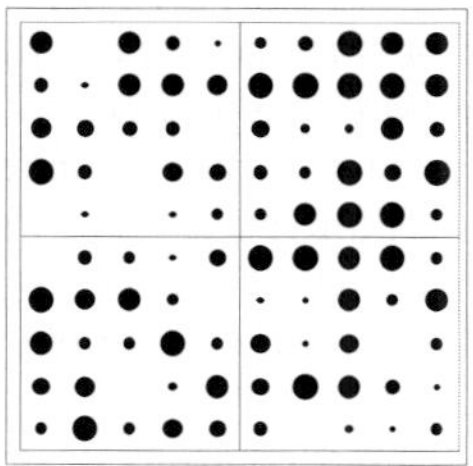	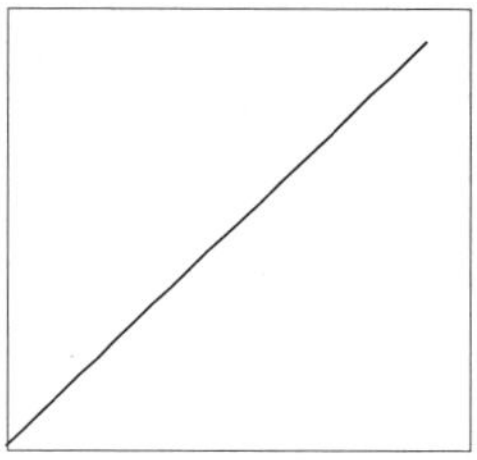	Perfect aggregation: Mean-field
b	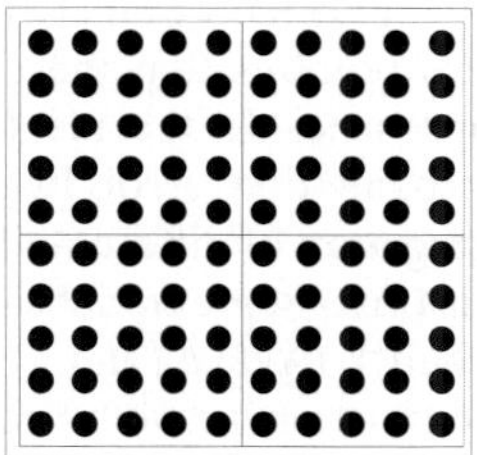	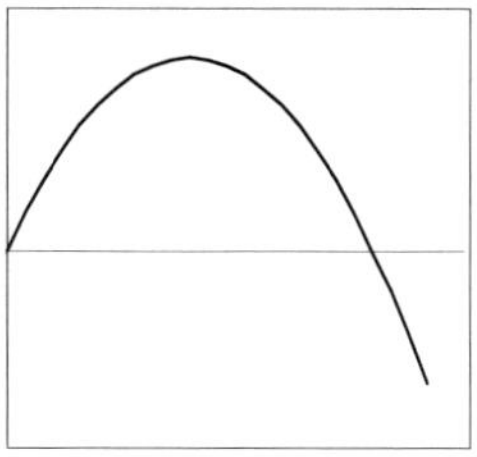	Perfect aggregation: Mean-field
c	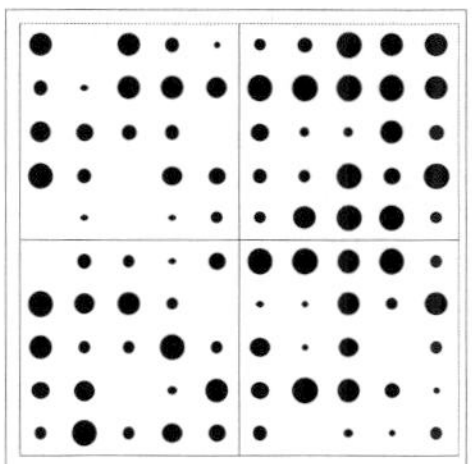	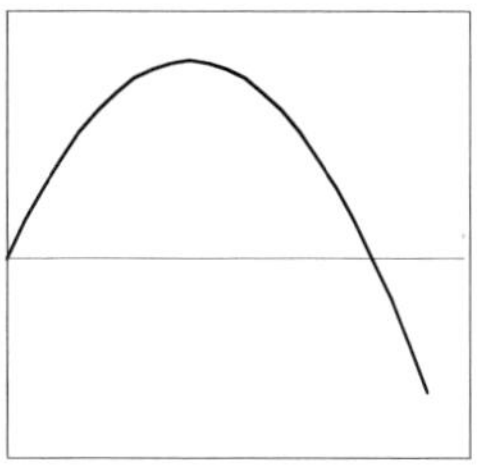	Direct evaluation Evaluation by expected value Scale transition Meta-modeling
d	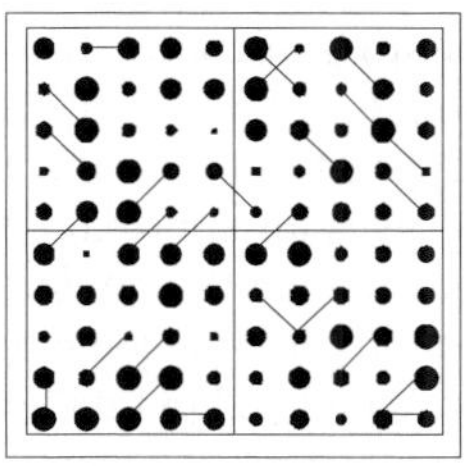	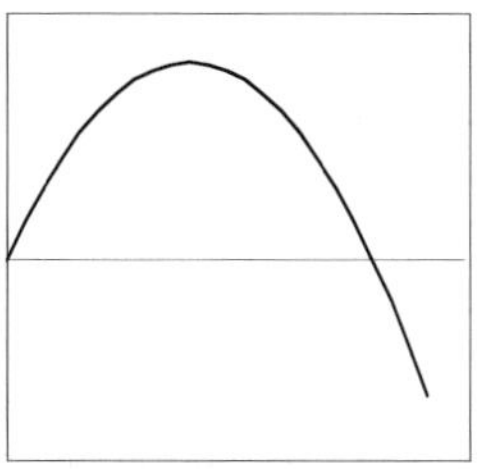	Moment equations Scale separation Meta-modeling
e	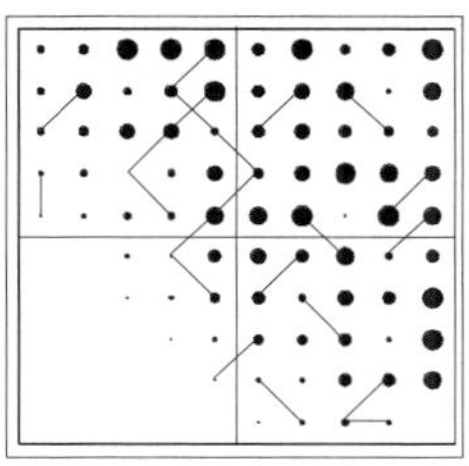	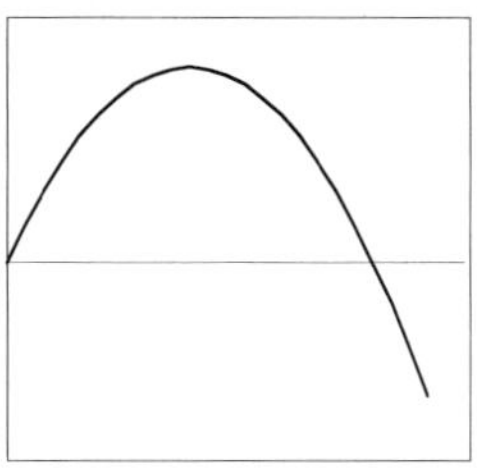	Adaptive grid Estimation of internal pattern Metamodeling

Another computationally intensive approach is to construct aggregated functions by **meta-modeling**, which means to fit the parameters of a simple model to simulations of a complex model (Urban 2005; Urban *et al.* 1999). Here, the fitting corresponds to the transformation T. It is achieved by generating artificial sets of large scale state and driving variables $Y_{s,k}$ and $Y_{d,k}$ by simulations with the source scale model, driven by various sets of small scale driving variables $y_{d,k}(x)$, $k=1,\ldots,n$. Next, the simple meta-model, $y_{d,k}(\Xi) = \widetilde{M}(Y_{s,k}(X), Y_{d,sk}(X))$, $k = 1,\ldots,n$, (e.g. a multivariate regression model or a matrix-population-model) is fitted to the $Y_{s,k}$ and $Y_{d,k}$. Once calibrated from exhaustive simulations of the finer scale model, such meta-models $\widetilde{M}$ can be applied very efficiently to huge regions and long periods. For example, Acevedo *et al.* (2001; 1995; 1996) ran simulations with a forest gap model under various environmental condition. They categorized the simulated stands into forest types, and noted under which combination of environment and surrounding forest the types changed into others. This yielded the transition probabilities of the forest types used in a simple type transition model that could be applied efficiently to large areas.

Since computing power is (still) limited, such approaches cannot be applied to very complex models, consisting of many interacting entities, and involving possibly numerical solution of differential equations. Up-scaling of such complex models requires that the aggregated variables and functions are derived with mathematical approaches. In the ideal case, the aggregated functions can be derived directly, without approximation error. Iwasa (1987) gives formal conditions and examples for such a **perfect aggregation** of differential equation models. For our common special case of averaging, (7) demonstrates some trivial cases where perfect aggregation is possible. If $y(x) = c =$ const. for all x, then $Y(X) = c$ (Tab. 1b), and the source scale function and its average, the target scale average function, are identical.

In this case $M^{ideal}(Y(\Xi)) = \frac{1}{h_X} a \int_{x=\Xi}^{\Xi+h_X} m(c)dx = m(c)$. If $y(x)$ is variable with mean $\bar{y}$ and the model m is linear (Tab. 1a), i.e. $m(y(x)) = a\,y(x) + b$, then

$$M^{ideal}(Y(\Xi)) = \frac{1}{h_X} a \int_{x=\Xi}^{\Xi+h_X} y(x)dx + b = a\bar{y} + b = m(\bar{y}).$$

In this case, the **mean-field approach**, i.e. applying the model to the average variables (lumping, King 1991), is a perfect aggregation.

However, four properties common to ecological systems usually impede a perfect up-scaling: **nonlinearity, variability, heterogeneity** and **within-holon-interactions.** If a model m m is nonlinear and the properties y of the entities are variable (e.g. Tab. 1c), the average of the model m (the aggregated model M) is different from m applied to the average property $\bar{y}$ (mean-field approach). Thus, the mean-field approach involves an approximation error. For example, Pierce and Running (1995) demonstrated that simulating the nonlinear carbon and water dynamics with driving variables averaged over cells of 0.5° latitude/longitude can differ significantly from the averaged dynamics over the same area simulated in smaller grid cells. The involved approximation errors are particularly large in complex terrains, whereas on a flat terrain the variability of the drivers is so low that little information is lost by the averaging.

Several approaches based on approximations address the problem of nonlinearity and variability. The first group of approaches works when the entities within a holon do not interact, i.e. $y_s(\xi) = m(y(\xi))$.

Some approaches (extrapolation by expected value, King 1991) **evaluate** the **expected value** of $m(y)$ (Tab. 1c) by integrating over the source scale model function $m(y)$ of all possible values of the source scale variables and their distribution densities $p_{\Xi,\Xi+h_X}$ within the holon

$$M(Y(\Xi)) = \int_{y=-\infty}^{\infty} m(y) p_{\Xi,\Xi+h_X}(y)dy \tag{8}$$

The integral (8) will often be impossible to solve. Thus, the integral, or its integrands must be approximated. For the approximation of p, empirical frequency distributions can be used. These can be shares of stratification units or theoretical distributions, e.g. the Normal distribution, and their parameters such as mean or variance. Again, estimating the distributions and their parameters involves an approximation error. Also m is sometimes approximated (Lischke *et al.* 1997b). An example is the scaling of an ecological process rate m, depending nonlinearly on the strongly variable temperature y, over the course of a year. The mean yearly process rate can be calculated e.g. by piece-wise linearizing m and describing the temporal temperature variability by a Normal distribution p over the year based on mean and variance (Lischke *et al.* 1997a).

Another possibility is to calculate the expected value (8) stochastically, e.g by Monte-Carlo-simulations. Here, values of y are generated stochastically, with frequencies determined from the distribution $p_{\Xi,\Xi+h_X}$. The $m(y)$ are then calculated and averaged. Depending on the number of replicates used, this approach requires much computing time, and will be more efficient than the explicit evaluation only for very large holons.

The **scale transition theory** (Tab. 1c, Chesson 1998; Melbourne and Chesson 2005) is based on the Taylor-series of the model functions around the mean values of the y. This Taylor-series is given in (9), for the case that where y consists of a single driving or state variable only. The aggregated function (10) then consists of terms including the nth derivative of m, and the nth central moment of the source scale variables.

$$m(y(x)) = \sum_{n=0}^{\infty} \frac{1}{n!} \left. \frac{d^n m(y)}{dy^n} \right|_{y=\bar{y}} (y(x) - \bar{y})^n \tag{9}$$

$$\Rightarrow M^{ideal}(Y(\Xi)) = \sum_{n=0}^{\infty} \frac{1}{n!} \left. \frac{d^n m(y)}{dy^n} \right|_{y=\bar{y}} \underbrace{\frac{1}{h_X} \int_{x=\Xi}^{\Xi+h_X} (y(x) - \bar{y})^n \, dx}_{n^{th} \text{ central moment of } y} \tag{10}$$

If (10) is truncated after the second term, we obtain a quadratic approximation (11)

$$M(Y(\Xi)) = m(\bar{y}) + \frac{1}{2} \left. \frac{d^2 m(y)}{dy^2} \right|_{y=\bar{y}} Var(y) \tag{11}$$

Eqn. (11) approximates the ideally up-scaled model by the mean-field function corrected by the second derivative of m (a measure of the model's nonlinearity) and by the variance of y, (a measure for its variability within the holon). If m depends on several y's, further terms containing the covariances are included in (11) (Chesson 1998; Melbourne and Chesson 2005). Yet, the problem remains to calculate the mean and (co)variances.

The scale transition and the expected value approaches rely on statistical distributions and their parameters. If the distributions (e.g. of soil pH) are constant during a simulation, they must be estimated only once before the simulation. If the distributions (e.g. of fluctuating temperature) change with time, then they have to be estimated for each simulation step, which can also be done before a simulation. However, if the small scale model depends on the state variables y_s (i.e. is dynamic), then the distribution of the y_s within a holon is dynamic, i.e. not known beforehand and has to be estimated in each time step. This results in further approximation errors, e.g by assuming a fixed type of the distribution and only deriving the parameters from the mean (see example "life cycles of trees").

Another promising idea is to use additional moments as aggregated variables, instead of only using the mean. For very simple models, using these moments can even yield a perfect aggregation (see below "Two logistic populations"). Yet, usually perfect aggregation is not possible for more complex models, particularly, when the entities within a holon interact (Tab. 1d, $y_s(\xi) = m(y(x))$). Such interactions can consist e.g. of movements between spatial grid-cells. One approach of dealing with dynamically varying and potentially interacting entities is to use **spatial moment equations** (for continuous space) or **pair approximations** (for discrete space) (see introduction in Bolker and Pacala 1997; Bolker and Pacala 1999; Dieckmann and Law 2000; Murrell *et al.* 2004). The central idea here is that a spatial pattern of $y(x)$, e.g. of a population, can be described by its spatial moments $C_1, \ldots, C_n$. The first spatial moment (C_1) is the mean density of y_s, the second spatial moment (C_2) the density of a pair of y_s as a function of the distance between them (closely related to spatial autocorrelation), and the nth spatial moment (C_n) is the density of n y_s as a function of n-1 distances. In this approach, the aggregated variables not only consist of the mean (as in the scale transition approach), but also of the higher spatial moments. Up-scaling, for example the differential equation model $\dot{y}(\xi) = m(y(x))$, implies finding the moments, and deriving the aggregated model $\dot{C}_i = M(C_j)$ depending on these moments. For example, Bolker and Pacala (1997; 1999) aggregated a stochastic individual based model to such a spatial moment differential equation model.

The following approach can be applied, if the source scale model operates on different time scales, meaning that the model processes differ significantly in speed. Such differences in process rates can be used in the process of up-scaling as criteria to group small scale variables into holons. Furthermore, such a **separation of time scales** is the basis of up-scaling methods by **variable aggregation** (Auger and de la Parra 2000; Auger and Lett 2003; Luckyanov 1995) or **relaxation projections** (Dieckmann and Law 2000). The idea behind is to express the model dynamics by a sum of fast processes with rate $m_{fast,i}$ and slow processes with rate $m_{slow,i}$. In the following, we explain this method using a differential equation model with state variables y_i of discrete entities.

$$\dot{y}_{s,i} = m_{fast,i}(y_{s,1},\ldots, y_{s,n}) + m_{slow,i}(y_{s,1},\ldots, y_{s,n}) \tag{12}$$

Using the aggregated variable $Y_s = \frac{1}{n}\sum_{i=1}^{n} y_{s,i}$, the aggregated model becomes

$$\dot{Y} = \frac{1}{n}\sum_{i=1}^{n}(m_{fast,i}(y_{s,1},\ldots, y_{s,n}) + m_{slow,i}(y_{s,1},\ldots, y_{s,n})) \tag{13}$$

On the fast time scale, the slow processes are nearly constant, i.e. the process rates $m_{slow,i}(y_{s,1},\ldots,y_{s,n}) \approx 0$. In this case the dynamics only consists of the fast processes, and the equilibria $(y^*_{s,1},\ldots,y^*_{s,n})$ can be determined by $m_{fast,i}(y^*_{s,1},\ldots,y^*_{s,n}) = 0$. The transformation T is now used to get the y^*_i as a function of the aggregated variable Y. In our example we thus get $Y_s = \frac{1}{n}\sum_{i=1}^{n} y^*_{s,i}$, yielding the fast process' equilibrium values $y^*_i(Y)$. In turn, on the slow time scale the fast processes can now be assumed to be in equilibrium. Substituting the fast scale equilibrium values into (13) makes the $m_{fast,i}$ terms disappear and we finally get the aggregated model, which only depends on the aggregated variable:

$$\dot{Y}_s = \frac{1}{n}\sum_{j=1}^{n} m_{slow,i}(y^*_{s,1}(Y),\ldots, y^*_{s,n}(Y)) \tag{14}$$

In all the methods described above, the variability within the holons is assumed to be homogenous (Tab. 1a-1d). For example, the moment methods assume the pattern characteristics (spatial moments) to be the same everywhere within a target-scale cell (translation invariance, isotropy). **Large scale heterogeneity** (Tab. 1e), as expressed e.g. by gradients or steep steps of y, is not explicitly taken into account. Such large scale heterogeneity may be the consequence of the heterogenous environment, e.g. topography, or of dynamically changing state variables, e.g. expanding waves of migrating organisms. Heterogeneities may have a tremendous impact on simulation outcomes of the target model. Let's imagine a population invasion front entering a given area at one side. All methods discussed so far assume that a certain share of the population immediately reaches the opposite edge of the area, and will thus overestimate the expansion speed.

To cope with large scale heterogeneities one can apply the model on the target scale, but **estimate the pattern within the cells** from additional information. In specific cases, the pattern can be roughly estimated from the model itself. For instance, when simulating an invasion process, the position of the front can be determined by simulating the spread of the outmost individuals in detail (e.g. including seed rain for plants), but in the area behind the front model processes are simulated in the up-scaled mode (Ellner *et al.* 1998).

Alternatively, we could estimate the within cell patterns from the larger scale patterns of the surrounding target cells, e.g. by using geospatial interpolation approaches from spatial statistics.

Finally, the **adaptive mesh approach** (e.g. Lee 1999; Pitman *et al.* 2003) is focused on the target scale patterns. In this approach, the grain is reduced automatically, where the gradient between neighboring cells (pixels or time steps) and thus presumably also within the larger cells is steep. Where gradients are flat, the mesh size (grain) can be kept or even increased without much loss of information. A prerequisite for this approach is that the aggregated functions are either independent of the grain or formulated explicitly dependent of the grain. This approach is rarely used in ecology, but seems to be common in other disciplines. Note that in all approaches that deal with large scale heterogeneity, the up-scaling is combined with model parts which remain on the source scale.

Examples

Two logistic populations

Some of the mentioned approaches are demonstrated with a very simple model. Imagine two populations of a species on two sites. Model (15) describes the dynamics of their population densities y_1 and y_2. The population growth follows the logistic model, with the parameters a (net reproduction rate) and b *(a/carrying capacity)*. We first assume that the populations do not interact, e.g. because the sites are too far apart. They differ only by the initial values $y_{1,0}$ and $y_{2,0}$,

$$\frac{d\,y_i(t)}{dt} = \dot{y}_i = m(y_i) = a\,y_i - b\,y_i^2, \quad y_i(0) = y_{i,0}, \quad i = 1,2 \tag{15}$$

As target scale variable Y we choose the mean of both populations. The aggregated model M then is the right hand side of the differential equation of Y.

$$Y := \bar{y} = \frac{1}{2}(y_1 + y_2) \tag{16}$$

$$\frac{d\,Y}{dt} = \dot{Y} = M(Y) \tag{17}$$

The mean field approach is used by applying the model M:= m with the mean initial value

$$\frac{d\,Y}{dt} = \dot{Y} = aY - bY^2, Y_0 = \frac{1}{2}(y_{1,0} + y_{2,0}).$$

The scale transition approach (18) is a perfect aggregation for model (15), because there are no higher derivatives with respect to y than the second one. However, the variance of the small scale variables has to be calculated from at least one of the small scale variables.

$$M(Y) = m(Y) + \frac{1}{2}\frac{d^2 m}{dy^2} Var = m(Y) - b(Y - y_1)^2 \tag{18}$$

To also track the dynamics of the variance, we use a simple moment equation approach with mean and variance as aggregated variables. After some reordering and substitutions we get the differential equations (19, 20). This is again a perfect aggregation.

$$\frac{d\,Y}{dt} = \frac{1}{2}(m(y_1) + m(y_2)) = a\frac{1}{2}(y_1 + y_2) - \frac{b}{2}(y_1^2 + y_2^2) = aY - b(Y^2 + Var) \tag{19}$$

$$\frac{d\,Var}{dt} = \frac{1}{4}(2y_1\frac{d\,y_1}{dt} + 2y_2\frac{d\,y_2}{dt} - 2\frac{d\,y_1}{dt}y_2 - 2\frac{d\,y_2}{dt}y_1) = 2Var(a - 2bY) \tag{20}$$

$$Var = \frac{y_1^2 + y_2^2 - 2y_1y_2}{4} \rightarrow 2y_1y_2 = -4Var + (y_1^2 + y_2^2)$$

$$Y^2 = \frac{y_1^2 + y_2^2 + 2y_1y_2}{4} \rightarrow 2y_1y_2 = +4Y^2 - (y_1^2 + y_2^2)$$

$$\rightarrow 4(Var + Y^2) = 2(y_1^2 + y_2^2) \rightarrow \frac{1}{2}(y_1^2 + y_2^2) = Var + Y^2$$

$$\frac{1}{4}(2y_1\frac{d\,y_1}{dt} + 2y_2\frac{d\,y_2}{dt} - 2\frac{d\,y_1}{dt}y_2 - 2\frac{d\,y_2}{dt}y_1) = \frac{1}{2}(\frac{d\,y_1}{dt}(y_1 - y_2) + \frac{d\,y_2}{dt}(y_2 - y_1))$$

$$= \frac{1}{2}(\frac{d\,y_1}{dt} - \frac{d\,y_2}{dt})(y_1 - y_2) = \frac{1}{2}(y_1 - y_2)(a(y_1 - y_2) - b\,(y_1^2 - y_2^2))$$

$$= 2a\frac{(y_1 - y_2)^2}{4} - b\frac{1}{2}(y_1^3 - y_2y_1^2 - y_1y_2^2 + y_2^3) = 2aVar - b\frac{1}{2}(y_1^3 - 2y_2y_1^2 - 2y_1y_2^2 + y_2y_1^2 + y_1y_2^2 + y_2^3)$$

$$= 2aVar - b\frac{8}{2}\frac{y_1(y_1^2 - 2y_2y_1 + y_2^2) + y_2(y_1^2 - 2y_2y_1 + y_2^2)}{2 \cdot 4} = 2aVar - 4b\frac{(y_1 + y_2)(y_1 - y_2)^2}{2 \cdot 4}$$

$$= 2aVar - 4bYVar = 2Var(a - 2bY)$$

Figure 2 demonstrates the accuracy of the approaches in a specific simulation for the model (16). The scale transition and the moment equation approach yield perfect aggregation, whereas the mean-field approach deviates from the direct evaluation, i.e. the average of the simulated y_1 and y_2. This deviation is large in the transient phase, but disappears in the equilibrium.

Now we extend the model (15) by adding an interaction between the populations, e.g. a movement between the two sites with dispersal rate c (proportion of the population at site i moving to site j per unit time) (21). We assume movement to occur much faster than the population growth, i.e. $a, b \ll c$. Thus, there are two separate time scales and we can apply the scale separation method.

$$\begin{aligned} \dot{y}_1 &= a\,y_1 - b\,y_1^2 + c(y_2 - y_1), \quad y_1(0) = y_{1,0} \\ \dot{y}_2 &= \underbrace{a\,y_2 - b\,y_2^2}_{m_{slow}} + \underbrace{c(y_1 - y_2)}_{m_{fast}}, \quad y_2(0) = y_{2,0} \end{aligned} \tag{21}$$

Again the mean is the target scale variable Y, and the aggregated model is

$$\dot{Y} = \frac{d}{dt}\left[\frac{y_1 + y_2}{2}\right] = \frac{1}{2}(a(y_1 + y_2) - b(y_1^2 + y_2^2)) + c\,\underbrace{(y_2 - y_1 + y_1 - y_2)}_{=0} \tag{22}$$

First we look at the up-scaled model on the fast time scale. The terms with a and b are so small that we assume them to be 0. The equilibrium of the remaining fast part of model (21) is then

$$c(y_1^* - y_2^*) = 0,\; c(y_2^* - y_1^*) = 0 \rightarrow y_1^* = y_2^*$$

Thus, every pair of equal population densities is a fast scale equilibrium. By substituting these equilibria into the aggregation rule (16) we get the equilibria depending on Y.

$$Y = \frac{1}{2}(y_1^* + y_2^*) \rightarrow y_1^* = y_2^* = Y$$

In this special case the equilibria are equal to the mean Y. It means that after some time of moving between the sites, the population densities in the two sites have leveled out.

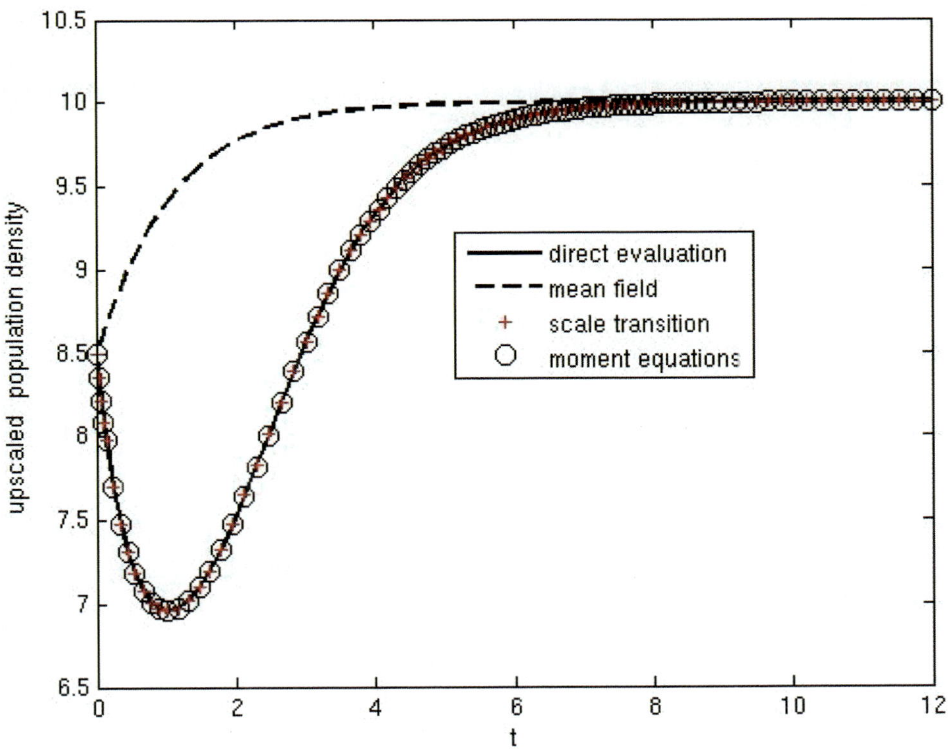

Fig. 2. Comparison of different up-scaling approaches with the directly evaluated model (15) of two logistic populations. a=1, b=0.1, $y_{1,0}$=1, $y_{2,0}$=16.

Now we change the viewpoint to the slow time scale. There, we assume the slow processes to be in the equilibrium, namely Y. Substituting the equilibrium into the slow part of the aggregated model (22) yields the final up-scaled model

$$\dot{Y} = \frac{1}{2}(a(Y + Y) - b(Y^2 + Y^2)) = aY - bY^2.$$

Note that this corresponds to the mean-field solution of the model without interaction. Thus, the interaction decreases the influence of the initial differences between the populations. The accuracy of this approach depends on the separability of the scales; the more the process rates differ, the better the assumptions of constancy hold, and the more accurate is the up-scaling. This is demonstrated in Figure 3, which shows the accuracy of this approach depending on the parameter c with fixed parameters a and b. For example, if the two populations consist of birds only a kilometer apart, their dispersal between the sites (measured in proportion of the population reaching the other site per unit time) is much faster than their reproduction; they will probably level out before their reproduction period. However, if the sites are very far apart, c is small: only few individuals will reach the other site before the next breeding period, and the populations at the sites will change before they can level out.

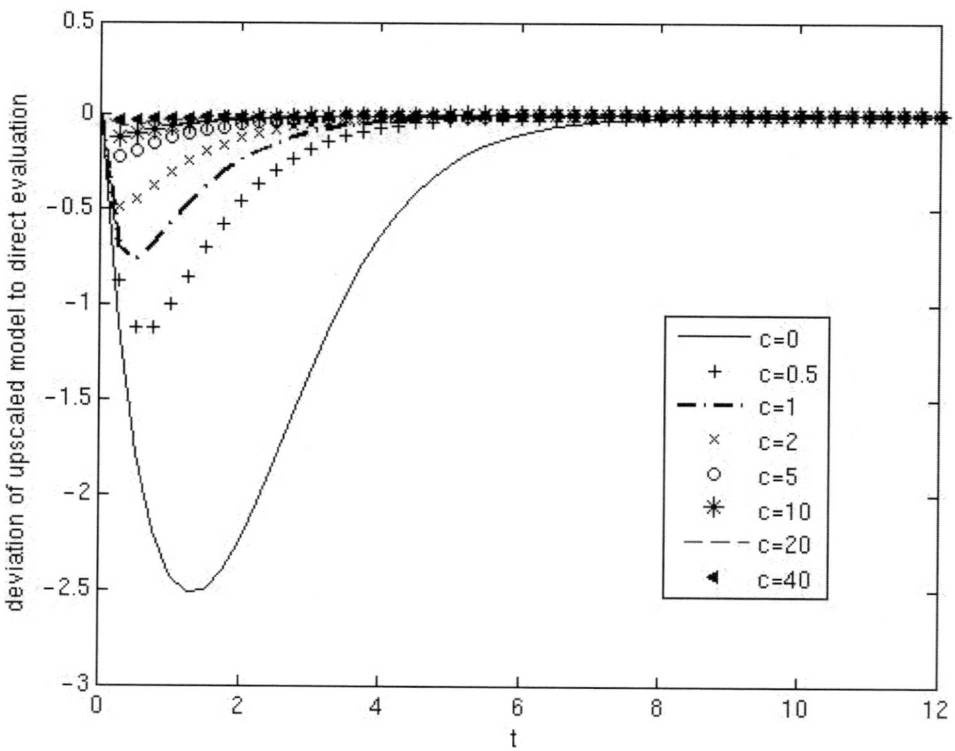

Fig. 3. Deviation (direct evaluation minus up-scaled) of model (21) up-scaled by scale separation from the direct evaluation. Source model: two logistic interacting populations. $a = 1$, $b = 0.1$, c: speed of movement between populations, varying between 0 (no interaction) and 40 (very fast movement).

Chloroplast to landscape – scaling models of photosynthesis

The estimation of gross primary production (GPP) at the landscape scale is a necessary step in modeling the overall dynamics of the terrestrial carbon cycle at regional to global scales (extents). The principles described above find a very natural expression in the analysis of landscape-scale carbon assimilation and photosynthesis necessary to simulate GPP. This is due to the strongly nested hierarchy of landscape, stands, individual plants, leaves, cells, and finally chloroplasts, where photosynthesis takes place (Fig. 4). Although photosynthesis is intrinsically a biochemical process operating at the level of the chloroplast, there are environmental constraints operating at the cell, leaf, and canopy (stand) levels that propagate to impact the biochemical reactions in the chloroplast. Therefore, these constraints have to be formulated depending on the respective hierarchical levels, which result from explicit or implicit up-scaling from the lower levels. The final level of scaling, from stand to landscape, is generally accomplished through a linear mixture model of parallel stands. Here we describe the quantitative mechanisms that operate between levels in this hierarchy, starting with the basic biochemical reaction inside the chloroplast, and tracing the dependencies up in scale through multiple environmental constraints.

The net biochemical fixation of CO_2 within the chloroplast, or assimilation rate *(A)*, depends in part on the intracellular concentration of dissolved CO_2 (C_i), the concentration of carboxylating enzyme *(Rubisco)*, and the flux of photosynthetically active radiation (PAR) (Farqhuar and von Caemmerer 1982).

$$A=f(C_{i,Rubisco},PAR)$$

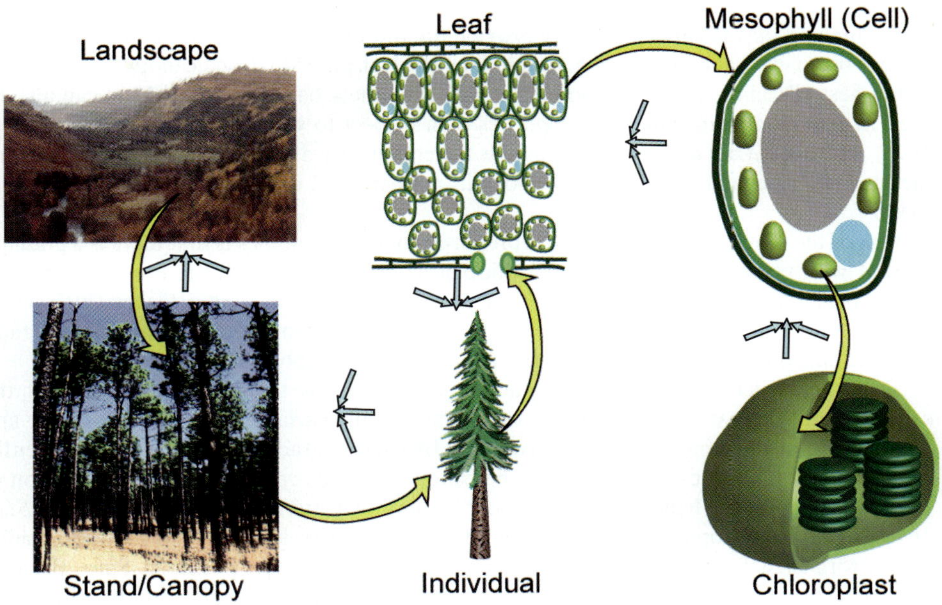

Fig. 4. Organization of a nested hierarchy for scaling estimates of photosynthesis from the chloroplast to the landscape level. Groups of 3 arrows indicate up-scaling, bent arrows environmental constraints imposed from higher levels onto lower levels.

The dependence of A on Ci can be written in terms of the diffusion-limited rate of delivery of CO_2 from the air outside the leaf (C_a) to the sites of carboxylation, as:

$$A=g(C_i - C_a),$$

where g is the total conductance of CO_2 across the gradient, including conductance through the leaf boundary layer, the stomatal pore, the intercellular air space, and the mesophyll cell wall. This relationship provides an implicit up-scaling of the transport processes of the CO_2 through three nested scales, from chloroplast, to cell, to leaf.

The dependence of A on *Rubisco* concentration is expressed as $A \propto V_{Cmax}$, where V_{Cmax} is the maximum rate of carboxylation at optimum CO_2 and light supply. V_{Cmax} depends on the available *Rubisco*. *Rubisco* is limited by the nitrogen supply, and can be related directly to measurable quantities at the leaf scale, as:

$$V_{Cmax} \propto 1/(SLA \cdot CN_{Leaf}),$$

where SLA is the specific leaf area (projected leaf area per unit mass of leaf carbon), and CN_{leaf} is the leaf mass ratio of carbon:nitrogen (Thornton *et al.* 2002) both aggregated from single leaves to the canopy.

The dependence of A on PAR is affected by the three-dimensional structure of the canopy; leaf shape and the orientation and arrangement of leaves in a canopy directly impact the transmission of PAR through the canopy, producing a light environment for each leaf that is a function of its position with respect to other nearby leaves, its position within the canopy, the total canopy leaf area, the solar zenith angle, and sky conditions dictating the fractions of direct and diffuse radiation. Radiative transfer (RT) models can produce very detailed representations of the canopy radiation environment (e.g. Myneni *et al.* 1997) by representing the details of leaf arrangement between and within the crowns of individual plants making up the canopy. So they provide quantitative information about the radiation pattern by an explicit extrapolation, thus linking the leaf, individual, and canopy (stand) levels of the nested scaling hierarchy. Simpler up-scaled models, based on Beers Law, can also be used to estimate mean radiation properties with respect to vertical canopy position (Jones 1992). The Beers Law representation avoids the complexity of individual crowns by ignoring the leaves' angles and spatial distributions in a mean-field approach, albeit with less precision.

One of the most commonly observed properties of plant canopies is that there is a gradient in SLA with vertical position, with the thickest leaves (low SLA) at the top of the canopy where they receive high PAR fluxes from direct sunlight, and thinner leaves (high SLA) lower in the canopy where they receive lower PAR fluxes from predominantly diffuse radiation (Monserud and Marshall 1999; Niinemets *et al.* 1998; Roberts *et al.* 1999).

The basic biochemical mechanism at the chloroplast is therefore constrained at the canopy scale with a nonlinear function of spatially highly variable drivers: the quantity and quality of light reaching individual leaves, and by canopy gradients in SLA. Consequently, the formulation and up-scaling of A to the canopy depends critically on the description of the distribution of the leaves. Thornton and Zimmermann (submitted) showed that the SLA of leaves within canopies varies from top (sunlit leaves with low SLA) to bottom (shaded leaves with high SLA) in a linear manner along vertical canopy positions. Since the $C{:}N$ ratio does not change along the SLA gradient, V_{Cmax} can now be easily scaled to the whole canopy using the above cited relationship.

Scaling of A from the canopy (stand) to higher hierarchical levels can generally be accomplished through simulating landscapes as averages of independent stands weighted linearly according to the stand areas. This corresponds to an extrapolation by expected value with empirical frequency distributions (8). Similarly, regions, continents and the global domain can be treated as simple numerical averages above the stand level. Considered in the broader context of a complete model of terrestrial energy, water, and biogeochemical cycles, other quantitative mechanisms operate at larger spatial scales. Examples include the mechanistic impact of landscape pattern on stand-level disturbance processes such as wildfire and wind damage, or the influence of regional topography on surface weather and climate at the landscape scale.

Individuals to stands – life cycles of trees

An example of a hierarchical up-scaling by a heuristic, approximative moment method is the aggregation of tree individuals to forest stands. The source scale of the aggregation was an individual based forest gap dynamics model (ForClim v2.6, Bugmann and Cramer 1998). In gap models, horizontal and species structures are taken into account by stochastically simulating establishment, growth and death of individual trees of different species on a set of small patches (Fig. 5, left). The output variables of the model are tree populations, i.e. up-scaling by *direct evaluation* is performed by summing the simulated individuals to population densities, biomasses or height structures. Because gap models are stochastic, the results have to be averaged over many replicates of single patch simulations. Thus, generally gap models are computationally very intensive and hence not applicable to certain research fields.

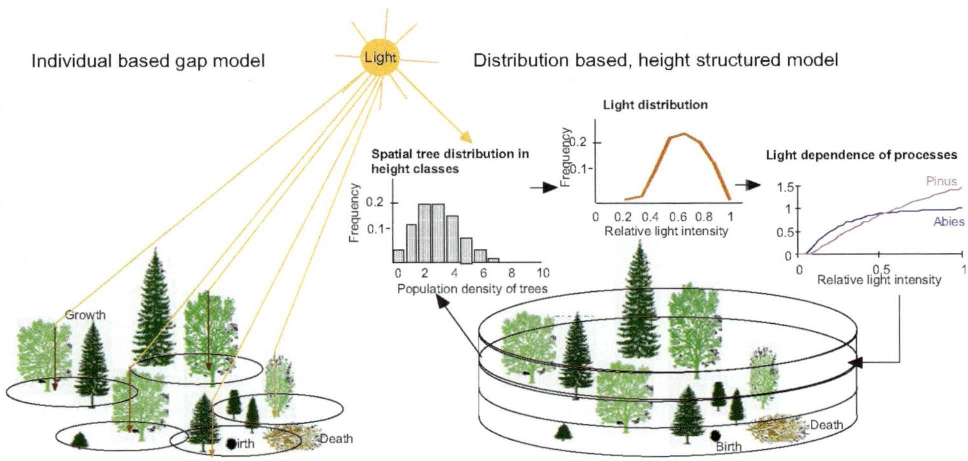

Fig. 5. DisCForM: In an individual based gap model (left), spatial variability is induced by stochastic birth and death event and the resulting differences between the simulated gaps with respect to shading and growth rates. Light reaching a certain level below canopy in a gap depends on the LAI (distribution) therein. In the aggregated form – i.e. the distribution based, height structured model – population densities per height classes are simulated. A theoretical distribution of tree densities/height class is calculated across the entire stand based on the average population density per height class. The resulting light distribution determines the process rates and the dynamics of all trees within this height class.

The up-scaling of the gap model to the model DisCForM (Fig. 5, right, Lischke *et al.* 1998; Löffler and Lischke 2001) was performed by aggregating the individuals to structured populations. Instead of describing the stochastic dynamics of several properties of each individual tree, the aggregated model describes the dynamics of the population densities per species and height class. By doing so the effects of vertical variability can be included. The fundamental approach is to describe the variability of tree population densities from patch to patch by theoretical frequency distributions. The trees per height class and species are assumed to be always randomly distributed over the patches, which yields a Poisson distribution for the population densities. The parameters of this distribution are dynamic; they are defined by the mean population densities, which are the state variables of the model. From the tree density distributions in the height classes, the distribution of the leaf area index (LAI) above each specific height class can be calculated, from the LAI distribution the distribution of the light intensity reaching the height class, and from the light distribution those of the light-dependent, height and species specific process rates such as growth, birth and mortality. This produces a purely deterministic description of the dynamics, which still reflects the variability in a forest affected by stochastic mortality and regeneration effects (per patch). This distribution based approach ressembles the scale transition and moment methods in that it tracks the dynamically changing variability of the tree density. However, by prescribing the Poisson-distribution, it uses a different approximation of the variability. An advantage of this approach compared to the gap model is that it is computationally much more efficient at only slightly reduced precision. Its precision largely depends on the number of height and light classes, rendering more precision at increased number of height classes (decreased class size), which in turn reduces the computational efficiency. However, above ~15 height classes, no more precision can be gained.

Scaling active seed dispersal – from individual behavior to general dispersal kernels

Dispersal of plant populations has received considerable attention in the scientific literature (Cain *et al.* 1998; Cain *et al.* 2000; Clark *et al.* 1999; Nathan *et al.* 2001; Nathan *et al.* 2000; Portnoy and Willson 1993). Much of this debate is related to understanding migration rates and invasion speed, particularly with respect to expected global change (Clark 1998; Clark *et al.* 2001a; Clark *et al.* 2001b; Eriksson 2000; Higgins and Richardson 1999; Kot *et al.* 1996; Lewis 2000; Melillo *et al.* 1996; Solomon and Kirilenko 1997). Dispersal is most often characterized by dispersal kernels, i.e. the probabilities of the seeds to travel to a certain distance from the source tree. These kernels can be derived empirically for small distances to the source tree, however for species spread, the long distance behavior (the tail of the kernel) is crucial (Nathan 2005).

Particularly for animal-dispersed seeds, the up-scaling of the local animal behavior to a large scale kernel, and particularly to its tail, is difficult. The up-scaling is even more complicated if the behavior is not spatially homogenous, for example if animals bury seeds only in preselected cache sites. Powell and Zimmermann (2004) have demonstrated one way of up-scaling such migration behavior to a general seed dispersal kernel by "calibrating the movements of animals with seeds in a landscape containing caching sites". They used a mathematical homogenization technique, which is based on scale separation and perturbation analysis. The source scale dispersal model consists of a partial differential equation of animal movement and caching of the seeds. For the homogenization, the spatial and temporal scales of the fast, short-distance movement of the dispersers between and in the cache sites (l and L in Fig. 6) are separated from the overall seed transport over larger distances (x in Fig. 6). The model is formulated depending on the resulting small and large temporal and spatial scales, where a parameter ($0<\varepsilon<<1$) describes the link between the small and large scales. The basic idea then is to express the state variable of the model by a power series of this ε, to substitute it in the partial differential equation, to sort the terms by powers of ε and to evaluate the terms with identical ε. Terms with powers larger than ε^2 are ignored. In the results, the parts depending on the small scales disappear or reach quickly an equilibrium. This leads to a dispersal kernel that depends on the large scales only and on the means of the movement and caching parameters.

The mathematical homogenization corroborates the intuitive reasoning that on the large scale, the fine scale spatial structure and its effect on the behavior of the dispersers can be aggregated ("smeared out"). In the special case of Figure 6, the shares of the seeds reaching the ground are calculated by taking into account the effects of the differing rates of seed movement and seed caching (λ, Fig. 6) over the dimension L (average distance between seed cache sites) and the dimension l (average size of a cache site). Thus, the spatially averaged dispersal kernel and the mean seed dispersal distance (an important number for modeling migration rates of plants) has to be corrected by the "caching scale ratio" $\gamma = l / L$, for realistic determination of effective dispersal distances.

By this, the homogenization approach is an alternative to estimating dispersal kernels from order statistics, where fat-tailed kernels are calibrated in order to come up with realistic simulations of plant migration rates.

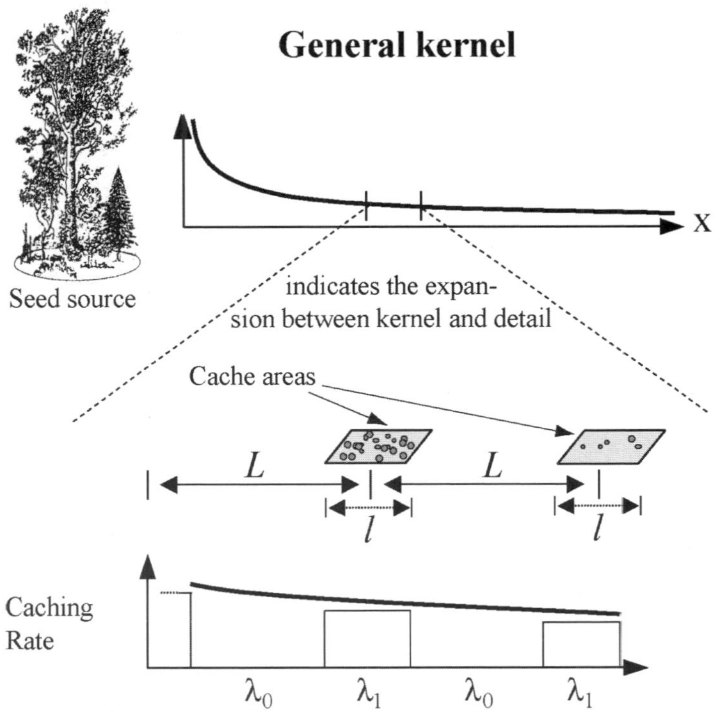

Fig. 6. Idealized spatial structure of active seed dispersal with L being the average distance between seed cache sites and l being the size of cache sites. The rate of caching differs significantly between cache sites (λ_1) and non-cache sites (λ_0). The slowly decreasing caching rate along any distance x results in decreasing seed availability for regeneration away from a seed source. Simplified from Powell and Zimmermann (2004).

Discussion

The various up-scaling methods shown in this chapter differ in how the aggregated functions are derived or approximated and how the variability within a holon is formulated. The appropriate up-scaling method for a given source scale model is determined by its nonlinearity, by the interactions, and by the variability/heterogeneity within the holons. The perfect up-scaling, where the aggregated function is derived directly from the aggregated variable is seldom possible. Usually, up-scaling involves different kinds of approximations: (1) approximating the model functions by empirical functions or series expansions, (2) describing the heterogeneity within the holon (e.g. a grid cell) by statistical parameters, (3) assessing the large scale heterogeneity within a holon, and (4) approximating much faster and slower processes by constant values. Consequently, the up-scaling method should keep the approximation error sufficiently small, and the up-scaled model should be validated against data or simulations with the source scale model. Furthermore, the approximation error increases together with the target scale grain size. The overall computing time however decreases with increasing grain size. With the appropriate trade-off between computing time and acceptable approximation error, up-scaling can yield target models which are at the same time efficient and still reflect the essential behavior of the source model (e.g. demonstrated in the DisCForM example).

Beyond this technical gain, up-scaling simplifies ecological models in a controlled way. Such simplified models contain less of the uncertainties usually associated with complex ecological models and are easier to deal with. The mechanisms of an up-scaled model are based on the source scale processes. Thus, an up-scaled model yields more trustworthy forecasts than an empirical model, because "with understanding of mechanisms one has predictive capacity that is impossible with correlations alone" (Levin 1992).

Up-scaling from a lower-level can also corroborate an existing target scale model, or determine under which conditions it is valid. For example, the well established logistic growth model could by up-scaling be traced back to individual behavior (Poggiale 1998)

In general, successful up-scaling can help to better understand a given system, because it reduces complexity to the necessary, and identifies how target scale mechanisms emerge from source scale processes. In the example of forest dynamics for instance, the up-scaling revealed that distributions of tree densities in height and space are sufficient to describe forest dynamics, and knowing detailed positions and heights of individual trees is not necessary. In the seed dispersal example the fine scale pattern of the cache sites turned out to be irrelevant for the overall dispersal, yet the relative amount of cache area had a strong influence on it.

Conclusions

Up-scaling is a central part of landscape research. It is involved explicitly or implicitly in activities where properties or processes are to be predicted at landscape scales and is used for both technical reasons (e.g. efficient simulations) and scientific understanding (e.g. emergent properties).

A successful up-scaling is based on four prerequisites: (1) a sound ecological understanding, (2) an abstract framework, such as hierarchy theory, which helps to identify, understand, order and abstract levels, entities, mechanisms, and relationships among and between levels, (3) the description of e.g. spatio-temporal patterns (a central topic in landscape ecology in general) and (4) mathematical methods from statistics, numerics and analysis to formulate the model transformations.

Landscape ecology could benefit from the large experience in up-scaling physical systems. This is especially true for the problem of large-scale within holon heterogeneity, which is currently one of the biggest challenges in landscape ecological up-scaling.

References

Acevedo M.F., Ablan M., Urban D.L. and Pamarti S. 2001. Estimating parameters of forest patch transition models from gap models. Environmental Modelling and Software 16, 7: 649–658.

Acevedo M.F., Urban D.L. and Ablan M. 1995. Transition and Gap Models of Forest Dynamics. Ecological Applications 5, 4: 1040–1055.

Acevedo M.F., Urban D.L. and Shugart H.H. 1996. Models of forest dynamics based on roles of tree species. Ecological Modelling 87, 1–3: 267–284.

Allen T.F.H. and Hoekstra T.W. 1992. The principles of ecological integration. In T. F. H. Allen and T.W. Hoekstra (eds.). Toward a unified ecology, pp. 13–53. Columbia University Press, New York a.o.

Auger P. and de la Parra R.B. 2000. Methods of aggregation of variables in population dynamics. Comptes Rendus De L Academie Des Sciences Serie Iii-Sciences De La Vie-Life Sciences 323, 8: 665–674.

Auger P. and Lett C. 2003. Integrative biology: linking levels of organization. Comptes Rendus Biologies 326, 5: 517–522.

Beerling D.J. and Woodward F.I. 1996. Palaeo-ecophysiological perspectives on plant responses to global change. Trends Ecol. Evol. 11, 1: 20–23.

Bolker B. and Pacala S.W. 1997. Using moment equations to understand stochastically driven spatial pattern formation in ecological systems. Theoretical Population Biology 52, 3: 179–197.

Bolker B.M. and Pacala S.W. 1999. Spatial moment equations for plant competition: understanding spatial strategies and the advantages of short dispersal. American Naturalist 153, 6: 575–602.

Bolliger J., Kienast F. and Zimmermann N.E. 2000. Risks of global warming on montane and subalpine forests in Switzerland – a modeling study. Regional Environmental Change 1: 99–111.

Brown D.G. 1994. Predicting vegetation types at treeline using topography and biophysical disturbance variables. Journal of Vegetation Science 5: 641–656.

Bugmann H. and Cramer W. 1998. Improving the behaviour of forest gap models along drought gradients. Forest Ecology and Management 103, 2–3: 247–263.

Bugmann H., Lindner M., Lasch P., Flechsig M., Ebert B. and Cramer W. 2000. Scaling issues in forest succession modelling. Climatic Change 44, 3: 265–289.

Cain M.L., Damman H. and Muir A. 1998. Seed dispersal and the holocene migration of woodland herbs. Ecological Monographs 68, 3: 325–347.

Cain M.L., Milligan B.G. and Strand A.E. 2000. Long-distance seed dispersal in plant populations. American Journal of Botany 87, 9: 1217–1227.

Chesson P. 1998. Making sense of spatial models in ecology. In: Bascompte J. and Solé R. (eds.). Modelling Spatiotemporal Dynamics in Ecology, pp. 151–166. Springer-Verlag and Landes Bioscience.

Clark J. 1998. Why trees migrate so fast: Confronting theory with dispersal biology and the paleorecord. American Naturalist 152, : 204–224.

Clark J., Horvath L. and Lewis M. 2001a. On the estimation of spread rate for a biological population. Statistics & Probability Letters 51, 3: 225–234.

Clark J.S., Lewis M. and Horvath L. 2001b. Invasion by extremes: Population spread with variation in dispersal and reproduction. American Naturalist 157, 5: 537–554.

Clark J.S., Silman M., Kern R., Macklin E. and HilleRisLambers J. 1999. Seed dispersal near and far: Patterns across temperate and tropical forests. Ecology 80, 5: 1475–1494.

Dieckmann U. and Law R. 2000. Relaxation projections and the method of moments. In Dieckmann U., Lawand R., Metz J.A.J. (eds.). The geometry of ecological interactions: Simplifying spatial complexity, pp. 412–455. Cambridge University Press, Cambridge.

Dieckmann U., Law R. and Metz J.A.J. 2000. The geometry of ecological interactions: Simplifying spatial complexity, pp. 564. Cambridge University Press, Cambridge.

Ellner S.P., Sasaki A., Haraguchi Y. and Matsuda H. 1998. Speed of invasion in lattice population models: pair-edge approximation. Journal of Mathematical Biology 36, 5: 469–484.

Eriksson O. 2000. Seed dispersal and colonization ability of plants – Assessment and implications for conservation. Folia Geobotanica 35, 2: 115–123.

Farqhuar G.D. and von Caemmerer S. 1982. Modelling of photosynthetic response to environmen-mental conditions. In: Lange O.L., Nobel P.S., Osmondand C.B., Ziegler H. (eds.). Physiological Plant Ecology. II. Water Relation and Carbon Assimilation. Encyclopedia of Plant Physiology, pp. 549–587. Springer, Berlin.

Higgins S.I. and Richardson D.M. 1999. Predicting plant migration rates in a changing world: The role of long-distance dispersal. American Naturalist 153, 5: 464–475.

Holdridge L.R. 1947. Determination of world plant formations from simple climatic data. Science 105: 367–369.

Iverson L.R. and Prasad A.M. 1998. Predicting abundance of 80 tree species following climate change in the eastern United States. Ecological Monographs 68: 465–485.

Iwasa Y., Andreasen V. and Levin S.A. 1987. Aggregation in model ecosystems. I. Perfect aggregation. Ecological Modelling 47: 287–302.

Jones H.G. 1992. Plants and microclimate. A quantitative approach to environmental plant physiology. Cambridge University Press, Cambridge, UK.

King A.W. 1991. Translating models across scales in the landscape. In: Turner M.G. and Gardner R.H. (eds.). Quantitative methods in landscape ecology: the analysis and interpretation of landscape heterogeneity, Vol. 82, pp. 470–517. Springer, New York.

Kirilenko A.P. and Solomon A.M. 1998. Modelling dynamic vegetation responses to rapid climate change using bioclimatic classification. Climatic Change 28: 16–49.

Kot M., Lewis M.A. and vandenDriessche P. 1996. Dispersal data and the spread of invading organisms. Ecology 77, 7: 2027–2042.

Lee C.K. 1999. Automatic adaptive mesh generation using metric advancing front approach. Engineering Computations 16, 2–3: 230–263.

Levin S.A. 1992. The problem of pattern and scale in ecology. Ecology 73, 6: 1943–1967.

Lewis M.A. 2000. Spread rate for a nonlinear stochastic invasion. Journal of Mathematical Biology 41, 5: 430–454.

Lischke H., Löffler T.J. and Fischlin A. 1997a. Calculating temperature dependence over long time periods: A comparison and study of methods. Agric. For. Meteorol. 86: 169–181.

Lischke H., Löffler T.J. and Fischlin A. 1997b. Calculating temperature dependence over long time periods: Derivation of methods. Ecol. Model. 98, 2–3: 105–122.

Lischke H., Löffler T.J. and Fischlin A. 1998. Aggregation of individual trees and patches in forest succession models – Capturing variability with height structured random dispersions. Theoretical Population Biology 54, 3: 213–226.

Löffler T.J. and Lischke H. 2001. Incorporation and influence of variability in an aggregated forest model. Natural Resource Modeling 14, 1: 103–137.

Luckyanov N.K. 1995. Model aggregation: Mathematical perspectives. In: Patten B.C. and Joergensen S.E. (eds.). Complex ecology: the part -whole relation in ecosystems, pp. 242–260. Prentice Hall, Englewood Cliffs.

Melbourne B.A. and Chesson P. 2005. Scaling up population dynamics: integrating theory and data. Oecologia 145, 2: 179–187.

Melillo J.M., Prentice I.C., Farquhar G.D., Schulze E.D. and Sala O.E. 1996. Terrestrial biotic responses to environmental change and feedbacks to climate. In: Houghton J.T., Meira Filho L.G., Callander B.A., Harris N., Kattenbergand A. and Maskell K. (eds.). Climate Change 1995: The Science of Climate Change, pp. 445–481. Cambridge University Press, Cambridge, UK.

Monserud R.A. and Marshall J.D. 1999. Allometric crown relations in three northern Idaho conifer species. Canadian Journal of Forest Research 29: 521–535.

Murrell D.J., Dieckmann U. and Law R. 2004. On moment closures for population dynamics in continuous space. Journal of Theoretical Biology 229, 3: 421–432.

Myneni R.B., Nemani R.R. and Running S.W. 1997. Estimation of global leaf area index and absorbed par using radiative transfer models. IEEE Transactions on Geoscience and Remote Sensing 35, 6: 1380–1393.

Nathan R. 2005. Long-distance dispersal research: building a network of yellow brick roads. Diversity and Distributions 11, 2: 125–130.

Nathan R., Safriel U.N. and Noy Meir I. 2001. Field validation and sensitivity analysis of a mechanistic model for tree seed dispersal by wind. Ecology 82, 2: 374–388.

Nathan R., Safriel U.N., Noy Meir I. and Schiller G. 2000. Spatiotemporal variation in seed dispersal and recruitment near and far from Pinus halepensis trees. Ecology 81 (8): 2156–2169.

Niinemets U., Kull O. and Tenhunen J.D. 1998. An analysis of light effects on foliar morphology, physiology, and light interception in temperate deciduous woody species of contrasting shade tolerance. Tree Physiology 18, 10: 681–696.

Nikora V.I., Pearson C.P. and Shankar U. 1999. Scaling properties in landscape patterns: New Zealand experience. Landscape Ecology 14: 17–33.

O'Neill R.V., DeAngelis D.L., Waide J.B. and Allen T.F.H. 1986. A hierarchical concept of ecosystems. Princeton University Press, Princeton, N.J.

Pacala S.W., Canham C.D. and Silander J.A.J. 1993. Forest models defined by field measurements: I. The design of a northeastern forest simulator. Canadian Journal of Forest Research 23, 10: 1980–1988.

Pattee H.H. 1973. Hierarchy theory: the challenge or complex systems. G. Braziller Inc., New York, NY, U.S.A.

Peterson D.L. and Parker V.T. 1998. Ecological Scale. Columbia University Press, New York, NY, U.S.A.

Pierce L.L. and Running S.W. 1995. The Effects of Aggregating Subgrid Land-Surface Variation on Large-Scale Estimates of Net Primary Production. Landscape Ecology 10, 4: 239–253.

Pitman E.B., Nichita C.C., Patra A., Bauer A., Sheridan M. and Bursik M. 2003. Computing granular avalanches and landslides. Physics of Fluids 15, 12: 3638–3646.

Poggiale J.C. 1998. From behavioural to population level: Growth and competition. Mathematical and Computer Modelling 27, 4: 41–49.

Portnoy S. and Willson M.F. 1993. Seed dispersal curves: behavior of the tail of the distribution. Evolutionary Ecology 7: 25–44.

Powell J.A. and Zimmermann N.E. 2004. Multi-scale analysis of seed dispersal and the resolution of Reid's Paradox. Ecology 85, 2: 490–506.

Prentice I.C., Cramer W., Harrison S.P., Leemans R., Monserud R.A. and Solomon A.M. 1992. A global biome model based on plant physiology and dominance, soil properties and climate. J. Biogeogr. 19: 117–134.

Rastetter E.B., King A.W., Cosby B.J., Hornberger G.M., Oneill R.V. and Hobbie J.E. 1992. Aggregating Fine-Scale Ecological Knowledge to Model Coarser-Scale Attributes of Ecosystems. Ecological Applications 2, 1: 55–70.

Roberts J., Hopkins R. and Morecroft M. 1999. Towards a predictive description of forest canopies from litter properties. Functional Ecology 13: 265–272.

Salthe S.N. 1985. Evolving Hierarchiscal Systems: their structure and representation. Columbia University Press, New York, NY, U.S.A.

Schneider D.C. 2001. The rise of the concept of scale in ecology. Bioscience 51, 7: 545–553.

Schulze E.D., Kelliher F.M., Körner C., Lloyd J. and Leuning R. 1994. Relationships among Maximum Stomatal Conductance, Ecosystem Surface Conductance, Carbon Assimilation Rate, and Plant Nitrogen Nutrition – a Global Ecology Scaling Exercise. Annual Review of Ecology and Systematics 25: 629–690.

Solomon A.M. and Kirilenko A.P. 1997. Climate change and terrestrial biomass: what if trees do not migrate! Global Ecology and Biogeography Letters 6, 2: 139–148.

Thornton P.E. 1998. Regional ecosystem simulation: combining surface- and satellite-based observations to study linkages between terrestrial energy and mass budgets. Ph.D. Dissertation, No., School of Forestry, University of Montana, Missoula, MT.

Thornton P.E., Law B.E., Gholz H.L., Clark K.L., Falge E., Ellsworth D.S., Goldstein A.H., Monson R.K., Hollinger D., Falk M., Chen J. and Sparks J.P. 2002. Modeling and measuring the

effects of disturbance history and climate on carbon and water budgets in evergreen needleleaf forests. Agricultural and Forest Meteorology 113: 185–222.
Thornton P.E. and Zimmermann N.E. submitted. An improved canopy integration scheme for a land surface model with prognostic canopy structure. Journal of Climate.
Urban D.L. 2005. Modeling ecological processes across scales. Ecology 86, 8: 1996–2006.
Urban D.L., Acevedo M.F. and Garman S.L. 1999. Scaling fine-scale processes to large-scale patterns using models derived from models: meta-models. In: Mladenoff D.J. and Baker W.L. (eds.). Spatial modeling of forest landscape change: approaches and applications, pp. 70–98. Cambridge University Press, Cambridge.
Wiens J.A. 1989. Spatial Scaling in Ecology. Functional Ecology 3, 4: 385–397.
Wu J. 1999. Hierarchy and scaling: extrapolating information along a scaling ladder. Canadian Journal of Remote Sensing 25, 4: 367–380.
Wu J. 2004. Effects of changing scale on landscape pattern analysis: scaling relations. Landscape Ecology 19, 2: 125–138.
Wu J. and David J.L. 2002. A spatially explicit hierarchical approach to modeling complex ecological systems: theory and applications. Ecological Modelling 153, 1–2: 7–26.
Wu J. and Qi Y. 2000. Dealing with scale in landscape analysis: an overview. Geograpical Information Science 6, 1: 1–5.
Zimmermann N.E. and Kienast F. 1999. Predictive mapping of alpine grasslands in Switzerland: species versus community approach. Journal of Vegetation Science 10, 4: 469–482.

Dynamic Spatio-temporal Landscape Models

Heike Lischke[1*], Janine Bolliger[1] and Ralf Seppelt[2]

[1] Swiss Federal Research Institute WSL, Zürcherstrasse 111, CH-8903 Birmensdorf, Switzerland
[2] Dept. for Applied Landscape Ecology, Environmental Research Centre (UFZ) Leipzig Halle, Germany
[*] Corresponding author: e-mail: heike.lischke@wsl.ch

Abstract
Modeling at the landscape scale is most relevant to quantify and understand landscape structure and dynamics. The existing landscape models encompass static empirical models, dynamic point and area models, dynamic regionalized and spatially linked spatio-temporal (SLST) models. The latter take into account both local dynamics and spatial interactions.

This chapter discusses various SLST model concepts and approaches that are applied in landscape research. SLST models are typically used to (1) advance general ecological theory, (2) test specific landscape-ecological hypotheses, (3) run scenario-simulations, and (4) derive decision support for landscape management. We present three case studies that illustrate the use and limitations of generic and complex SLST models. In the first case study we employ SLST models to explain the formation of forest-landscape patterns. The second and third case study highlight the use of SLST models to analyze the spread of tree-species during the Holocene and to identify landscape functions for management purposes. Finally, major challenges in the field of spatio-temporal landscape modeling, scaling issues and model testing are discussed.

Keywords: landscape models, spatio-temporal modeling, spatially linked spatio-temporal models; SLST models

F. Kienast, O. Wildi & S. Ghosh (eds.), A Changing World. Challenges for Landscape Research, 273–296.

Modeling Landscape Patterns and Dynamics

Landscape patterns (landscape heterogeneity) are the result of various drivers acting on the landscape. Drivers include exogenous factors, endogenous processes, or both (Bolliger 2005; Bolliger *et al.* 2003; Patten *et al.* 1997). In most natural systems both endogenous processes and exogenous factors influence landscape dynamics. Exogenous landscape drivers are climate, soil, or disturbances (e.g., fire, floods). Endogenous processes involve interactions between landscape elements and are usually of biotic nature (e.g., trophic interactions, competition between or within species). These interactions generate small variations between time steps and locations that finally accumulate and lead to pattern formation (Bak 1996; Bak *et al.* 1987). In environmentally extreme habitats (e.g., desert, arctic, extreme pH values in soils) landscape dynamics are likely to be primarily driven by exogenous factors. Habitats with relatively constant environmental heterogeneity between and within years (e.g., rainforests), are primarily driven by endogenous interactions (Solé *et al.* 2002; Solé and Manrubia 1995).

Interactions between the various drivers of landscape patterns and dynamics can be linear, non-linear, unidirectional or form positive (mutuality, self-reinforcing) or negative feedbacks (e.g., self-inhibition). Nonlinear and feedback interactions have been reported to be a primary source of structure or patterning in many kinds of natural systems (Bascompte *et al.* 2003; Bascompte and Sole 1995; Farkas *et al.* 2002; Green 2000). These complex interactions between endogenous and exogenous drivers of landscape patterns make intuitive understanding or direct assessments of likely cause-and-effect relationships difficult. One way to quantify and predict spatially dynamic patterns and their underlying processes at the landscape scale is to use mathematical or computer models (cf. Richter *et al.* 2002). These are implementations of conceptual models on the basis of empirical observations and experiments using the generic and uniform language of mathematics (see Seppelt 2003). Models integrate current knowledge about interactions and influences of drivers, rank it, point to inconsistencies and uncertainties, and make simplifications explicit.

Landscape models formulate interactions within and between landscape elements and/or with environmental factors in space and time. By doing so, landscape models are able to relate spatial and temporal pattern to exogenous and endogenous drivers. Depending on the research question, the extents to which landscape models are applied range from one square meter to entire continents. Beyond interpretation of data, landscape models allow scenario testing by assessing various degrees of changes on a particular landscape. This may lead to confirmation, rejection, or generation of hypotheses and support environmental decisions and policy making.

In this chapter, we discuss mathematical models and computer programs that generate quantitative descriptions of landscapes in space, time or both. The chapter is an overview of a variety of landscape model approaches that are based on different model concepts, temporal and spatial resolutions, and levels of complexity. Special emphasis is given to the spatially linked spatio-temporal landscape (SLST) models.

Types of Landscape Models

Although there is no generally applicable model-classification scheme, a number of model types can be distinguished based on various aspects of the modeling approach, ranging from purely conceptual, descriptive word and graphic models to semi-quantitative graphical schemes, mathematically formalized models to computer programs yielding quantitative descriptions. These model types differ particularly in the way landscape heterogeneity is taken into account.

Often, **static modeling approaches** are the basis of large-scale spatial predictions. These approaches assume that the landscape is in equilibrium with the environment, as they do not account for transient adaptation phases. A static model $Y(s_j) = f(X(s_j))$ links observed state variables Y, (biotic units, e.g., trees), to exogenous factors X (e.g., climate) at positions s_j in the landscape. The simulated landscape heterogeneity is thus a simple mapping of the heterogeneity of the exogenous factors. The link between state variables and exogenous factors is often performed using various regression approaches (reviews in Bakkenes *et al.* 2002; Guisan and Zimmermann 2000) ranging from logistic regression (Bolliger *et al.* 2000) to CART models (De 'Ath and Fabricius 2000), or General Additive Models (Yee and Mitchell 1991). Applied at discrete timestep(s) t_i, the model yields $Y(t_i,s_j) = f(X(t_i,s_j))$.

Simulations from static models thus represent the spatially detailed distributions of the biotic unit at individual timestep(s). Applications of this type of model include risk assessments of global climatic change on vegetation distribution (Bolliger 2002; Bolliger *et al.* 2000; Guisan and Theurillat 2000; Guisan *et al.* 1998), habitat suitability models for individual species (Akçakaya *et al.* 1995; Guisan and Hofer 2003; Lindenmayer *et al.* 1991), for species groups (Bonn and Schröder 2001), for communities (Peppler-Lisbach and Schröder 2004), cascades of landscape filters (Poff 1997; Schröder and Reineking 2004), or biogeographic models (Haxeltine and Prentice 1996; Holdridge 1947; Leemans and van-den-Born 1994; Neilson 1995; Prentice *et al.* 1992; Woodward and Smith 1994) for species, communities, or biomes. In these models, the variable describing vegetation, is fitted to variables expressing ecophysiological constraints, e.g., yearly day degree sum, maximum net ecosystem production (NEP) or leaf area index which can be attained under the given moisture and nutrient conditions. These vegetation models are sometimes combined with dynamic nutrient cycling models.

Advantages of static models include that they allow quick and easy calculations (regressions). Since the predictions are given in a geographically explicit form, they are interpretable as maps, e.g., in Geographical Information Systems (GIS). Spatial interactions may be accounted for by applying methods that consider spatial autocorrelation (e.g., chapter 4.2 Spatial dynamics, Augustin *et al.* 1996). Static statistical models allow assessment of factors and factor combinations that are relevant for a given landscape pattern, and are therefore in many situations a good starting point for further modeling approaches. Static models imply, however, that the ecosystems are in quasi-equilibrium, i.e., the transient behavior, the way in which the equilibrium is attained, is not accounted for. Additionally, the basic mechanisms of the spatio-temporal patterns are not explicitly included in statistical models. This limits cause-and-effect analysis and restricts extrapolations to the range of the factors (in space and time) where the model was calibrated (Lischke 2001; Lischke *et al.* 1998a; Peng 2000), as in empirical models in general.

Dynamic models emerge from the concept that the landscape state is defined by change, driven by exogenous factors and endogenous processes and interactions. Thus, dynamic models take the transient nature of systems into account. They are based on assumptions about the underlying processes and describe for specific localities the temporal change ($\frac{d}{dt}Y(t)$ in continuous time or $Y(t_{i+1}) - Y(t_i)$ in discrete time) of the state variables Y (e.g., the biomasses oftree species): ($\frac{d}{dt}Y(t) = f(Y(t),X(t),\varepsilon)$. These models can be deterministic ($\varepsilon = 0$) or stochastic, i.e., take into account random influences ($\varepsilon \neq 0$).

A broad variety of dynamic landscape models are available. Among these, *dynamic point models* describe the dynamics of the state variables for specific locations, thus points in space through time. Dynamic point models include e.g., forest gap models (Botkin *et al.* 1972; Bugmann 2001), which simulate the establishment, growth and death of single trees on small patches and take into account the shifting mosaic of these patches created by stochastic death and birth processes of small subpopulations (e.g., Botkin *et al.* 1970). Other approaches involve models that account for the spatial variability by using theoretical descriptions (distributions or moment equations) (DisCForM; Lischke *et al.* 1998b; Picard and Franc 2001; Bolker and Pacala 1997). The obvious disadvantage of dynamic point models is that they do not consider space explicitly. This, however, has the advantage that their simulations are usually fast, require only small computer storage capacity and are able to simulate state variables in a highly detailed form.

Dynamic area models simulate larger areas in a single simulation. However, spatial heterogeneity is not taken into account, resulting in average representations of the properties of the area under consideration. In other models the state variables are structured with regard to essential properties, e.g., areas of stand age and stand volume (see matrix model EFISCEN, Nabuurs *et al.* 2000).

Advantages of dynamic models in general include that the temporal course of the state variables is interpretable. A major disadvantage of dynamic point and area models is that the simulations do not allow spatial interpretations of the results. In addition, highly resolved temporal input data for larger spatial scales may not be readily available.

Dynamic regionalized (distributed) models incorporate landscape heterogeneity by applying dynamic point or area models in parallel at many locations s_j, e.g., on a grid: ($\frac{d}{dt} Y(t,s) = f(Y(t,s_j), X(t,s_j), \varepsilon)$. Thus, this model type combines spatial and temporal aspects of the landscape. However, the simulated locations are not spatially linked, i.e., do not communicate with each other. Applications of dynamic distributed models include e.g., the evaluation of global change phenomena such as predicting the behavior and properties of ecosystems at large scales under scenarios of possible future land use and climate. In the following, several examples of dynamic, distributed models are discussed.

An example of an *empirically derived forest model* is MASSIMO (Kaufmann 2001; Thürig *et al.* 2005). The model simulates the growth of individual trees at any of the sample plots of the Swiss National Forest Inventory on a 1.4*1.4 km grid, under different forest management scenarios. Since the sample plots are very small (1ha) and thus not representative for the surrounding areas, the simulation results are lumped into geographical and ecological strata. An example of a *big-leaf-model* (Lexer 1995) is the biogeochemical model BIOME-BGC (Thornton 1998; Thornton *et al.* 2002). The model enables the calculation of fluxes and pools of carbon, water and nitrogen. These pools are extended over large areas; e.g., all leaves in a region form one "big leaf". In contrast to this coarse spatial and organizational resolution, BIOME-BGC includes relatively detailed processes of photosynthesis, nutrient and carbon re-allocation and partitioning, litter decomposition and a variety of mortality functions that allow realistic simulations of ecosystem properties over decades. *DGVMs* (Dynamic global vegetation models) (see comparison in Cramer *et al.* 2001) simulate the development of the vegetation composition across the whole globe, based on ecophysiological processes and nutrient cycling. Vegetation dynamics are based on annual net primary production and biomass growth; they include competition among plant functional types, disturbances and succession. *Frame-based models* (Starfield and Chapin 1996) simulate transitions between vegetation types ("frames") in grid cells on the regional to continental scale. The transition probabilities are based on intrinsic state variables and on rules that depend on environmental conditions. In some cases the transition probabilities are fitted to the results of gap models (Acevedo *et al.* 2001). This approach assumes biomes to be fixed entities, thereby neglecting the individualistic responses of species to changing environmental conditions (Davis 1986; Kittel *et al.* 2000). Finally, *patch models at many locations* have been applied to assess succession details in a larger spatial context, either explicitly on a geographical transect (e.g., ForClim, Bugmann and Fischlin 1996) or in bioclimatic classes spanned by drought and day degree sum (Löffler and Lischke 2001). Advantages of dynamic large-scale models and of dynamic, distributed models include that the dynamic simulations can be interpreted spatially. The lack of spatial communication between the dynamic simulations on the discrete landscape elements is, however, a disadvantage.

Dynamic, Spatially Linked Spatio-temporal Models (SLST models)

Introducing spatial communication in regionalized models leads to the group of dynamic, *spatially linked spatio-temporal* (SLST) models, the focus of this chapter. Landscape elements influence each other dynamically, not only locally, but also across larger spatial scales, revealing that local dynamics and spatial interactions are crucial in assessing landscape dynamics. Spatial interactions in natural systems and the models describing them occur at a variety of scales. On smaller spatial scales, interactions occur by flow of resources, e.g., water (Jakeman and Letcher 2003), or by competition for habitat or for resources, e.g., by lateral shading as in some individual-based, position-dependent forest models (e.g., SORTIE Pacala *et al.* 1993; Picard *et al.* 2001; SILVA Pretzsch 2002). On intermediate spatial scales, interactions may be represented by active movement or passive dispersal (Clark and Ji 1995; Neilson *et al.* 2005), of animals, plants, (e.g., by the formation of tillers) or (e.g., pathogens). Movement or dispersal determines the rate and direction of spread. For example, different seed dispersal mechanisms (e.g., ballistic, wind, animals) identify (together with generation time and a variety of other factors) a plant species' migration rate and direction. Variation in migration rates is one process that generates landscape heterogeneity.

Dynamic spatio-temporal models that account for such spatial interactions between locations z and the simulated location s (SLST models) take the form:

$$\frac{d}{dt} Y(t,s) = f(Y(t,z_j), X(t,s), \varepsilon), i = ,..., n_s$$

Same as all dynamic models, SLST models include time in the temporal change of the state variables, either in discrete time steps (e.g., years, generations) or continuously. Spatial dependencies (and thus spatial interactions) in SLST models are coded by making f depend on a set of state variables $Y(t,z_i)$ from the neighborhood $i = ,..., n_s$. The approaches to deal with these spatial interactions vary considerably (Caswell and Etter 1993). Classical, spatially implicit metapopulation models (Gilpin and Hanski 1991; Hanski 1999; Levins 1970) or network models (Green 1995) deal with subpopulations on patches or network-nodes without explicit positions. Distances are implicitly contained in transfer rates between the patches. In some position-dependent models (e.g., Pacala *et al.* 1993; Picard *et al.* 2001; Pretzsch 2002; Prevosto *et al.* 2003), explicit coordinates of the simulated biota are recorded. Other models simulate the state variables on each location in the simulation domain, either continuously (partial differential equation, reaction-diffusion, integral equation, integro-difference models, see Renshaw 1991), or in the cells or on the nodes of grids (coupled map-lattices Bjørnstad *et al.* 1999; Kaneko 1992). A specific type of lattice models are cellular automata (see case study 1 and Bolliger 2005; Bolliger *et al.* 2003; Syphard *et al.* 2005; With and King 1999; With *et al.* 2002). In cellular automata, each cell can take several discrete states (e.g., one single individual of different species) defined by rules that depend on the states of the surrounding cells.

Comparable to other ecological models, SLST models differ in their degree of complexity (generic/complex), in the organizational level (cells, individuals, biomes), in the relative importance of endogenous processes and exogenous drivers, and in the general model approach (deterministic vs. stochastic) (Bolliger *et al.* 2005).

Applications of SLST Models

SLST models are applied in various fields of landscape research where the processes and interactions that generate a landscape pattern are of interest. They are used to answer different types of research questions that can be subsumed under the following headings:

- Models to develop theories: What are the general mechanisms behind an observed landscape phenomenon?
- Generating and testing hypotheses: Why are landscapes as we observe them?
- Scenarios: What might happen, if...?
- Projections: What might be in the future?
- Optimization and decision support for management: What is the best way to achieve a goal?

These questions are used in the following to give an overview of typical SLST model applications. Note that, although usually models are initially developed for certain applications, in many cases they can be attributed to several of these questions.

Models to develop theories: what are the general mechanisms behind an observed landscape phenomenon?

SLST models are often applied to enhance the understanding of general mechanisms or laws that drive landscape processes and patterns. Certain observable phenomena in landscapes may be related to specific conditions and processes of one particular landscape, but may also be viewed as the result of general, even fundamental and universal mechanisms applicable to various independent systems. The search for universality or generality is fundamental to the development of ecological theory and is one of the most important aims of modeling studies (Green and Sadedin 2005; Jorgensen 1992). The models involved often belong to the group of SLST models and are usually parsimonious, i.e., simulate landscapes with very few generic and abstract variables and relationships.

The search for general phenomena in systems, including landscapes, often involves complex systems theory (Milne 1998; Strogatz 2001; Wu and Marceau 2002). One aspect of complex systems theory is self-organization. Self-organization originates from dynamic interactions between system constituents which spontaneously lead to order and organization in multi-component (complex) systems (Bak 1996; Perry 1995). A system self-organizes to a critical state (SOC) if its dynamics lead to a state characterized by scale invariance (Bak *et al.* 1987; Gisiger 2001; Solé and Manrubia 1995). Mathematically, scale invariance is expressed by power laws (straight lines on a log-log scale) (see also Bolliger *et al.* 2005), indicating that no particular scale (spatial, temporal) is singled out. This means in a spatial context that large-scale patterns may be predicted from small-scale patterns and vice versa. Scale invariance in a temporal context indicates that no particular time scale is singled out. The phenomenon of scale invariance has been observed in many research disciplines, including landscape ecology (Bolliger 2005; Bolliger *et al.* 2003; Cousens *et al.* 2004; Lennon *et al.* 2001; Milne 1998; Storch *et al.* 2002). However, empirical evidence for processes leading to scale invariance is still largely missing (Levin 1998) and the observation of power laws does not automatically imply that they have been produced by SOC (Allen *et al.* 2001; Li 2000a; b) However, it has been stated that scaling relationships may offer clues and hypotheses to how the fundamental processes of biology give rise to emergent diversity (Brown *et al.* 2002b). There is a wide range of modeling approaches to investigate complex systems, e.g., network models (Green 2000), partial differential equations with diffusion (Deutschman *et al.* 1993), or coupled map lattices (Green and Sadedin 2005). An example of a parsimonious cellular automaton is presented in case study 1.

Other ecological theories where SLST models have played a crucial role include the concepts of patch-dynamics (Levin *et al.* 1993; Steele 1993). Population movements, e.g., the surprisingly fast migration of some plant species during the Holocene, have fascinated researchers since the 19th century (Reid 1899) and initiated a series of analytical spatially dynamic modeling approaches (reaction-diffusion, integro-difference, integral-equations). These are based on different hypotheses about the redistribution function for the propagules, e.g., Gaussian, negative-exponential, fat-tailed, (Clark 1998; Kot *et al.* 1996; Powell and Zimmermann 2004; Skellam 1951; van den Bosch *et al.* 1990). Another example for the contribution of spatially dynamic models to ecological theory are modeling studies of the survival and coexistence of species for understanding the effect of space on biodiversity and its maintenance (e.g., Chesson 2000; Gurney *et al.* 1998).

In contrast to the mostly parsimonious models used for general theories, SLST models tend to be more complex if applied to more specific questions of landscape research. Such specific questions are addressed in the next paragraphs.

Generating and testing hypotheses: why are landscapes as we observe them?

One motivation for the application of SLST models is to understand observed pattern or development of a specific landscape to assess the relative importance of various exogenous and endogenous drivers. This requires analysis of past landscape dynamics, because the landscape state as observed in the present is the result of the landscape development in the past.

One topic for modeling such past landscape developments is the effect of disturbance history on landscape development. Coffin and Lauenroth (1989) used a gap-model to assess the effect of disturbances on grassland dynamics, and Wimberly (2002) investigated the influence of wildfire history on spatial forest composition using a combined forest-fire model. Wagner *et al.* (2006) used a cellular automaton for continuous tree growth combined with a lattice model for lichen (sub)population development and spread to explain abundance and genetic diversity of lichens in forest stands with different disturbance history. Moravie and Robert (2003) used a position dependent, individual based forest model to assess whether forest structure can be used to assess past forest dynamics, including disturbances.

The history of natural migrations is another area for the application of SLST models for the study of past landscape dynamics. Case study 2 illustrates the modeling of plant species migration during past climatic changes. Also the spread of herbivores and the resulting effect on plant distributions has been studied by SLST models (Lewis 1994; Maron and Harrison 1997; Pastor *et al.* 1999). A range of different models has been developed to study invasion of exotic plant species (see review in Higgins and Richardson 1996), including individual-based or lattice models (Higgins *et al.* 2000), cellular automata (Cannas *et al.* 2003), or reaction-diffusion models (Frappier *et al.* 2003).

The influence of heterogeneity on landscape or community patterns is for example studied with the individual-based, position-dependent forest model SORTIE that acts at small spatial scales (Pacala *et al.* 1996). Deutschman *et al.* (1999) found that the very fine scale light variability did not have any significant influence on the community structure. Land-cover change affected by biotic, abiotic and anthropogenic factors has been studied with semi-empirical frame-based type models (Brown *et al.* 2002a; Irwin and Geoghegan 2001).

Scenarios: what might happen if … ?

Once hypotheses about the mechanisms behind an observed landscape are generated and tested, models can be applied in an "if-then"-modus, evaluating either the model outputs' sensitivity to different parameter or input variables, different forms of process functions or even different variants of model structure (sensitivity analysis). Alternatively scenario outcomes can be estimated with models whose control parameters' or input variables are set according to a previously defined scenario (scenario studies).

A prominent application of SLST models in "what-if" studies is the unintended spatial spread of genetic information, either of genetically manipulated organisms or of resistant pests and weeds. Such complex systems require an integrated view of population dynamics, genetics, and physical transport processes. Different model approaches have been used to analyze this complexity, focusing on pollen dispersal (Baker and Preston 2003; Klein *et al.* 2003; Ma *et al.* 2004; Meagher *et al.* 2003; Tufto *et al.* 1997), population dynamics, and/or seed dispersal and genetics (Cresswell *et al.* 1995; Richter and Seppelt 2004; VanRaamsdonk and Schouten 1997). Pollen dispersal is taken into account either empirically (Baker and Preston 2003; Ma *et al.* 2004; Meagher *et al.* 2003) or by mechanistically describing pollen transport by wind (Loos *et al.* 2003; Tufto *et al.* 1997) or insect vectors (Cresswell *et al.* 1995). Also the effect of different types of landscape fragmentation on the persistence and migration of species has been studied with SLST models, e.g., with a single species spatial logistic model (Collingham and Huntley 2000), with a lattice model of plant functional types (Cousins *et al.* 2003), with a spatial gap model (Malanson and Cairns 1997), and with a model for the flight of individual butterflies in a structured landscape (Kindlmann *et al.* 2005).

Projections: what might be in the future?

Probably the most tempting application of modeling is to develop scenarios for the outcome of changes (projections). The basic assumption behind projections is that all functional relationships that depend on time t whether directly or indirectly, e.g., by definition of rates and initial conditions, remain true. Furthermore, the intrinsic uncertainty, that might be acceptable at present, is assumed to be tolerable in the future.

One prominent field of SLST model projections is the ongoing and anticipated climate change. In this context it is crucial, whether species or ecosystems are resilient, adapt, or respond with diebacks to the changed environmental conditions, or can follow the climate change induced latitudinal and altitudinal shifts in the local site conditions (see e.g., Kirschbaum and Fischlin 1996). To study vegetation change and migration on the global scale, first steps are made towards including migration in dynamic global vegetation models (Neilson *et al.* 2005). On the continental to regional scale, various vegetation models have been extended to accommodate seed dispersal processes and have been applied in climate change studies: e.g., the frame-based model ALFRESCO (Rupp *et al.* 2000), the landscape model LANDIS (He *et al.* 1999), cellular automata (Iverson *et al.* 2004), lattice models (Dullinger *et al.* 2004), and forest and grassland-shrubland patch-models (Lexer *et al.* 2000; Peters 2002). Additionally, there is a wealth of models which study the combined effect of spatial-temporal fire dynamics, vegetation dynamics and succession, which are influenced by climate change (see classification in Keane *et al.* 2004).

Optimization and decision support for management: what is the best way to achieve a goal?

An extension of "what-if" studies is the use of SLST models as tools for decision support in environmental management (Seppelt 2003). The aim is to facilitate decisions about whether intervention in environmental systems is desirable or likely to be necessary, or which interventions might yield the best results. This kind of application requires – besides the assumption that an extrapolation in time is correct – that the models used are robust regarding variation of the driving variables. In the decision-support mode SLST models can by applied in two ways:

Scenario analysis: SLST models compare the outcome of a given set of scenarios. Each scenario is a representation of a possible management strategy. As environmental processes are complex and highly interacting, scenario definition is itself a difficult task, requiring consensus within the group defining the scenario, e.g., consensus on the input variables of the model. Scenarios are frequently set up by discussing management options and future developments with a group of scientists, stakeholders and people involved in the process of interest (cf. Millennium Ecosystem Assessment 2005).

Optimization: Optimization procedures systematically search over all combinations of values of the input variables until given management goals are satisfied. Such procedures support consensus finding by providing transparent evaluations of scenarios. The major drawback is the high computational complexity that depends on two factors: the complexity of the process model (number of state variables, degree of nonlinearity etc.) and the spatial complexity (size of study area, grid cell size, number of spatially interacting processes). The more complex the simulation model and the larger the number of spatial relationships, the lower are the chances of success in the optimization. For such complex models, scenario analysis is usually the only feasible (Fig. 1 in Seppelt and Voinov 2003). Case study 3 gives an example for this kind of analysis.

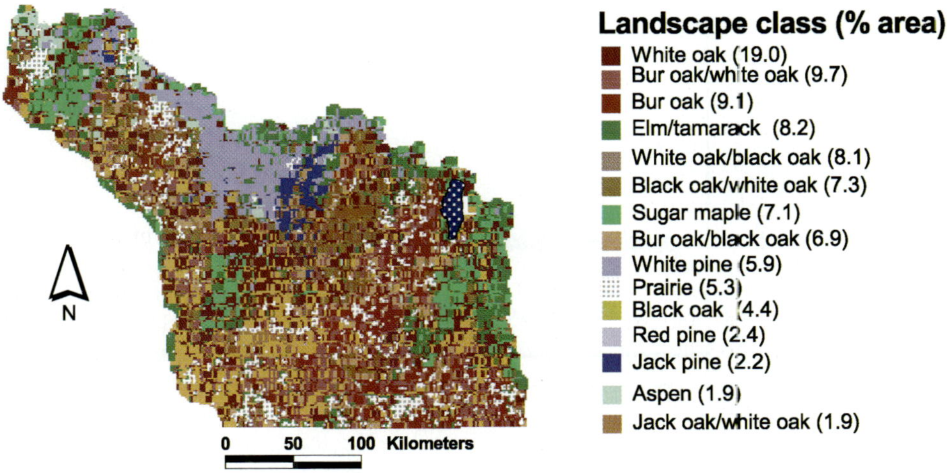

Fig. 1. Southern Wisconsin (U.S.A) represented by a quantitative (fuzzy) landscape classification. ϕ=1.1 indicates the degree of fuzziness with which the landscape was classified (for details see Bolliger 2005)

Case Studies

Case study 1: assessing forest-landscape patterns using a cellular automaton model

This case study applies a generic cellular automaton model to statistically reproduce the spatio-temporal patterns of an empirical landscape. The model is generic with respect to the model structure. The model development rules define interaction distances for landscape patterns without accounting for any ecological detail. The interaction distances are identified by a circular radius *r* within which a cell on the landscape is chosen according to various stochastic rules (details cf. Bolliger 2005). The so chosen cell is then replaced by a randomly selected cell within a circular neighborhood of radius *r* ($1< r < 10$), where *r* represents the interaction distances. Small interaction distances are defined by $r = 1$ encompassing 4 cells on the landscape. Intermediate interaction distances are represented by $r = 3$, accounting for 27 cells. Large interaction distances are identified by $r = 10$, encompassing 314 cells (details cf. Bolliger 2005; Bolliger *et al.* 2003; Sprott *et al.* 2002).

Input to the model is a historical landscape of southern Wisconsin (U.S.A). The landscape is represented by the US General Land Office Surveys, conducted during the 19th century, at a time prior to Euro-American settlement (Manies *et al.* 2001; Schulte and Mladenoff 2001; Schulte *et al.* 2002). Major landscape patterns (Fig. 1) are assessed using fuzzy classification (details cf. Bolliger 2005).

Comparisons between model simulations and the empirical landscape include temporal dynamics and spatial patterns. The temporal dynamics are assessed using cluster probability. Cluster probability is defined as the proportion of cells that are part of a cluster on the cell array. A cell is part of a cluster if its value is identical with the four nearest neighbors (e.g., bur oak). Cluster probability is thus a measure of aggregation of cells with identical properties and a measure to identify landscape patterns. The spatial development of landscape patterns is assessed using fractals (Bolliger 2005; Bolliger *et al.* 2003). Fractals provide measures to quantify spatial characteristics at a variety of scales based on algorithms that quantify the proportion of the geometrical space occupied by the fractal (Mandelbrot 1983; Milne 1988; 1991; Milne *et al.* 1992; Sprott *et al.* 2002).

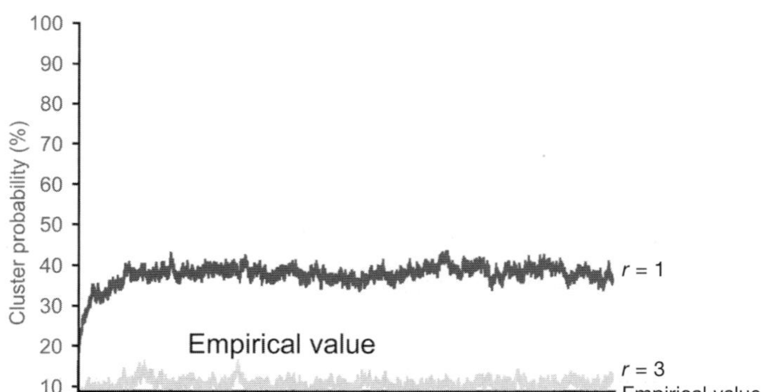

Fig. 2. Self-organizing landscapes: the simulated landscape self-organizes to the empirical value for intermediate neighborhoods for r = 3 where r is the interaction distance (neighborhood) and the only parameter of the model. Cluster probability is defined as the proportion of cells that are part of a cluster on the cell array. A cell is part of a cluster if its value is identical with the four nearest neighbors (e.g., bur oak).

For the temporal development, large neighborhoods (r = 10) result in simulated landscapes that do not self-organize (Fig. 2) as the likelihood that cells with identical properties interact is decreased. For small neighborhoods (r = 1) over-organization is observed because landscape properties are more likely to be similar within small neighborhoods (spatial autocorrelation). If intermediate neighborhoods (r = 3) are chosen, however, the model simulation self-organize to the respective empirical values. Interaction distances (connectivity) across the landscape are thus important to generate organized heterogeneity.

For the spatial development, the fractal dimensions D for both simulated and empirical landscapes are within comparable ranges for intermediate neighborhoods of r = 1 and 3 (Fig. 3). Similarly, for the temporal development of neighborhoods of r = 1 and 3 no particular

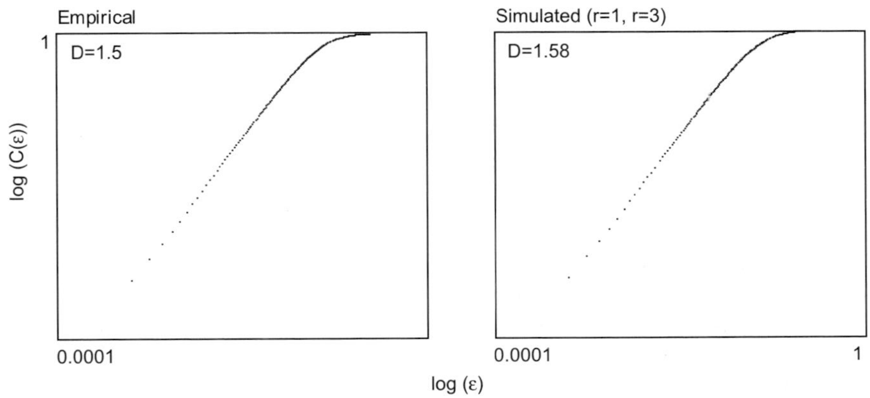

Fig. 3. Spatial characterization of model simulations: the fractal dimension D (slope of straight part of function) is similar for both the modeled (r = 1 and r = 3) and the empirical landscape. Scale invariance (power laws) are observed for r = 1 and r = 3. Power laws indicate that no particular space scale is singled out. ε: distance. $C(\varepsilon)$: probability that two identical entities ar distance ε apart.

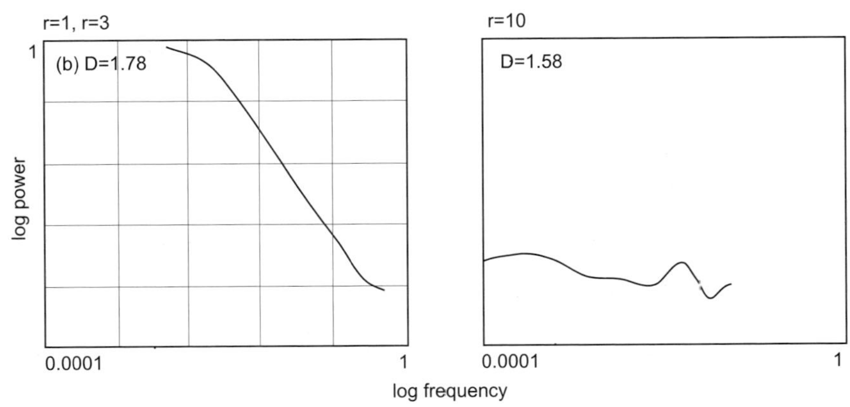

Fig. 4. Temporal characterization of model simulations: power laws are observed for simulations with r = 1 and r = 3. Temporal scale invariance is observed, i.e. no particular time-scale is characteristic for the simulations for r = 1 and r = 3. The neighborhood of r = 10 does not exhibit scale invariance since specific time scales are singled out and no straight line can be observed.

time-scale is characteristic for the simulations (Fig. 4). Thus, no particular space or time scales are singled out for small and intermediate neighborhoods, indicating that the patterns are self-similar and scale invariant with no characteristic scale length. Scale-free behavior reflects long-range correlations between space and time scales, indicating that processes may act similarly on landscapes across wide ranges of spatial and temporal scales.

Consistent with the empirical evidence, the simulated landscape shows self-organization and spatio-temporal scale invariance by relying exclusively on the internal variation of the patterns themselves since the model accounts for no ecological detail. It is therefore concluded that a generic model calibrated independently from specific ecological processes may suffice to statistically replicate a complex landscape. Other advantages of generic models include that simulations can be directly interpreted as a function of its few parameters, thus do not require detailed assessments of potential interactions with other model parameters. Limitations of the approach emerge from the fact that the simulated patterns are not spatially explicit since environmental constraints or gradients are not accounted for.

Case study 2: simulating tree species spread with a migration model

To assess the potential of tree species adaptation to a changing climate, the SLST model TreeMig has been developed. It is suitable to simulate the succession and migration of tree species in a structured environment and under changing environmental conditions.

TreeMig (Fig. 5, Lischke *et al.* in press) is based on the distribution based, height structured population model DisCForM (see chapter upscaling, Lischke *et al.* 1998b; Löffler and Lischke 2001), in which the spatial within-stand variability is taken into account by theoretical distributions of tree densities and light intensities. To obtain the spatial model TreeMig, DisCForM has been implemented on a 1 km x 1 km grid. In each grid cell, trees in different height classes – tracked as population densities per height class and light availability class – germinate, grow, and die. TreeMig simulates explicitly seed production, seed dispersal to other grid cells, seed bank dynamics including density regulation and the development of seedlings/saplings.

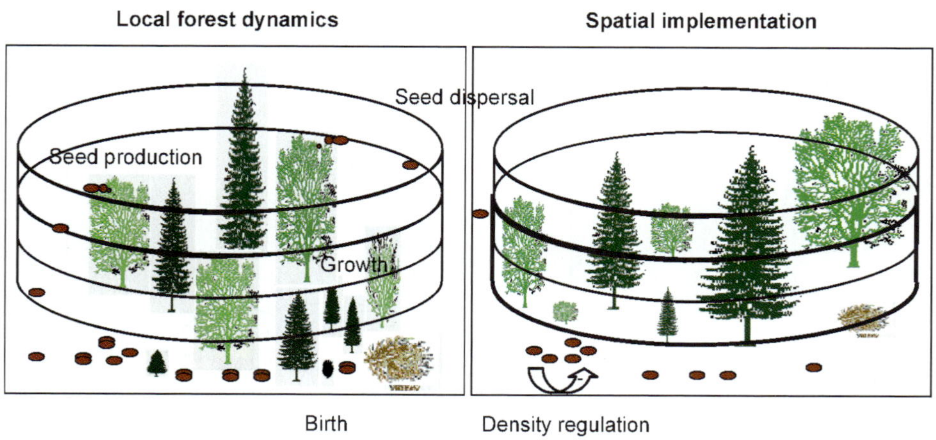

Fig. 5. Processes of the model TreeMig. The model is implemented on a grid. State variables are the densities of trees in different height classes. Trees germinate from the seed bank, grow, die and produce seeds. The seeds are dispersed to other grid cells, where they enter the seed bank. Seed bank dynamics follows a species-specific density regulation.

TreeMig was used to study the factors and processes driving the spatio-temporal vegetation development under past changing environmental conditions. The model was applied to the region of Valais (Switzerland) for the period since the last glacial during the Holocene (Lischke 2005). Figure 6 shows the spatial species composition at different time points

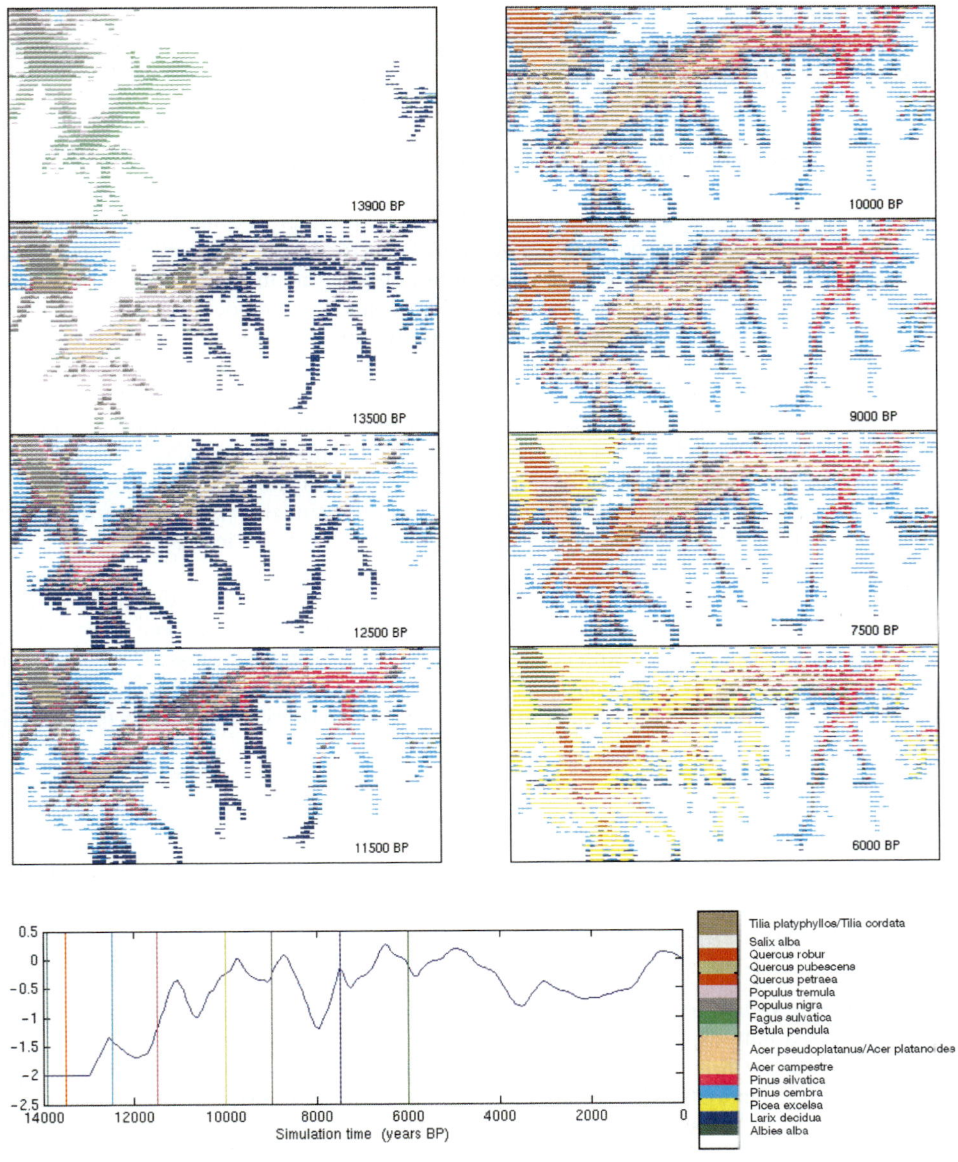

Fig. 6. TreeMig simulation of tree species spread on a 1 km x 1 km grid over 100 km x 50 km in the region of Valais, Switzerland. In each grid cell individual species biomass (t/ha) is drawn as a stacked column. A cell that is completely filled corresponds to 450 t/ha total biomass, resulting in white stripes in regions of low biomass. The graph at the bottom indicates the assumed temperature anomaly, the vertical lines refer to the times of the maps shown above.

during the simulation. The simulation is driven by time series of day degree sum, drought stress, and minimum winter temperature, based on a temperature anomaly scenario and current temperature and precipitation values. Years and sites of species immigration were derived from pollen records. The results show a vivid pattern of species spread, changes of dominance, and up and down shifts of the timberline. These phenomena are triggered by the variability of the external factors. Drastic changes of the boundary conditions, such as immigration of species into the simulation area or strong climate changes, initiate endogenous dynamics, i.e., migration and succession.

Case study 3: using optimization for identification of landscape functions

Spatially explicit ecosystem models allow calculation of water and matter dynamics in a landscape as functions of spatial location of habitat structures and matter input. Seppelt and Voinov (2002; 2003) studied in a mainly agricultural region the nitrogen balance as a function of different land use and land cover schemes. The landscape model uses a grid structure to calculate water- and matter-dynamics in a spatially explicit way, e.g., flow of matter is calculated from cells to neighboring cells for surface, subsurface and groundwater according to the flow network and conductivities, to soil properties and land use. The task of the case study was to calculate optimum land use and fertilizer application for three different nested investigation areas, i.e., Partuxent watershed (2365 km^2), Hunting Creek (77,8 km^2) and a sub-watershed of Hunting Creek (20,5 km^2), all in the Chesapeake Bay region, U.S.A (for details cf. Figure 12.1 in Seppelt 2003). Here, we focus on optimizing nitrogen loss out of the watershed; different aspects such as NPP or base flow are discussed in (Seppelt and Voinov 2003).

In a first step optimization tasks were formulated. This required the definition of performance criteria, which compare economic aspects, such as farmer's income from harvest "*A*", costs for fertilization "*B*", with ecological aspects, such as nutrient loss "*C*". As "*A*" and "*B*" can be quantified by monetary units and "*C*" is given for instance by mass per area, a weight λ (shadow price) is introduced in the performance criterion J, which is to be maximized:

$$J = A - B - \lambda C \tag{1}$$

These variables aggregate the processes of the entire study area. If J is defined for each grid cell (z_j) separately, the optimization task can be simplified by

$$J(z_j) = A(z_j) - B(z_j) - \lambda C(z_j),\ i = 1, ncells \tag{2}$$

The maximum values of the criterion J have been calculated based on numerical optimization in spatially explicit dynamic ecosystem simulation models. Tests have been performed by Monte-Carlo simulations and gradient-free optimization procedures. The core idea of the investigation is to study optimum habitat patterns as a function of the unknown weight λ in Eqn. (1,2). Increasing means that nutrient loss out of the watershed is increasingly "punished" (compared to economic income by agriculture). Consequently the number of agricultural habitats decreases when λ is increased. The interesting results of these findings are i) although Eqn (1,2) describes a linear performance criterion, the resulting patterns are not linear, and ii) these nonlinear pattern are similar for all study areas of differing sizes, see Figure 7.

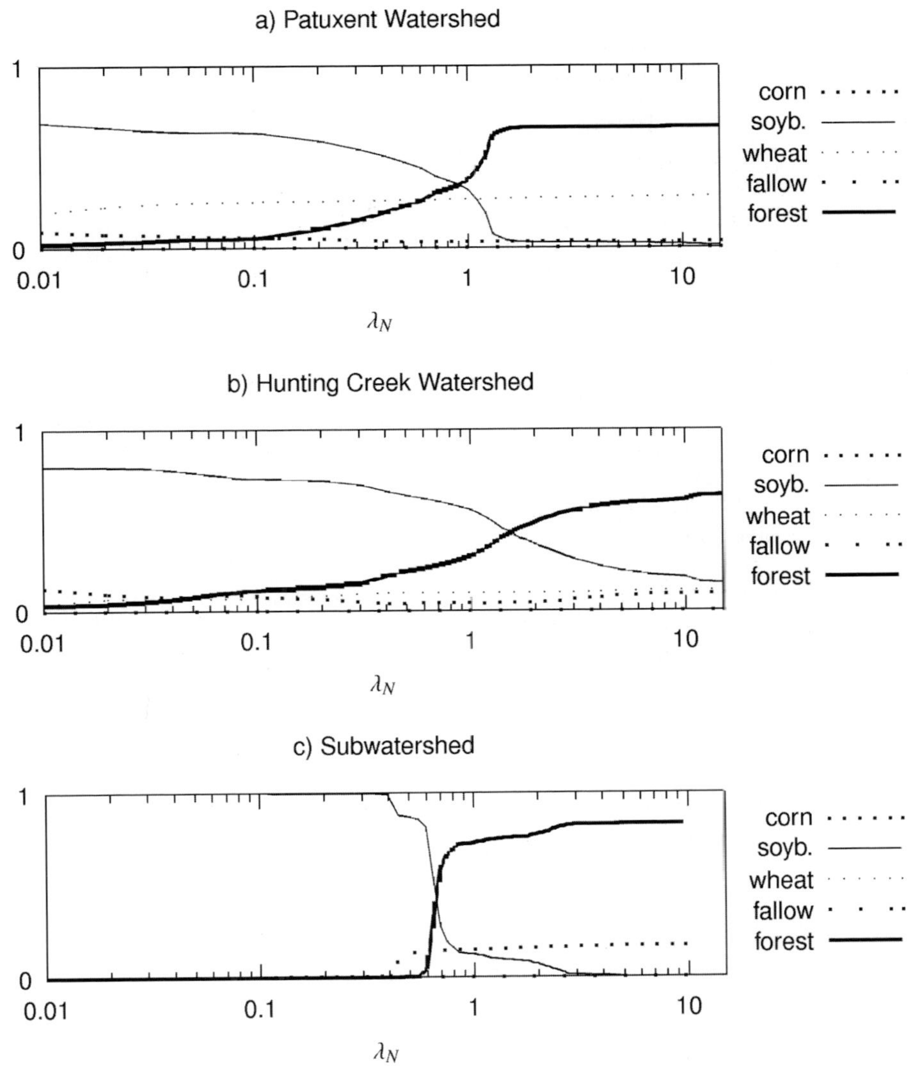

Fig. 7. Comparing optimum land use distribution patterns. In the present example yield is optimized against nutrient loss for three study areas, namely (a) Patuxent study area, (b) Hunting Creek watershed, and (c) the sub-watershed.

In addition to the aggregated results, one is interested in the spatially explicit results. Figure 8 displays optimum land use patterns for different λ-values. By doing so sensitive regions can be identified. For instance, the maps with higher values of λ show that forest habitat near the rivers and creeks supports retention capability of the ecosystem in terms of nutrients. As a result important areas with high retention capacity can be identified and fertilizer schemes can be evaluated depending on soil properties and topological relations to neighboring cells. The example shows that such optimization techniques can be a useful tool for a systematic analysis of management strategies.

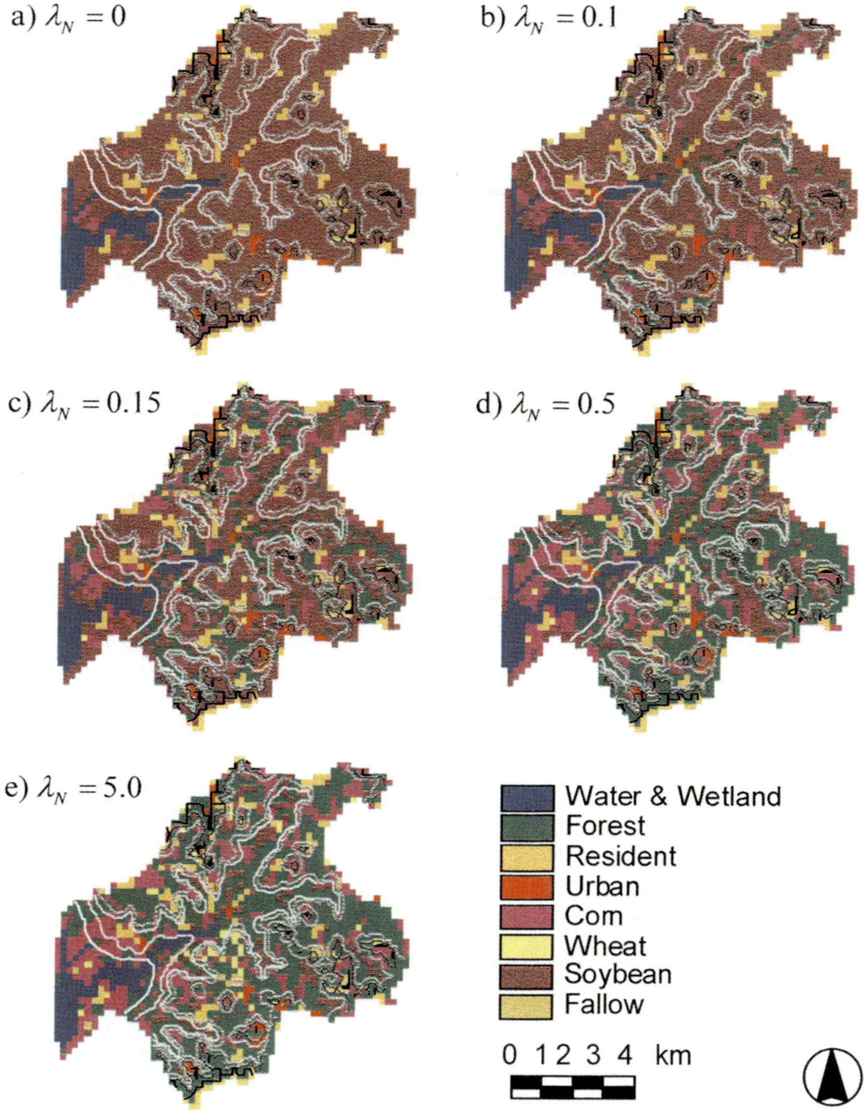

Fig. 8. Optimum land use maps derived from an SLST model. The models use different λ values (weights), namely $\lambda = 0$ (a), $\lambda = 0.1$ (b), $\lambda = 0.13$ (c), $\lambda = 0.5$ (d) and $\lambda = 5$ (e) (see text for more details). Lower areas near rivers and creeks can be identified by contour lines.

Challenges

The presented applications have illustrated that SLST models are helpful tools for basic and applied landscape research as well as environmental management. However, there are major challenges involved in the development and application of SLST models, which include data availability, choice of appropriate model complexity and scale and model evaluation.

Modeling, simulation and testing model performance requires data and information on the processes of interest. A major challenge in landscape modeling is data availability. There is, on the one hand, a fast development in earth observation and monitoring technologies (esp. remote sensing), which support large-scale spatio-temporal modeling. On the other hand, a well-known experience in modeling studies is data scarcity: e.g., because modeling was not planned when a sampling strategy was developed and the data do not suit the modeling task or because they do not cover the spatial and temporal scale of interest.

Tests of models which are to be used to assess long-term consequences of environmental change are often hindered due to a lack of appropriate long-term data and there is clearly a lack of field experiments that run for more than 5–10 years. Alternatively, validation might be performed with historical archives, but has then to rely on data quality that never reaches the quality of recently measured data. For example, one important data source for testing simulations of vegetation change over long time periods are pollen assemblages. Yet, these are connected with a whole chain of uncertainties (Lischke *et al.* 1998a), concerning the interpretation of the pollen data, their temporal resolution and the input data scenarios required for such comparisons. Taking into account the above mentioned problems of model testing we recommend that (i) modelers should be integrated in field surveys from the very beginning and (ii) concerted actions for the collection of environmental data should be strongly encouraged (Jørgensen *et al.* 2000).

Another challenge is to find model approaches, whose complexity (in terms of number of state variables, parameters and processes) suits best the research question and the available data (Bolliger *et al.* 2005). Although this holds for ecological models in general, it is particularly true for SLST models, because the concept of interacting entities involves additional processes and parameters, and may produce a qualitatively new behavior. The aforementioned lack of appropriate information is one reason to adapt the model to the available information, which means to keep it simple (parsimonious). Such parsimonious models have the advantage that they are easy to manage and robust because they consist of only few highly aggregated and abstracted model components, i.e., state variables, process functions, parameters and driving variables. Sometimes they are even simple enough to be analyzed mathematically, e.g., by stability or bifurcation analysis. If the components of such a parsimonious model are chosen appropriately, it can be generic, i.e., can give insights into the general behavior of a landscape (see case study 2). However, generic models are usually not valid for a specific landscape or cannot represent specific landscape properties such as locations of landscape elements. Only if the model captures all essential basic processes, resulting in more complex models, projections or scenario studies of the specific landscape can be reliable. Complex models furthermore allow more detailed assessments of the spatio-temporal dynamics of landscapes and their drivers. However, due to the high number of parameters and potentially non-linear interactions, complex models can lose their robustness. Additionally, complex spatial models can require excessive computing time. Although this is only a technical constraint, it encumbers model development and testing and currently restricts applications either to small regions, coarse resolutions or very general formulation of the dynamics. Thus, the decision on which model complexity to choose relies on trade-offs between model generality, performance, accuracy, robustness, and manageability. Model complexity is partly associated with the scale at which the landscape dynamics is depicted.

Landscape processes operate and generate patterns across a range of hierarchical, spatial, and temporal scales. The choice of the adequate scale, the integration of model parts acting on different scales, and the change between scales are considerable challenges in landscape research in general. Scale changes of SLST models are even more complicated, due to the spatial interactions.

Modeling involves generalizations and thus introduction of uncertainties on a variety of levels. It is thus important that the model simulations are thoroughly tested for their robustness and validity (see Oreskes *et al.* (1994) for discussion on terms like "validation", "verification" etc.) within the range of conditions in which they are to be applied. Furthermore, landscape models are often optimized for clearly defined study areas and for specific species that are simulated based on abiotic factors relevant for the study area. If they are to be applied under different conditions, validity needs to be tested for changing environments or a variety of situations (model generality). Due to the complexity of many landscape models it is evident that there is no standard procedure that provides appropriate tests. Rather, a variety of model aspects, e.g., model structure and behavior, can be evaluated using different approaches and scenarios. Model testing is often complicated by the lack of appropriate data. An alternative to comparing the models to empirical data is model comparisons, i.e., qualitative or quantitative comparisons within or between different model types.

Outlook

Despite all these challenges, SLST landscape models will play an even more important role in future landscape ecology. Apart from the classical use in system understanding and scenario-testing, models will be increasingly used to interprete large-scale data (e.g., remote sensing signals), or for temporal and spatial interpolation of observations, e.g., in landscape inventories. Additionally, models will help to quantify state and trends of landscapes, and to identify critical thresholds where abrupt changes may occur. Thus, landscape models are and will be helpful tools for theoretical and applied research and environmental management, as they allow hypothesis testing for theory building and scenario testing for applications.

References

Acevedo M.F., Ablan M., Urban D.L. and Pamarti S. 2001. Estimating parameters of forest patch transition models from gap models. Environmental Modelling and Software 16, 7: 649–658.

Akçakaya H.R., McCarthy M.A. and Pearce J.L. 1995. Linking landscape data with population viability analysis: management options for the helmeted honeyeater Lichenostomus melanops cassidix. Biological Conservation 73: 169–176.

Allen P.A., Li B.-L. and Charnov E.L. 2001. Population fluctuations, power laws, and mixtures of lognormal distributions. Ecological Letters 4: 1–3.

Augustin N.H., Mugglestone M.A. and Buckland S.T. 1996. An autologistic model for the spatial distribution of wildlife. Journal of Applied Ecology 33: 339–347.

Bak P. 1996. How nature works. Springer, New York.

Bak P., Tang C. and Wiesenfeld K. 1987. Self-organized criticality: an explanation of 1/f noise. Physical Review Letters 59: 381–384.

Baker J. and Preston C. 2003. Predicting the spread of herbicide resistance in Australian canola fields. Transgenic Research 12, 6: 731–737.

Bakkenes M., Alkemade J.R.M., Ihle F., Leemans R. and Latour J.B. 2002. Assessing effects of forecasted climate change on the diversity and distribution of European higher plants for 2050. Global Change Biology 8: 390–407.

Bascompte J., Jordano P., Melian C.J. and Olesen J.M. 2003. The nested assembly of plant-animal mutualistic networks. Proceedings of the National Academy of sciences of the United States of America 100: 9383–9387.

Bascompte J. and Sole R.V. 1995. Rethinking complexity – Modeling spatiotemporal dynamics in ecology. Trends in Ecology & Evolution 10, 9: 361–366.

Bjørnstad O.N., Ims R.A. and Lambin X. 1999. Spatial population dynamics: Analyzing patterns and processes of population synchrony. Trends in Ecology and Evolution 14: 427–432.

Bolker B. and Pacala S.W. 1997. Using moment equations to understand stochastically driven spatial pattern formation in ecological systems. Theoretical Population Biology 52, 3: 179–197.

Bolliger J. 2002. Schweizer Wälder und Klimaveränderungen: Vergleich von Simulationen quantitativer Vegetationsmodelle. Schweizerische Zeitschrift für das Forstwesen 153: 167–175.

Bolliger J. 2005. Simulating complex landscapes with a generic model: Sensitivity to qualitative and quantitative classifications. Ecological Complexity 2: 131–149.

Bolliger J., Kienast F. and Zimmermann N.E. 2000. Risks of global warming on montane and subalpine forests in Switzerland – a modeling study. Regional Environmental Change 1: 99–111.

Bolliger J., Lischke H. and Green D.G. 2005. Simulating the spatial and temporal dynamics of landscapes using generic and complex models. Ecological Complexity 2, 2: 107–116.

Bolliger J., Sprott J.C. and Mladenoff D.J. 2003. Self-organization and complexity in historical forest landscape patterns. OIKOS 100: 541–553.

Bonn A. and Schröder B. 2001. Habitat models and their transfer for single- and multi-species groups: a case study of carabids in an alluvial forest. Ecography 24: 483–496.

Botkin D.B., Janak J.F. and Wallis J.R. 1970. A simulator for northeastern forest growth. No. 3140, IBM Thomas J. Watson Research Center, Yorktown Heights, N.Y.

Botkin D.B., Janak J.F. and Wallis J.R. 1972. Rationale, limitations and assumptions of a northeastern forest growth simulator. IBM Journal of Research and Development 16: 101–116.

Brown D.G., Goovaerts P., Burnicki A. and Li M.Y. 2002a. Stochastic simulation of land-cover change using geostatistics and generalized additive models. Photogrammetric Engineering and Remote Sensing 68: 1051–1061.

Brown J.H., Gupta V.K., Li B.-L., Milne B.T., Restrepo C. and West B.G. 2002b. The fractal nature of nature: power laws, ecological complexity and biodiversity. Philosophical Transactions of the Royal Society of London 357: 619–626.

Bugmann H. 2001. A review of forest gap models. Climatic Change 51, 3–4: 259–305.

Bugmann H. and Fischlin A. 1996. Simulating forest dynamics in a complex topography using gridded climatic data. Clim. Change 34: 289–313.

Cannas S.A., Marco D.E. and Paez S.A. 2003. Modelling biological invasions: species traits, species interactions, and habitat heterogeneity. Mathematical Biosciences 183, 1: 93–110.

Caswell H. and Etter R.J. 1993. Ecological intercations in patchy environments: From patch-occupancy models to cellular automata. In: Levin S.A., Powelland T.M. and Steele J.H. (eds.). Patch dynamics, Vol. 96: 90–92. Springer, Berlin Heidelberg New York.

Chesson P. 2000. Mechanisms of maintenance of species diversity. Annual Review of Ecology and Systematics 31: 343–366.

Clark J. 1998. Why trees migrate so fast: Confronting theory with dispersal biology and the paleorecord. American Naturalist 152, 2: 204–224.

Clark J.S. and Ji Y. 1995. Fecundity and dispersal in plant populations: implications for structure and diversity. Am. Nat. 146, 1: 72–111.

Coffin D. and Lauenroth W.K. 1989. Disturbances and gap dunamics in a semiarid grassland: A landscape-level approach. Landscape Ecol. 3, 1: 19–27.

Collingham Y.C. and Huntley B. 2000. Impacts of habitat fragmentation and patch size upon migration rates. Ecological Applications 10, 1: 131–144.

Cousens R., Wallinga J. and Shaw M. 2004. Are the spatial patterns of weeds scale-invariant? Oikos 107: 251–264.

Cousins S.A.O., Lavorel S. and Davies I. 2003. Modelling the effects of landscape pattern and grazing regimes on the persistence of plant species with high conservation value in grasslands in south-eastern Sweden. Landscape Ecology 18, 3: 315–332.

Cramer W., Bondeau A., Woodward F.I., Prentice I.C., Betts R.A., Brovkin V., Cox P. M., Fisher V., Foley J.A., Friend A.D., Kucharik C., Lomas, M.R. Ramankutty N., Sitch S., Smith B., White A. and Young-Molling C. 2001. Global response of terrestrial ecosystem structure and function to CO_2 and climate change: results from six dynamic global vegetation models. Global Change Biology 7, 4: 357–373.

Cresswell J.E., Bassom A.P., Bell S.A., Collins S.J. and Kelly T.B. 1995. Predicted pollen dispersal by honey-bees and three species of bumble-bees foraging on oil-seed rape: A comparison of three models. Functional Ecology 9, 6: 829–841.

Davis M.B. 1986. Climatic instability, time lags and community disequilibrium. In: Diamond J. and Case T. (eds.). Community Ecology, pp. 269–284. Harper & Row Publishers Inc., Cambridge a.o.

De 'Ath G. and Fabricius K.E. 2000. Classification and regression trees: a powerful yet simple technique for ecological data analysis. Ecology 81: 3178–3192.

Deutschman D.H., Bradshaw G.A., Childress W.M., Daly K.L., Grünbaum D., Pascual M., Schumaker N.H. and Wu J. 1993. Mechanisms of patch formation. In: Levin S.A., Powelland T. M. and Steele J.H. (eds.). Patch dynamics, Vol. 96: 184–212. Springer, Berlin Heidelberg New York.

Deutschman D.H., Levin S.A. and Pacala S.W. 1999. Error propagation in a forest succession model: The role of fine-scale heterogeneity in light. Ecology 80, 6: 1927–1943.

Dullinger S., Dirnbock T. and Grabherr G. 2004. Modelling climate change-driven treeline shifts: relative effects of temperature increase, dispersal and invasibility. Journal of Ecology 92, 2: 241–252.

Farkas I., Derenyi I., Jeong H., Neda Z., Oltvai Z.N., Ravasz E., Schubert A., Barabasi A.-L. and Vicsek T. 2002. Networks in life: scaling properties and eigenvalue spectra. Physica A 314: 25–34.

Frappier B., Lee T., Olson K. and Eckert R. 2003. Small-scale invasion pattern, spread rate, and lag-phase behavior of Rhamnus frangula L. Forest Ecology and Management 186, 1–3: 1–6.

Gilpin M.E. and Hanski I. 1991. Metapopulation dynamics. Academic Press, London.

Gisiger T. 2001. Scale invariance in biology: coincidence or footprint of a universal mechanism? Biological Reviews 76: 161–209.

Green D.G. 1995. Connectivity and the evolution of biological systems. Journal of Biological Systems 2: 91–103.

Green D.G. 2000. Self-Organization in complex systems. In: Bossomaier T.J. and Green D.G. (eds.). Complex Systems, pp. 7–41. Cambridge University Press, Cambridge.

Green D.G. and Sadedin, S. 2005. Interactions matter – complexity in landscapes and ecosystems. Ecological Complexity 2: 117–130.

Guisan A. and Hofer U. 2003. Predicting reptile distributions at the mesoscale: relation to climate and topography. Journal of Biogeography 30: 1233–1243.

Guisan A. and Theurillat J.-P. 2000. Assessing alpine plant vulnerability to climate change: a modeling perspective. Integrated Assessment 1: 307–320.

Guisan A., Theurillat J.-P. and Kienast F. 1998. Predicting the potential distribution of plant species in an alpine environment. Journal of Vegetation Science 9: 65–74.

Guisan A. and Zimmermann N.E. 2000. Predictive habitat distribution models in ecology. Ecological Modelling 135, 2–3: 147–186.

Gurney W., Veitch A., Cruickshank I. and McGeachin G. 1998. Circles and spirals: Population persistence in a spatially explicit predator-prey model. Ecology 79, 7: 2516–2530.

Hanski I. 1999. Metapopulation ecology. Oxford Univ. Press, Oxford.

Haxeltine A. and Prentice I.C. 1996. BIOME3: An equilibrium terrestrial biosphere model based on ecophysiological constraints, resource availability, and competition among plant functional types. Global Biogeochemical Cycles 10, 4: 693–709.

He H.S., Mladenoff D.J. and Crow T.R. 1999. Linking an ecosystem model and a landscape model to study forest species response to climate warming. Ecological Modelling 114: 213–233.

Higgins S.I. and Richardson D.M. 1996. A review of models of alien plant spread. Ecological Modelling 87, 1–3: 249–265.

Higgins S.I., Richardson D.M. and Cowling R.M. 2000. Using a dynamic landscape model for planning the management of alien plant invasions. Ecological Applications 10, 6: 1833–1848.

Holdridge L.R. 1947. Determination of world plant formations from simple climatic data. Science 105: 367–369.

Irwin E.G. and Geoghegan J. 2001. Theory, data, methods: developing spatially explicit economic models of land use change. Agriculture, Ecosystems and Environment 85: 7–24.

Iverson L.R., Schwartz M.W. and Prasad A.M. 2004. How fast and far might tree species migrate in the eastern United States due to climate change? Global Ecology and Biogeography 13, 3: 209–219.

Jakeman A.J. and Letcher R.A. 2003. Integrated assessment and modelling: features, principles and examples for catchment management. Environmental Modelling and Software 18: 491–501.

Jørgensen L.A., Jørgensen S.E. and Nielsen S.N. 2000. ECOTOX – Ecological Modelling and Ecotoxicology. Elsevier Science B.V., Amsterdam, The Netherland.

Jorgensen S.E. 1992. Integration of ecosystem theories: a pattern. Kluwer, Dordrecht.

Kaneko K. 1992. Overview of coupled map lattices. Chaos 2: 279–282.

Kaufmann E. 2001. Prognosis and management scenarios. In: Brassel P. and Lischke H. (eds.). Swiss National Forest Inventory: Methods and Models of the Second Assessment., pp. 336. Swiss Federal Research Institute WSL, Birmensdorf.

Keane R.E., Cary G.J., Davies I.D., Flannigan M.D., Gardner R.H., Lavorel S., Lenihan J.M., Li C. and Rupp T.S. 2004. A classification of landscape fire succession models: spatial simulations of fire and vegetation dynamics. Ecological Modelling 179, 1: 3–27.

Kindlmann P., Aviron S. and Burel F. 2005. When is landscape matrix important for determining animal fluxes between resource patches? Ecological Complexity 2: 150–158.

Kirschbaum M. and Fischlin A. 1996. Climate change impacts on forests. In: Watson R.T., Zinyoweraand M.C. and Moss R.H. (eds.). Climate change 1995 – Impacts, adaptations and mitigation of climate change: Scientific-technical analyses: Contribution of working group II to the second assessment report of the intergovernmental panel on climate change, pp. 95–131. Cambridge University Press, Cambridge.

Kittel T.G.F., Steffen W.L. and Chapin F.S. 2000. Global and regional modelling of Arctic-boreal vegetation distribution and its sensitivity to altered forcing. Global Change Biology 6: 1–18.

Klein E.K., Lavigne C., Foueillassar X., Gouyon P. H. and Laredo C. 2003. Corn pollen dispersal: Quasi-mechanistic models and field experiments. Ecological Monographs 73, 1: 131–150.

Kot M., Lewis M.A. and vandenDriessche P. 1996. Dispersal data and the spread of invading organisms. Ecology 77, 7: 2027–2042.

Leemans R. and van-den-Born G.J. 1994. Determining the potential distribution of vegetation, crops and agricultural productivity. Water, Air, & Soil Pollution 76: 133–161.

Lennon J.J., Koleff P., Greenwood J.J.D. and Gasto K.J. 2001. The geographical structure of British bird distributions: diversity, spatial turnover and scale. Journal of Animal Ecology 70: 966–979.

Levin S.A. 1998. Ecosystems and the bioshpere as complex adaptive systems. Ecosystems 1: 431–436.

Levin S.A., Powell T.M. and Steele J.H. 1993. Patch dynamics, Lecture Notes in Biomathematics Vol. 96, pp. 307. Springer, Berlin Heidelberg New York.

Levins R. 1970. Extinction. Lecture Notes in Mathematics 2: 75–107.

Lewis M.A. 1994. Spatial Coupling of Plant and Herbivore Dynamics – the Contribution of Herbivore Dispersal to Transient and Persistent Waves of Damage. Theoretical Population Biology 45, 3: 277–312.

Lexer M. 1995. Anwendung eines "big-leaf"-Modelles zur Simulation des Bodenwasserhaushaltes in Fichtenbeständen (*Picea abies*(L.) Karst.). Applying a "big-leaf"-model to simulate seasonal sooil water dynamics in Norway spruce stands. Centralblatt für das gesamte Forstwesen 112, 4: 209–225.

Lexer M.J., Hönninger K., Scheifinger H., Matulla C., Groll N. and Kromp-Kolb H. 2000. The sensitivity of central European mountain forests to scenarios of climatic change: Methodological frame for a large-scale risk assessment. Silva Fennica 34, 2: 113–129.

Li B.L. 2000a. Fractal geometry applications in description and analysis of patch pattern and patch dynamics. Ecological Modelling 132: 33–50.

Li B.L. 2000b. Why is the holistic approach becoming so important in landscape ecology? Landscape Urban Plan. 50: 27–41.

Lindenmayer D.B., Cunningham R.B., Tanton M.T., Nix H.A. and Smith A.P. 1991. The conservation of arboreal marsupials in the montane ash forests of the central highlands of Victoria, South East Australia: III. The habitat requirements of Leadbeater's Possum Gymnobelideus leadbeateri and models of the diversity and abundance of arboreal marsupials. Biological Conservation 56: 295–315.

Lischke H. 2001. New developments in forest modeling: Convergence between applied and theoretical approaches. Natural Resource Modeling 14, 1: 71–102.

Lischke H. 2005. Modelling of tree species migration in the Alps during the Holocene: What creates complexity? Ecological Complexity 2, 2: 159–174.

Lischke H., Guisan A., Fischlin A., Williams J. and Bugmann H. 1998a. Vegetation responses to climate change in the Alps – Modeling studies. In: Cebon P., Dahinden U., Davies H., Imbodenand D. and Jaeger C. (eds.). A view from the Alps: Regional perspectives on climate change, pp. 309–350. MIT Press, Boston.

Lischke H., Löffler T.J. and Fischlin A. 1998b. Aggregation of individual trees and patches in forest succession models – Capturing variability with height structured random dispersions. Theoretical Population Biology 54, 3: 213–226.

Lischke H., Zimmermann N.E., Bolliger J., Rickebusch S. and Löffler T.J. in press. TreeMig: A forest-landscape model for simulating spatio-temporal patterns from stand to landscape scale. Ecological Modelling.

Löffler T.J. and Lischke H. 2001. Incorporation and influence of variability in an aggregated forest model. Natural Resource Modeling 14, 1: 103–137.

Loos C., Seppelt R., Meier-Bethge S., Schiemann J. and Richter O. 2003. Spatially explicit modelling of transgenic maize pollen dispersal and cross-pollination. Journal of Theoretical Biology 225: 241–255.

Ma B.L., Subedi K.D. and Reid L.M. 2004. Extent of cross-fertilization in maize by pollen from neighboring transgenic hybrids. Crop Science 44, 4: 1273–1282.

Malanson G.P. and Cairns D.M. 1997. Effects of dispersal, population delays, and forest fragmentation on tree migration rates. 131, 1: 67–79.

Mandelbrot B. 1983. The fractal geometry of nature. Freeman, New York.

Manies K.L., Mladenoff D.J. and Nordheim E.V. 2001. Assessing large-scale surveyor variability in the historic forest data of the Original U.S. Public Land Surveys. Canadian Journal of Forest Research 17: 1719–1730.

Maron J.L. and Harrison S. 1997. Spatial pattern formation in an insect host-parasitoid system. Science 278, 5343: 1619–1621.

Meagher T.R., Belanger F.C. and Day P.R. 2003. Using empirical data to model transgene dispersal. Philosophical Transactions of the Royal Society of London Series B-Biological Sciences 358, 1434: 1157–1162.

Millennium Ecosystem Assessment 2005. Ecosystems and Human Well-being: Synthesis. Island Press, Washington, DC.

Milne B.T. 1988. Measuring the fractal geometry of landscapes. Appl. Math. Computation 27: 67–79.

Milne B.T. 1991. Lessons from applying fractal models to landscape patterns. In: T.M.G. and G.R.H. (eds.). Quantiative methods in landscape ecology, pp. 199–235. Springer.

Milne B.T. 1998. Motivation and benefits of complex systems approaches in ecology. Ecosystems 1, 5: 449–456.

Milne B.T., Turner M.G., Wiens J.A. and Johnson A.R. 1992. Interactions between the fractal geometry of landscapes and allometric herbivory. Theoretical Population Biology 41: 337–353.

Moravie M.A. and Robert A. 2003. A model to assess relationships between forest dynamics and spatial structure. Journal of Vegetation Science 14, 6: 823–834.

Nabuurs G.J., Schelhaas M.J. and Pussinen A. 2000. Validation of the European Forest Information Scenario Model (EFISCEN) and a projection of Finnish forests. Silva Fennica 34, 2: 167–179.

Neilson R.P. 1995. A model for predicting continental-scale vegetation distribution and water balance. Ecol. Applications 5, 2: 362–385.

Neilson R.P., Pitelka L.F., Solomon A., Fragoso J., Lischke H., Nathan R. and Thompson K. 2005. Forecasting regional to global plant migration in response to climate change. Bioscience 55, 9: 749–759.

Oreskes N., Shrader-Frechette K. and Belitz K. 1994. Verification, validation and confirmation of numerical models in the earth sciences. Science 263: 641–646.

Pacala S.W., Canham C.D., Saponara J., Silander J.A., Kobe R.K. and Ribbens E. 1996. Forest Models Defined By Field Measurements: Estimation, Error Analysis and Dynamics. Ecological Monographs 66, 1: 1–43.

Pacala S.W., Canham C.D. and Silander J.A.J. 1993. Forest models defined by field measurements: I. The design of a northeastern forest simulator. Canadian Journal of Forest Research 23, 10: 1980–1988.

Pastor J., Cohen Y. and Moen R. 1999. Generation of spatial patterns in boreal forest landscapes. Ecosystems 2, 5: 439–450.

Patten B.C., Straškraba M. and Jørgensen S.E. 1997. Ecosystems emerging: 1. conservation. Ecological Modelling 96: 221–284.

Peng C.H. 2000. From static biogeographical model to dynamic global vegetation model: a global perspective on modelling vegetation dynamics. Ecological Modelling 135, 1: 33–54.

Peppler-Lisbach C. and Schröder B. 2004. Predicting the species composition of mat-grass communities (Nardetalia) by logistic regression modelling. Journal of Vegetation Science 15: 623–634.

Perry D.A. 1995. Self-organizing systems across scale. Trends in Ecology and Evolution 10: 241–244.

Peters D.P.C. 2002. Plant species dominance at a grassland-shrubland ecotone: an individual-based gap dynamics model of herbaceous and woody species. Ecological Modelling 152, 1: 5–32.

Picard N., Bar-Hen A. and Franc A. 2001. Spatial pattern induced by asymmetric competition: A modelling approach. Natural Resource Modeling 14, 1: 147–175.

Picard N. and Franc A. 2001. Aggregation of an individual-based space-dependent model of forest dynamics into distribution-based and space-independent models. Ecological Modelling 145: 69–84.

Poff N.L. 1997. Landscape filters and species traits: towards mechanistic understanding and prediction in stream ecology. Journal of the North American Benthological Society 16: 391–409.

Powell J.A. and Zimmermann N.E. 2004. Multi-scale analysis of seed dispersal and the resolution of Reid's Paradox. Ecology 85, 2: 490–506.

Prentice I.C., Cramer W., Harrison S.P., Leemans R., Monserud R.A. and Solomon A.M. 1992. A global biome model based on plant physiology and dominance, soil properties and climate. J. Biogeogr. 19: 117–134.

Pretzsch H. 2002. Application and evaluation of the growth simulator SILVA 2.2 for forest stands, forest estates and large regions. Forstwissenschaftliches Centralblatt 121: 28–51.

Prevosto B., Hill D.R.C. and Coquillard P. 2003. Individual-based modelling of Pinus sylvestris invasion after grazing abandonment in the French Massif Central. Plant Ecology 168, 1: 121–137.

Reid 1899. The origin of the British flora. Dulau, London, UK.

Renshaw E. 1991. Modelling biological populations in space and time. Cambridge University Press, Cambridge.

Richter O. and Seppelt R. 2004. Flow of genetic information through agricultural ecosystems: a generic modelling framework with application to pesticide resistent weeds and genetically modified crops. Ecological Modelling 174: 55–66.

Richter O., Seppelt R. and Söndgerath D. 2002. Computer Modelling. Encyclopedia for Environmetrics 1: 402–411.

Rupp T.S., Starfield A.M. and Chapin F.S. 2000. A frame-based spatially explicit model of subarctic vegetation response to climatic change: comparison with a point model. Landscape Ecology 15, 4: 383–400.

Schröder B. and Reineking B. 2004. Modellierung der Art-Habitat-Beziehung – ein Überblick über die Verfahren der Habitatmodellierung. pp. 5–26. No. 9.

Schulte L.A. and Mladenoff D. J. 2001. The original U.S. Public Land Survey records: Their use and limitations in reconstruction pre-European settlement vegetation. Journal of Forestry 99: 5–10.

Schulte L.A., Mladenoff D.J. and Nordheim E.V. 2002. Quantitative classification of a historic northern Wisconsin (U.S.A.) landscape: mapping forests at regional scales. Canadian Journal of Forest Research 32: 1616–1638.

Seppelt R. 2003. Computer-based environmental management. Wiley-VCH, Weinheim.

Seppelt R. and Voinov A. 2002. Optimization methodology for land use patterns using spatially explicit landscape models. Ecological Modelling 151, 2–3: 125–142.

Seppelt R. and Voinov A. 2003. Optimization Methodology for Landuse Patterns: Evaluation based on Multiscale Habitat Pattern Comparison. Ecological Modelling 168, 3: 217–231.

Skellam J.G. 1951. Random dispersal in theoretical populations. Biometrika 38: 196–218.

Solé R.V., Alonso D. and McKane A. 2002. Self-organized instability in complex ecosystems. Philosophical Transactions of the Royal Society of London Series B-Biological Sciences 357, 1421: 667–681.

Solé R.V. and Manrubia S.C. 1995. Are rainforests self-organized critical? Journal of Theoretical Biology 173: 31–40.

Sprott J.C., Bolliger J. and Mladenoff D.J. 2002. Self-organized criticality in forest-landscape evolution. Physics Letters A 297, 3–4: 267–271.

Starfield A.M. and Chapin F.S. 1996. Model of transient changes in arctic and boreal vegetation in response to climate and land use change. Ecological Applications 6, 3: 842–864.

Steele J.H. 1993. Concepts and models: An overview. In: Levin S.A., Powelland T.M. and Steele J.H. (eds.). Patch dynamics, Vol. 96: 90–92. Springer, Berlin Heidelberg New York.

Storch D., Gaston K.J. and Cepák J. 2002. Pink landscapes: 1–f spectra of spatial environmental variability and bird community composition. Proceedings of the Royal Society of London/B 269: 1791–1796.

Strogatz S.H. 2001. Exploring complex networks. Nature 410: 268–276.

Syphard A.D., Clarke K. and Franklin J. 2005. Using a cellular automaton model to forecast urban growth patterns in a fire-prone Mediterranean landscape. Ecological Complexity 2: 185–203.

Thornton P.E. 1998. Regional ecosystem simulation: combining surface- and satellite-based observations to study linkages between terrestrial energy and mass budgets. Ph.D. Dissertation, No., School of Forestry, University of Montana, Missoula, MT.

Thornton P.E., Law B.E., Gholz H.L., Clark K.L., Falge E., Ellsworth D.S., Goldstein A.H., Monson R.K., Hollinger D., Falk M., Chen J. and Sparks J.P. 2002. Modeling and measuring the effects of disturbance history and climate on carbon and water budgets in evergreen needleleaf forests. Agricultural and Forest Meteorology 113: 185–222.

Thürig E., Kaufmann E., Frisullo R. and Bugmann H. 2005. Evaluation of the growth function of an empirical forest scenario model. Forest Ecology and Management 204: 53–68.

Tufto J., Engen S. and Hindar K. 1997. Stochastic dispersal processes in plant populations. Theoretical Population Biology 52, 1: 16–26.

van den Bosch F., Metz A.J. and Diekmann O. 1990. The velocity of spatial population expansion. Journal of Mathematical Biology 28, 5: 529–565.

VanRaamsdonk L.W.D. and Schouten H.J. 1997. Gene flow and establishment of transgenes in natural plant populations. Acta Botanica Neerlandica 46, 1: 69–84.

Wagner H.H., Werth S., Kalwij J.M., Bolli J.C. and Scheidegger C. 2006. Modelling forest recolonization by an epiphytic lichen using a landscape genetic approach. Landscape Ecology 21: 849–865.

Wimberly M.C. 2002. Spatial simulation of historical landscape patterns in coastal forests of the Pacific Northwest. Canadian Journal of Forest Research-Revue Canadienne De Recherche Forestiere 32, 8: 1316–1328.

With K.A. and King A.W. 1999. The use and misuse of neutral landscape models in ecology. Oikos 79: 219–229.

With K.A., Pavuk D.M., Worchuck J.L., Oates R.K. and Fisher J.L. 2002. Threshold effects of landscape structure on biological control in agroecosystems. Ecological Applications 12: 52–65.

Woodward F.I. and Smith T.M. 1994. Predictions and measurements of the maximum photosynthetic rate at the global scale. In: Schulze E.D. and Caldwell M.M. (eds.). Ecophysiology of Photosynthesis, Vol. 100: 491–509. Springer, New York.

Wu J. and Marceau D. 2002. Modeling complex ecological systems: an introduction. Ecological Modelling 153, 1–2: 1–6.

Yee T.W. and Mitchell N.D. 1991. Generalized additive models in plant ecology. Journal of Vegetation Science 2: 587–602.

Landscape Series

1. H. Palang and G. Fry (eds.): *Landscape Interfaces*. Cultural Herita;ge in Changing Landscapes. 2003 ISBN 1-4020-1437-6
2. I. Vanslembrouck and G. Van Huylenbroeck: *Landscape Amenities*. Economic Assessment of Agricultural Landscape. 2005 ISBN 1-4020-3134-3
3. F. Almo: *Principles and Methods in Landscape Ecology.* Towards a Science of Landscape. 2005 ISBN 1-4020-3327-3 Pb: ISBN 1-4020-3328-1
4. D.G. Green, N. Klomp, G. Rimmington and S. Sadedin: *Complexity in Landscape Ecology.* 2006 ISBN 1-4020-4285-X
5. T.S. Terkenli and A.-M. d'Hauteserre (eds.): *Landscapes of a New Cultural Economy of Space*. 2006 ISBN 1-4020-4095-4
6. Z. Naveh (ed.): *Transdisciplinary Challenges for Landscape Ecology and Restoration Ecology*. Anthology of Studies on Mediterranean and Global Issues. 2006 ISBN 1-4020-4420-8
7. H. Palang, H. Sooväli, A. Printsmann (eds.): *Seasonal Landscapes*. 2007 ISBN 1-4020-4982-X
8. F. Kienast, S. Ghosh, O. Wildi (eds.): *A Changing World.* Challenges for Landscape Research. 2007 ISBN 1-4020-4434-8